KW-361-047

THE NATIONAL FOOD CENTRE
A DIVISION OF TEAGASC

2 2 MAY 2006

LIBRARY

Acc. No. B2206 Class No. 061.1:631

OECD Agricultural Outlook
2004-2013

OECD

ORGANISATION FOR ECONOMIC CO-OPERATION AND DEVELOPMENT

ORGANISATION FOR ECONOMIC CO-OPERATION AND DEVELOPMENT

Pursuant to Article 1 of the Convention signed in Paris on 14th December 1960, and which came into force on 30th September 1961, the Organisation for Economic Co-operation and Development (OECD) shall promote policies designed:

- to achieve the highest sustainable economic growth and employment and a rising standard of living in member countries, while maintaining financial stability, and thus to contribute to the development of the world economy;
- to contribute to sound economic expansion in member as well as non-member countries in the process of economic development; and
- to contribute to the expansion of world trade on a multilateral, non-discriminatory basis in accordance with international obligations.

The original member countries of the OECD are Austria, Belgium, Canada, Denmark, France, Germany, Greece, Iceland, Ireland, Italy, Luxembourg, the Netherlands, Norway, Portugal, Spain, Sweden, Switzerland, Turkey, the United Kingdom and the United States. The following countries became members subsequently through accession at the dates indicated hereafter: Japan (28th April 1964), Finland (28th January 1969), Australia (7th June 1971), New Zealand (29th May 1973), Mexico (18th May 1994), the Czech Republic (21st December 1995), Hungary (7th May 1996), Poland (22nd November 1996), Korea (12th December 1996) and the Slovak Republic (14th December 2000). The Commission of the European Communities takes part in the work of the OECD (Article 13 of the OECD Convention).

Publié en français sous le titre :
Perspectives agricoles de l'OCDE : 2004-2013

© OECD 2004

Permission to reproduce a portion of this work for non-commercial purposes or classroom use should be obtained through the Centre français d'exploitation du droit de copie (CFC), 20, rue des Grands-Augustins, 75006 Paris, France, tel. (33-1) 44 07 47 70, fax (33-1) 46 34 67 19, for every country except the United States. In the United States permission should be obtained through the Copyright Clearance Center, Customer Service, (508)750-8400, 222 Rosewood Drive, Danvers, MA 01923 USA, or CCC Online: *www.copyright.com*. All other applications for permission to reproduce or translate all or part of this book should be made to OECD Publications, 2, rue André-Pascal, 75775 Paris Cedex 16, France.

Foreword

T he OECD Agricultural Outlook *provides a medium term assessment of future trends and prospects in the major agricultural commodity markets of OECD countries. The report is published annually, as part of a continuing effort to promote informed discussion of emerging policy issues. This tenth edition of the OECD Agricultural Outlook, 2004-2013, is set against the background of a world economy that is on the path to economic recovery, and where OECD agricultural policy is being influenced by changes taking place in the European Union with the 2003 reform of the CAP and enlargement of the Union as well as the multi-year provisions of the US Farm Act of 2002. The* Outlook *for agricultural markets is for a gradual strengthening in market conditions for all commodities over the period to 2013. Stronger global economic growth is expected to lead to increased consumption and trade and firmer agricultural product prices in nominal terms. But these outcomes are highly conditional on the geopolitical and global economic situation, as well as a continuation of domestic policies and policy settings, particularly in OECD countries. A restart of the stalled Doha round of multilateral trade discussions in the WTO and their successful conclusion in terms of further trade reform, would strengthen the prospects for agricultural markets beyond that contained in this assessment which is based on a continuation of existing policy reforms and URAA commitments.*

The projections to 2013, presented in the Outlook, *constitute a plausible medium-term future for the markets of key commodities. They are the result of close co-operation between the OECD Secretariat and experts in member countries, and some national co-operators in Non-member economies (NMEs), and hence, reflect their combined knowledge and expertise. This year's report takes account of the enlargement of the European Union, from fifteen to twenty-five countries, from 2004. The commodity projections are based on a number of assumptions relating to current or announced agricultural and trade policies in OECD countries, the outcome of the URAA multilateral trade negotiations in the WTO, the underlying macroeconomic environment and its expected evolution, as well as developments in major NMEs. The OECD's Aglink model is used to guarantee internal consistency in the projections. In addition, the model is employed to generate scenarios around the* Outlook *baseline so that sources of uncertainty in relation to key assumptions and selected policy issues can be analysed. Thus, the report includes – inter alia – an assessment of the market impacts of the 2003 CAP reform in the European Union, an evaluation of the implications for oilseed markets of different rates of growth in Brazilian oilseed production, the potential market implications of a rundown in the huge level of grain stocks held by China and the possible interaction between milk quotas and other instruments to achieve specific milk policy objectives. It also presents results of ongoing work on the introduction of stochastic elements in the baseline generation. Finally, the report includes a background section on the Indian agricultural sector covering the evolution of the main agricultural industries, policy settings, world trade integration and trade prospects. The fully documented outlook database, including historical data, projections and selected scenario results, is available through the OECD internet site.*

This publication is prepared by the Directorate for Food, Agriculture and Fisheries of the OECD with the active participation of all member countries. The policy assessments provided in this report is supported and extended by another annual report prepared by the Directorate: Agricultural Policies in OECD Countries: At a Glance (July 2004).

The OECD Agricultural Outlook is published under the responsibility of the Secretary-General of the OECD. The views expressed and conclusions reached in this report do not necessarily correspond to those of the governments of OECD member countries.

Acknowledgement. This edition of the OECD *Agricultural Outlook* was prepared by the following team of economic and market analysts from the OECD Directorate of Food, Agriculture and Fisheries. Loek BOONEKAMP (team leader), Garry SMITH (coordinator), Pete LIAPIS, Grégoire TALLARD, Wyatt THOMPSON, Pavel VAVRA, Martin von LAMPE, Andrew DEVLIN and David DOWEY.

Research and statistical assistance were provided by Armelle ELASRI, Andrew DEVLIN, David DOWEY and Claude NENERT. Secretarial services and coordination in report preparation was provided by Christine CAMERON. Technical assistance in the preparation of the Outlook database was provided by Eric ESPINASSE and Serge PETITEAU. Many other colleagues in the OECD Secretariat and member country delegations furnished useful comments on earlier drafts of the report. A theme section in the report on the Indian agricultural sector was prepared by the Secretariat and drew on information contained in a consultancy report provided by Mr. Neeraj SHARMA of Lucknow, India.

Table of Contents

Table of Contents

Acronyms and Abbreviations

ABARE	Australian Bureau of Agricultural and Resource Economics
ACP	African, Caribbean and Pacific countries
ALIC	Agriculture and Livestock Industry Corporation
AMAD	Agricultural Market Access Database
AMS	Agricultural Marketing Service
ASEAN	Association of Southeast Asian Nations
BSE	Bovine Spongiform Encephalopathy
CEEC	Central and Eastern European Countries
CAP	Common Agricultural Policy (EU)
CCP	Counter-Cyclical Payments (US)
CIS	Commonwealth of Independent States
CoOl	Country-of-Origin Labelling
CPI	Consumer Price Index
CMO	Common Market Organisation for sugar (EU)
CRP	Conservation Reserve Program (US)
CWT	Cooperatives Working Together
DPC	Direct Payments for Crops (US)
EBA	Everything-But-Arms Initiative (EU)
ECB	European Central Bank
EEP	Export Enhancement Program (US)
ERS	Economic Research Service of the US Department of Agriculture
EU	European Union
EU-15	Fifteen member states of the European Union
EU-10	Ten new member states of the European Union from May 2004
EU-25	Twenty five member states of the European Union from May 2004
EUROSTAT	Statistical Office of the European Communities
FAIR ACT	Federal Agriculture Improvement and Reform Act (US) of 1996
FAO	Food and Agriculture Organisation of the United Nations
FMD	Foot and Mouth Disease
FAS	Foreign Agricultural Service of the US Department for Agriculture
FOB	Free on board (export price)
FSRI ACT	Farm Security and Rural Investment Act (US) of 2002
FTAA	Free Trade Area of the Americas
GATT	General Agreement on Tariffs and Trade
GDP	Gross Domestic Product
GM	Genetically modified
GMO	Genetically engineered or modified plant, animal, micro-organism or virus
HFCS	High Fructose Corn Syrup
HS	Harmonised Commodity Description and Coding System
IMF	International Monetary Fund
MAF	Ministry of Agriculture and Forestry (New Zealand)
MAFF	Ministry of Agriculture, Forestry and Fisheries (Japan)
MERCOSUR	Common Market of the South
MLAP	Marketing Loan Assistance Program (US)
MLC	Meat and Livestock Commission (United Kingdom)

MFN	Most Favoured Nation
MPC	Milk protein concentrates
NAFTA	North American Free Trade Agreement
NME	Non-Member economies
NTBs	Non-Tariff Barriers
NZDB	New Zealand Dairy Board
OECD	Organisation for Economic Co-operation and Development
OIE	Office International des Epizooties
OMB	Office of Management and Budget (United States)
OTMS	Over Thirty Month Scheme
PFCP	Production Flexibility Contract Payments (US)
PSE	Producer Support Estimate
PSD	Production supply and distribution
R&D	Research and Development
RR	Roundup Ready seed varieties
RRAC	Relative Risk Aversion Coefficient
RTAs	Regional Trading Arrangements
SARS	Severe Acute Respiratory Syndrome
SFP	Single Farm Payment (EU)
SMP	Skim milk powder
SPS measures	Sanitary and phyto-sanitary measures
STE	State Trading Enterprises
TRQ	Tariff rate quota
UK	United Kingdom
UNCTAD	United Nations Conference on Trade and Development
UNESCO	United Nations Educational Scientific and Cultural Organisation
URAA	Uruguay Round Agreement on Agriculture
US	United States
USDA	United States Department of Agriculture
WMP	Whole milk powder
WTO	World Trade Organisation

Abbreviations and symbols

AUD	Dollars (Australia)		kt	thousand tonnes
ARS	Pesos (Argentina)		L	litre
Bn	Billion		lw	live weight
BRL	Real (Brazil)		mha	million hectares
CAD	Dollars (Canada)		mn	million
CNY	Yuan (China)		mt	million tonnes
cwe	carcass weight equivalent		MXN	Peso (Mexico)
c.i.f.	cost insurance freight		NZD	Dollars (New Zealand)
cts/lb	Cents per pound		PPP	Purchasing Power Parity
dw	dressed weight		pw	product weight
est	estimate		RUR	Ruble (Russia)
EUR	Euro (Europe)		rse	raw sugar equivalent
f.o.b.	Freight on board		rtc	ready-to-cook
Ha	Hectare		rw	retail weight
JFY	Japanese fiscal year beginning 1 April		t	tonnes
JPY	Japanese yen		t/ha	tonnes per hectare
Kg	Kilogram		USD	Dollars (United States)
KRW	Won (Korea)		VAT	Value added tax

OECD AGRICULTURAL OUTLOOK: 2004-2013 – ISBN 92-64-02008-X – © OECD 2004

Outlook in Brief

Broad-based income growth in both OECD and Non-member economies, moderate population growth and low inflation lead to higher per capita incomes and consumption gains world-wide. Consumption in the Non-member economies is expected to grow at rates much faster than those of the OECD area, especially for dairy products such as butter, cheese and whole milk powder as well as livestock products. Consumption gains for these products are faster than growth in population providing the potential to reduce malnutrition and hunger.

In the mature markets of the OECD area, where incomes are high and basic dietary needs have long been more than satisfied, consumption gains for commodities are expected to post only moderate growth rates as preferences shift towards products such as poultry meat, cheese and whole milk powder. Higher growth rates in the non-OECD region during the projection period imply an increasing share of agricultural produce and feedstuffs is consumed outside the OECD area, indicating increasing activity in animal production in the Non-member economies.

Global production for wheat, rice, coarse grains, beef, cheese and vegetable oils, expands faster than consumption. Most of the gains in production are through expected productivity improvements especially in crops where area expands at a much lower rate, and these tend to be concentrated in the Non-member economies. Production expansion in countries outside the OECD area outpaces that in OECD countries taken together. As a result, the OECD share in world production falls; more for butter and skim milk powder, less for pig meat and whole milk powder, with minor changes for the other products.

Global trade for wheat and coarse grains is expected to grow moderately with more substantial increases in rice trade. Trade in sugar is also expected to expand over the projection period with Brazil, the leading sugar exporting nation expected to increase its market share. World trade in dairy products continues to represent a small share of world milk production, is dominated by OECD countries and is not expected to expand significantly during the *Outlook* period. OECD countries continue to dominate world trade in these products. Net exports of dairy products from the OECD to the Non-member economies are expected to decline, other than for whole milk powder. But OECD countries remain big in meat trade, especially poultry meat.

Prices for almost all products covered in this *Outlook* are expected to strengthen over the projection period in nominal terms, but to continue to trend downwards in real terms.

Domestic and trade policies are important factors in the *Outlook* as they influence markets and the degree of integration and variability of domestic and world prices. For some commodities, these policies preserve large differences between domestic and world prices, imposing high costs on consumers and mitigating the responsiveness of domestic markets to changing scarcities on international markets. The persistence of large price gaps suggests that more needs to be done to liberalise sensitive sectors, in particular through redressing border measures and related domestic policies. Renewed progress in the agricultural negotiations underway as part of the WTO Doha Development Agenda would be an important contribution in this respect.

Overview

The main underlying assumptions

*Solid income growth prospects and low
inflation expected*

For the first time, this year's *Agricultural Outlook* contains projections over a ten-year period to 2013 and includes an expanded European Union of 25 member countries. The *Outlook* this year occurs against a macroeconomic background that is more optimistic than that of the last two years. Economic growth in most OECD member countries is expected to be higher, led by the resurgent growth in the United States and its NAFTA trading partners, Canada and Mexico. Japan too, is expected to post solid growth numbers in 2003, and a path of moderate growth is expected following years of stagnation, even though this is anticipated to lessen in the medium-term. Growth in the euro zone in 2003 is lagging that of other major OECD member countries, but prospects in future years are expected to improve from the current low level. Beyond the OECD, macroeconomic conditions are optimistic across the globe as growth prospects for Non-member economies are also strong. Of the Non-member economies included explicitly in this *Outlook*, Russia, Argentina, and especially China and Brazil are expected to post healthy income growth rates over the projection period.

Helping the economic recovery are the accommodating monetary policies followed by many OECD member countries. Interest rates in many countries continue to hover at historical lows providing investment incentives and fuelling consumer borrowing. This supports consumer spending, the driving force behind the optimistic income growth projections. And, even with higher expected income growth rates and low interest rates, core inflation conditions are expected to remain moderate for most OECD member countries. But, recent spikes in the price of energy and other industrial raw materials such as base metals, if they persist, may challenge the assumption of low inflation. This is particularly true in the case of China, a country with an economy that many compare to the early stages of the US industrial emergence. China is increasingly becoming the manufacturing centre of the world. Its income growth rate has been phenomenal for a long time and it is increasingly becoming more integrated into the world economy. Taking the EU as a single trader, China is now the fourth largest trader in the world. Since it has relatively low wages and its manufacturing sector is very competitive, it has become the price setter for many goods. If China passes on the possible higher costs of energy and other raw materials through its exports to the world, the assumption of low inflation may be at risk.

Population growth rates are expected to moderate across the globe over the projection period, potentially slowing demand growth for agricultural products, especially bulk commodities. The expected rate of expansion of population is lower in all regions for the next ten years compared to the last ten. Except for Africa, population growth is expected to average less than 2%, with most regions averaging around one per cent per year. Coupled with projected income growth rates of a little more than 3% per year over the next ten years, per capita income world-wide should expand, raising the standard of living of most people across the globe and potentially increasing demand for higher value-added agricultural products such as meats and dairy products.

Low U.S dollar could influence competitiveness

The US dollar, the currency in which most agricultural commodity trade is denominated, depreciated against most OECD currencies in 2003. The projections are conditioned on the US dollar remaining at its current level *vis-à-vis* other major currencies during the *Outlook* period. This adjustment should reduce price competitiveness amongst competing exporters within the OECD while increasing the purchasing power of importing countries such as Japan. The currencies of the group of developing countries in the *Outlook*, with the exception of China whose currency is basically tied to the US dollar, are expected to continue to depreciate in nominal terms relative to the US dollar, increasing their competitiveness in international markets while hindering their ability to import.

The *Outlook* is based on a relatively smooth path in the development of macroeconomic variables such as income growth and exchange rates. However, the world usually does not follow such a smooth path over an extended period. In order to incorporate some of the inherent randomness over the *Outlook* period, a special section in the Economic and Policy Assumptions chapter examines the implications of different developments in income growth and exchange rates and their effects on world and domestic prices directly and indirectly through market adjustments. For example, the world oilseed meals price in 2004 is projected at almost USD 190/t while at the end of the projection period it falls to USD 172/t, assuming that income growth and exchange rates evolve as in the *Outlook*. But, given the randomness in the income growth and exchange rates exhibited in the past, on average, two-thirds of the price observations could fall within the range of USD 178 to USD 199 in 2004 and two-thirds of the price observations could fall within the range of USD 163 and USD 188 in 2013.

World trade since 2001, partly as a result of the macro economic slowdown, has been stagnant compared to the growth in world trade during the 1990s when it grew much faster than world income. In the last half of 2003 however, merchandise trade in general, including trade in agricultural products, rebounded and is estimated to have grown at an annualised rate close to the increase in world income. One of the key developments is the growing presence and importance of developing countries in world trade. Not only is their trade with OECD member countries expanding, but trade among themselves is also increasing.

Domestic and border policies matter

In addition to the macroeconomic environment, domestic agricultural policies and levels of support, as well as border measures such as tariff levels and tariff rate quotas, or the use of export competition measures, also significantly influence production, consumption, prices and trade. On the international arena, the policy environment is in flux. After finally including agriculture in a multilateral trade agreement of the Uruguay Round negotiations and agreeing to the need for further reforms, discussions under the Doha Development Round failed to make progress in Cancún in September 2003. Prior to that time the agricultural discussions under the built in agenda repeatedly missed deadlines to agree to modalities to further reform trade in agricultural products along the three pillars of market access, export competition, and domestic support. Since Cancún, discussions have continued but, to this point in time, agreement on even the broad modalities for reform of agricultural trade policy remains elusive.

One important development in the continuing discussions is the emergence of the group of G-20, a grouping of developing countries led by Brazil, China, India and South Africa, which have demanded greater agricultural policy reforms and market liberalisation on the part of the developed world. Partly as a result of inabilities to reach a multilateral trade agreement, countries are continuing their pursuit of new regional trade agreements (RTA). At the end of 2002, 176 RTAs were in force and notified to the WTO, 17 more than at the end of the preceding year. Another development in Cancún was the emergence of the G-90 group of least developed countries who are concerned about maintaining their preferential access to developed country markets through the non-reciprocal preferential trading arrangements (PTAs). Their concern is that preference erosion may result from further trade liberalisation. RTAs may create trade and thus promote welfare. However, a latent problem with these agreements is their potential to divert trade from more competitive non-members which is becoming of greater importance as such agreements spread.

Outlook assumes current policies in place

The *Outlook* is conditioned on policies that are in place or have been announced within well-defined programmes. Thus, the *Outlook* assumes trade policies as agreed in the URAA and excludes any possible modifications that may result from the current discussions under the Doha Development Agenda. This means that the potential WTO accession of new members, such as Russia, are not considered, nor are the current discussions by Korea and its trading partners on the changes to Korea's current import regime for rice.

On the domestic front, the *Outlook* includes the various provisions and programmes of the US Farm Security and Rural Investment Act of 2002 (FSRI Act) which is assumed to continue through out the projection period even though its mandate ends after 2007. For the European Union, the main elements of the Common Agricultural Policy (CAP) 2003 Reform are included. One of the most significant changes is the new Single Farm Payment (SFP) which will enter into force in 2005 though member states can delay implementation up to the year 2007. The SFP provides direct payments to farmers, based on reference areas and livestock numbers over the 2000 to 2002 period. The 2003 reform continues the gradual shift in EU policies away from market price support, which is the most distorting and least effective means of providing income support to farmers, to more decoupled and less distorting payments.

In addition to the SFP, the reforms include reductions in the intervention price of rice by 50%, while the butter intervention price is 10% lower, and the milk quota is retained until 2014. Additionally, through modulation, direct payments that exceed EUR 5 000 per year for any farm will be reduced. The reductions are to be phased-in over a three-year period starting in 2005 when the cutback will be 3%, rising to 5% in 2007. More details of the recent EU reforms are provided in the section "Medium Term Market Impacts of the 2003 EU Common Agricultural Policy Reform". This section summarises the main findings of the market impacts for the commodities covered in this *Outlook* of a special analysis made by the Secretariat. It concludes that, although the impacts on the international markets depend on how the reforms will actually be implemented and on the degree of decoupling, they are nonetheless small. The largest impacts on world markets were found for butter and whole milk powder, markets in which the EU is a major actor. Domestically, the reforms tend to result in more extensive production, particularly of livestock products.

Although the level of support changes only slightly relative to previous policy measures, the composition shifts toward less coupled and hence less production distorting measures.

A summary of the main market trends and developments

Population and income growth result in broad-based consumption gains

Broad-based income growth and moderate increases in population should result in consumption gains world-wide. Increasing urbanisation and changing eating patterns lead to diet diversification, which generates increased demand for high value-added products such as dairy products and meats. The projections suggest consumption increases for all products and regions in the *Outlook*. Between 2003 and 2013, of the products that are for direct human consumption, the product with the highest growth rate is vegetable oil with an expected average compound annual growth of 2.9%. Butter, cheese, whole milk powder and poultry consumption are also expected to increase at rates above 2% per annum (see Table 1). Consumption of all products, except rice and skim milk powder, is expected to grow at rates greater than the growth in population, including in low income countries, providing the potential to reduce malnutrition and hunger.

Table 1. **Consumption and production growth rates, 2003-2013**

	CONSUMPTION			PRODUCTION		
	Total	OECD	NON-OECD	Total	OECD	NON-OECD
	%			%		
Wheat	1.2	0.8	1.4	1.8	1.5	2.0
Rice	0.8	0.8	0.8	1.3	1.1	1.3
Coarse grains	1.3	0.8	1.8	1.6	1.4	1.8
Coarse grains used for feed	1.5	1.0	2.1	n.a.	n.a.	n.a.
Oilseeds	n.a.	n.a.	n.a.	2.7	2.5	2.8
Oilseed meal	2.6	1.6	3.8	2.6	2.2	2.9
Beef	1.5	0.4	3.0	1.6	0.6	2.8
Pig meat	1.5	0.8	2.0	1.5	0.8	2.0
Poultry meat	2.0	1.7	2.5	1.9	1.7	2.1
Butter	2.3	0.4	3.3	2.2	0.0	3.8
Cheese	2.0	1.7	2.8	2.0	1.6	3.4
Skim milk powder	1.0	0.0	2.3	0.7	−0.7	5.6
Whole milk powder	2.6	1.7	2.8	2.6	1.9	3.4
Vegetable oils	2.9	1.7	3.8	3.0	2.0	2.9
Sugar	1.9	0.5	2.2	1.7	0.5	2.1

Source: OECD Secretariat.

The diversification of diets is also illustrated in the growth rate of oil meals and coarse grains, products used in the production of livestock and milk. Oil meal consumption is expected to grow at an annual compound growth rate of 2.6% while coarse grains used in animal feeding rather than food is expected to grow at 1.5% per annum.

Especially in the non-OECD region

Although consumption gains are broad based, the bulk of these occur in the Non-member economies. This is the area where most of the people live, where most of the population gains will occur, and where income growth is expected to remain strong. As Table 1 shows, consumption in the non-OECD area is expected to post solid gains and to grow at rates much faster than those of the OECD. Consumption of dairy products, butter, cheese and whole milk powder, as well as livestock products such as beef, pig meat and poultry meat are expected to grow considerably faster, not only than in the OECD but faster also than population in these regions. And, products that are consumed indirectly, such as coarse grains and especially oil seed meals used to feed animals, are projected to show robust gains.

In the OECD area where incomes are high and basic dietary needs have long been more than satisfied, consumers are increasingly looking for variety in their diets and for new taste experiences and convenience, while an increasing share of the meals consumed are prepared outside the home. In this area, consumption gains are less spectacular and for skim milk powder, consumption at the end of the period is about equal to 2003 levels as demand for this product does not expand. Consumption of beef and butter, although growing, is expected to post moderate growth rates as preferences shift towards products such as poultry meat, cheese and whole milk powder.

The higher consumption growth rates in the non-OECD area during the projection period imply that an increasing share of agricultural produce is consumed outside the OECD area. For example, the OECD's share of global skim milk powder consumption falls from 60% in 2003 to 55% at the end of the projection period. Similarly, slow growth rates for beef, butter and pig meat result in a falling share of consumption of these products in the OECD region. Beef consumption in the OECD falls from 61% in 2003 to 54% in 2013; that for butter falls to 31% from 37%; while that for pig meat falls to 39% from 42% in 2003. The consumption share in the non-OECD area increases even for products used in feeding animals, indicating increasing activity in animal production in the Non-member economies. The OECD consumption share of global coarse grains use for animal feed falls from 55% in 2003 to 52% while that for oilseed meals falls from 58% to 52% by the end of the period.

While production expands even faster

With normal weather conditions and continued productivity gains the growth rate in the production of most of the agricultural products considered in the *Outlook* is expected to be larger than that for consumption, leading to a continuation of the long-term decline in real prices (see Table 1).

Production growth in NMEs outpaces that in OECD countries

Out of the 14 commodities covered in Table 1, production growth is faster than that for consumption in the NME region for six of them: wheat, rice, butter, cheese, skim milk powder and whole milk powder. On a global basis, only for butter, poultry meat, sugar and skimmed milk powder is consumption growing marginally faster than production. For all products covered in the *Outlook*, the expansion of production in countries outside the OECD-area outpaces that in OECD member countries and by a large margin for dairy products and sugar though to a lesser extent for meat. As a result, the OECD share in global output for these products declines considerably. For instance, the OECD's share in world production falls by about 12 percentage points for skimmed milk powder, by 9 percentage points for butter and by 3 to 5 points for pig meat, beef, cheese and whole milk powder. The changes for other products are smaller, and for cereals they remain nearly unchanged (Table 2).

Table 2. **Consumption and production of OECD countries as a share of world total**

	CONSUMPTION			PRODUCTION		
	2003	2008	2013	2003	2008	2013
	%			%		
Wheat	33.0	32.0	31.0	44.0	43.0	42.0
Rice	4.0	4.0	4.0	4.0	4.0	4.0
Coarse grains	51.0	50.0	48.0	54.0	54.0	53.0
Coarse grains used for feed	55.0	54.0	52.0	n.a.	n.a.	n.a.
Oilseeds	n.a.	n.a.	n.a.	38.0	39.3	37.2
Oilseed meal	58.0	55.0	52.0	41.7	41.9	40.1
Beef	61.0	57.0	54.0	59.0	56.0	54.0
Pig meat	42.0	41.0	39.0	43.0	42.0	40.0
Poultry meat	64.0	64.0	63.0	64.0	64.0	64.0
Butter	37.0	34.0	31.0	46.0	41.0	37.0
Cheese	77.0	77.0	75.0	79.0	78.0	76.0
Skim milk powder	60.0	58.0	55.0	83.0	77.0	71.0
Whole milk powder	19.0	18.0	18.0	54.0	52.0	50.0
Vegetable oils	36.0	33.0	32.0	29.0	29.0	26.0
Sugar	27.8	26.0	24.5	28.8	27.3	25.8

Source: OECD Secretariat.

World cereal production is projected to expand 1.6% per year, with wheat and coarse grains production increasing at an annual compounded rate of 1.8% and 1.6% respectively while rice production lags behind, growing 1.3% per annum. Cereal production growth rates in the OECD area are expected to slightly trail those in the Non-member economies. Most of the gains in production are through expected productivity improvements as area harvested expands at a much lower rate (about 0.7% per annum for wheat area and less than 0.4% for coarse grains and rice area). Most of the additional growth in area harvested occurs outside the OECD-area.

Oilseed production is expected to post strong growth rates, averaging 2.7% annually during the 2003 to 2013 period. This is more than one percentage point higher than the rate of growth for cereal production. Production growth is a result of both productivity gains but also increases in area, especially outside the OECD region. Area devoted to oilseeds is expected to expand by 1.5% per year in the non-OECD region compared to about 0.4% in the OECD area. Output of derived products from oilseeds, oilseed meal and vegetable oils, is growing roughly in line with that of oilseeds, again with a relatively stronger expansion in the Non-member economies.

In particular for dairy products and certain types of meat

World milk production is expected to increase by 121 million tonnes with large gains in Argentina and especially China. OECD member countries, where 50% of production is constrained by quotas, only contribute 25 million tonnes of the total gain. Most of the gains in the OECD region occur in New Zealand and Australia. The wide margin, by which growth in non-OECD raw milk production exceeds that in the OECD region, as well as productivity gains through increased investment in processing capacities in NMEs, is reflected in much stronger growth in the output of dairy products outside the OECD as well. The starkest difference is for butter and skimmed milk powder, where OECD production shows a small negative growth, which compares to 3.8% and 5.6% annually, on average, in the Non-member economies as a whole. The differences in growth rates for meats are less substantial, but those for beef and pig meat in particular are expected to be substantially higher in Non-member economies. Globally, relatively low inflation and stable feed prices, with lower oilseed meal prices offsetting higher coarse grain prices, enable meat production to expand in an expected weak product price environment.

Moderate trade expansion for cereals and oilseed meals

Global trade* for wheat and coarse grains is expected to grow moderately over the projection period with more substantial increases in rice trade. Gross wheat exports from the OECD area are expected to grow 3.7% per year, compared to 3.3% for coarse grains. The bulk of the increased trade is from the OECD area to the non-OECD zone, mostly China. OECD exports of wheat to Non-member economies is expected to grow at an annual rate of 4.4% while that for coarse grains, mostly to be used as feed, expands at an annual rate of 10.1%.

Trade in oilseed meal among the countries/regions in the *Outlook* is expected to grow at an annul rate of 3.4%. Again, most of this growth occurs in the Non-member economies, with Argentina and Brazil remaining the leading exporters. Brazil is expected to export more than 6 million tonnes more at the end of the period compared to the beginning while Argentina exports about 4.7 million tonnes more. The United States is expected to resume exporting substantial volumes during the projection period, gaining market share, primarily at the expense of Argentina. Major importers are the enlarged EU, where imports increase by almost 5 million tonnes over the period, and China.

NMEs dominate sugar trade...

Trade in sugar is also expected to expand over the projection period. Brazil, among the world's lowest cost producers and the leading sugar exporting nation, is expected to further expand exports by 50%. Thailand is also projected to expand exports by 21% and Australia by 15% with assumed productivity improvements. Mexico is expected to become a consistent exporter, mainly to the US market. The import side of the world sugar market is much less geographically concentrated. Russia is expected to remain the world's largest importer, but rising domestic production replaces some imports.

... but OECD countries remain big in meat trade

Beef exports (live animals and meat) by the countries included in the *Outlook* is expected to increase by more than 2 million tonnes over the projection period, at an annual growth rate of almost 3.1%. Australia is expected to maintain its role as the world's leading beef exporter. Beef trade between the United States, Canada and Mexico is expected to resume from early 2004 following import bans due to BSE incidents. United States and Canadian beef exports to other markets are assumed to resume from early 2005 and total exports should recover to around their pre BSE levels. Pig meat trade is growing over the projection period with exports from OECD countries in 2013 some 673 000 tonnes greater than in 2003. Most of this

---

* The trade numbers reported here represent exports, imports or net trade (exports minus imports) of countries or regions explicitly considered in the *Outlook* and may not necessarily represent total exports or total imports reported elsewhere.

Figure 1. **Outlook for world crop prices to 2013**

(Index of nominal prices, 1993 = 1)

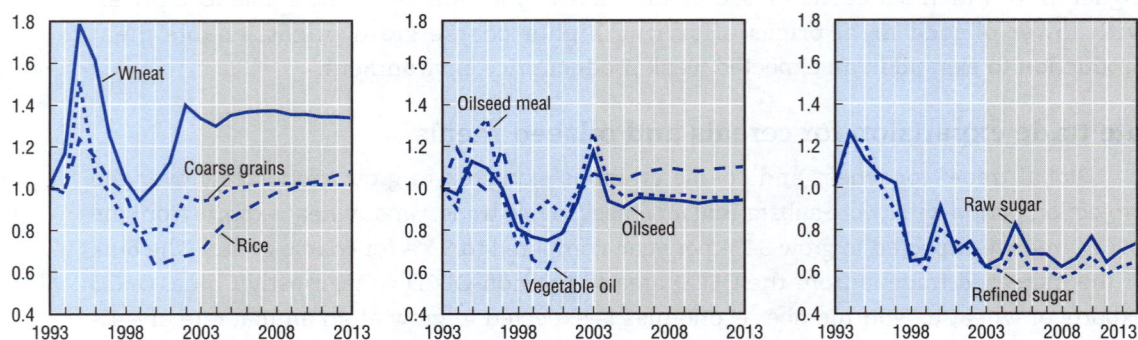

Source: OECD Secretariat.

trade is within the OECD area however, as exports to the Non-member economies are little changed. The EU continues to be the largest exporter of pig meat in the OECD area, with Canada a close second by the end of the *Outlook* period.

The United States continues to export significant volumes of poultry meat, reaching nearly 3 million tonnes by 2013, even as domestic consumption continues to expand, reaching 50 kg per person annually, the highest level in the OECD area. Exports of another major trading country, Brazil, however, are expected to fall as more of the domestic production is destined for the home market. The major importers of poultry meat are China, and especially Russia which remains the largest net importer.

Dairy markets remain thin and with little growth in trade

World trade in dairy products continues to represent a small share of world milk production and remains relatively regional, though imports of some dairy products represent a relatively large share of consumption. For instance, the share of imports in world consumption hovers around 30% for skimmed milk powder and 35% for whole milk powder. This compares with much lower shares of 7% for cheese and 8% for butter. Cheese imports are expected to grow at a compounded rate of 2.1% per year during the Outlook while that of whole milk powder expands at the rate of 1.3% per annum. Imports of skim milk powder are expected to grow at only 0.4% per year, while butter imports actually decline by about 0.4% per year over the Outlook period.

OECD countries continue to dominate world trade in dairy products. Dairy product exports by OECD countries represent more than 90% of the world's imports of cheese, butter and whole milk powder. However, most of the dairy product trade occurs within the OECD region. Net exports from the OECD to the non-OECD area are expected to decline during the projection period for all products other than whole milk powder, reflecting strong production growth in the latter group of countries.

Agricultural commodity prices continue their long term decline in real terms

In nominal terms, prices for almost all agricultural products covered in this *Outlook* are expected to strengthen over the projection period. The evolution of nominal prices of selected commodities is shown in Figures 1 and 2. The price development in real prices, i.e once nominal movements have been corrected for inflation, is different. In some cases,

Figure 2. **Outlook for world livestock product prices to 2013**

(Index of nominal prices, 1993 = 1)

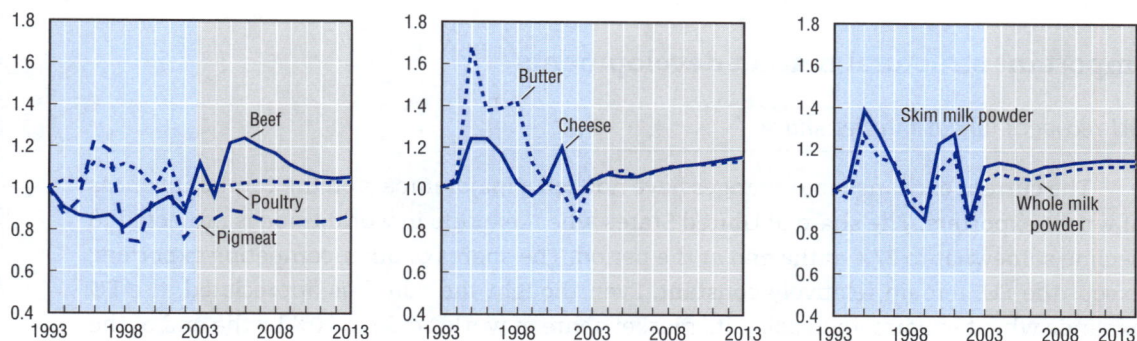

Source: OECD Secretariat.

notably for cereals, beef and lamb, real prices over the *Outlook* period are projected to be higher than the very depressed levels for these commodities in the most recent years. For all products, however, even for those mentioned earlier, real prices continue their declining trend when considered over the longer term. The various commodity chapters show nominal and real price developments graphically.

Stronger prices for cereals and vegetable oils

Even as cereal supply expands faster than demand, prices in nominal terms are expected to increase somewhat for wheat and coarse grains and especially rice. This is partly as a result of lowering global stock levels. Stock-to-use ratios for cereals are projected to fall to levels not seen in recent years. A significant contributing factor to the expected decline in global cereal stocks is the expected reduction in China and in other Non-member economies.

In contrast to the price evolution in cereals, oilseed prices, after an initial sharp drop, remain flat in nominal terms throughout the projections period, as production expands to meet increased demand. This stability in oilseed prices reflects declining oilseed meal prices over the projection period, and slightly increasing prices for vegetable oils.

Little scope for higher sugar prices

Fundamentals in the sugar market remain bearish. Sugar consumption world-wide is expected to increase at an average annual rate of 1.9%. Production is expected to exceed consumption in most years putting a break on the potential for prices to rise. The global stock-to-use ratio declines somewhat, but not enough to materially affect prices. Nominal world prices are expected to remain in a USD 7-9 cents/lb band over the projection period. As such, the long run pattern of falling prices in real terms is set to continue.

Prices for livestock and dairy products to increase

World dairy prices especially cheese are expected to increase in nominal terms during the *Outlook*, boosted by relatively large gains in consumption, reasonably strong growth in imports for cheese and whole milk powder, and the weaker US dollar. Following the initial BSE related drop, world beef prices resume their cyclical pattern over the *Outlook* period, peaking in 2006. Pig meat prices continue to show pronounced cycles as well, but remain, on average, below historical levels. Poultry prices are relatively flat over the *Outlook* period, but those of lamb remain much above historic levels (in nominal terms) due to lower export supplies from New Zealand and Australia.

Some important structural market developments

EU recovers wheat market share

The current exchange rate changes have not had a major impact on the competitiveness of wheat exporters. The share of United States wheat exports in world trade is projected to decline somewhat to 27% at the end of the period. The shares of other competing exporters, except the EU, remain relatively constant. Even thought the euro has appreciated, the EU expands wheat exports, increasing its market share of world trade to 20% by the end of the projection period from the drought induced low of 9% in 2003. This expansion in wheat exports is driven by lower domestic wheat prices which enable the EU to export within the WTO's subsidized export constraint, even with a stronger euro.

The pattern is somewhat different in the case of coarse grains. In this instance, the United States is the dominant exporter and increases its export share of world trade from 59% in 2003 to 62% by the end of the projection period. The other competing exporters manage to maintain their export shares, except the EU whose market share increases to 10% from the drought induced low of 3% in 2003. As in the wheat case, lower domestic prices enable the EU to export within its WTO's export limit constraint.

China becomes a large cereal importer

It was mentioned that a large share of the additional trade is due to China and her projected expansion in import demand for cereals. A continuing question in recent years has been China's role in world cereal markets and whether it will continue being a net exporter of grains or – as many expect – switch to being a net importer. In the recent past, many analysts expected China to become a net importer of grains, due to rising income and expectations of a dietary shift to more dairy and livestock products, thus demanding more cereals to feed an expanding livestock sector. But, in recent years, each time there were expectations of large imports, domestic supply increased through releases of stocks. The actual volume of Chinese stocks are not well known, raising a question as to how long it can continue to release them to meet domestic needs rather than import. This uncertainty is addressed in a special section entitled "Chinese Grain Import Surge – Market Implications of Alternative Stock Estimates and Trade Policies" found in the Cereals chapter. As discussed in this section, if Chinese stock levels are higher than those assumed in the *Outlook*, Chinese imports of cereals would decline as would domestic and world prices.

Another uncertainty explored in this section is the administration of China's tariff rate quotas (TRQs) for cereals. The *Outlook* assumes seamless administration of these TRQs. But, if this is not the case and imports fall short of those projected in the Outlook due to TRQ administration hindrance, then world cereal prices would be lower than those projected, while domestic prices in China would be higher.

Brazil becomes the largest oilseed exporter...

World import demand for oilseeds is expected to grow by 2.7% per annum; mostly coming from Non-member economies as OECD import demand is relatively flat, expanding by only 0.7% per annum. Most of this increased import demand is expected to be met by Non-member economies, primarily Argentina and Brazil. The United States is projected to lose market share and its past position as the world's largest oilseeds exporter. The United States share of world trade at the end of the period, at 31%, is substantially lower than that at the beginning of the period, having lost share to Argentina and Brazil. Argentina's share expands to 21% of the total at the end of the period while Brazil has become the world's largest exporter, representing 34% of global trade in oilseeds by 2013.

... but there is uncertainty regarding its role in world oilseed markets

An uncertainty in the oilseed markets is Brazil's ability to expand area sown to oilseeds, primarily soyabeans. The *Outlook* assumes that the oilseed area will continue to increase. But, what would happen if Brazil is not able to maintain this pace, or contrary if area expands at an even faster rate? This question is addressed in the section "The Impact of Further Area Expansion for Oilseeds in Brazil". It found that an effective 12% reduction in the oilseeds area in Brazil results in Brazilian exports falling dramatically so that in 2013 they are half the projected level in the *Outlook*. Falling exports result in higher world prices, on average 5% greater than prices in the *Outlook*. Given the links between oilseeds and oilseed meal and vegetable oil production, the price for these products changes as well. On average, oilseed meal prices would be nearly 5% higher than the *Outlook* in this scenario while vegetable oil prices would average about 1% more. And, if in contrast, Brazil is able to expand oilseed area at an even faster rate than that assumed in the *Outlook*, the opposite effect would occur. In this situation, world prices would average 5% lower than those reported in the *Outlook*. Price declines of this magnitude are sufficient to trigger the US marketing loan benefits for soyabean producers in most years of the *Outlook*, sheltering them from price declines and shifting the burden of adjustment to producers in other countries.

South America also big in meat trade

Argentina and especially Brazil are two of the world's major beef suppliers. Brazil's beef exports have risen dramatically since the late 1990s and further growth is expected during the *Outlook* period. However, as a greater share of domestic production is expected to be consumed locally, export growth should slow, resulting in a declining share of the world market. These countries, however, do not compete significantly in the lucrative Pacific beef market because of foot and mouth considerations which continues to segregate world beef trade.

And the EU becomes again, a net importer of beef

Apart from Argentina and Brazil, the European Union (as an importer and exporter) and Russia (as a large importer) are also major players in beef markets outside the Pacific region. The *Outlook* projects that the EU will become a net beef importer during the projection period, a status not seen since the late 1970s, partly due to a relatively strong euro and partly as a result of changes in domestic policies. The section "Analysis of Beef Imports by the European Union", found in the chapter on meat, traces the market and policy developments in the EU over the last thirty years illustrating the interaction between policies and markets that have contributed to the EU's net trade position.

More agricultural activity shifting to Non-member economies

The *Outlook* suggests that Non-member economies are playing an ever larger role in shaping world agricultural markets. For example, China is important in the cereal markets while Argentina and Brazil are major players in the oilseed and oilseed meal markets. Brazil also has a large role in sugar and beef markets. India is another large Non-member economy that could become increasingly important in world agricultural markets as domestic markets become more open to the world. Previous *Outlook* reports have reviewed in more detail agricultural markets in specific Non-member economies. This year's special focus section examines the agricultural market and policy situation and Outlook in India.

Uncertainty about India's future role

India is a very rural country with heavy dependence on the agricultural sector. About 59% of India's population depends on agriculture which employs 57% of the workforce. India's agricultural structure is dominated by small holdings as 80% of the farms have less than 2 hectares. But, as is the case for most other countries across the world, as they develop, agriculture contributes a smaller share to national income. Although much greater than the share of agriculture in the income of OECD countries, India's agricultural share of gross domestic product has declined from around 45% in the early 1970s to 27% at the turn of the 21st century.

Although dominated by small holdings, its massive area makes India a large agricultural producer. In the year 2000, India planted more area to rice, wheat, cotton and pulses than any other country in the world, while area planted to oilseeds and sugarcane was the second largest, and that planted to coarse grains the third largest in the world. The resulting production was also very large. India in 2000 was the second largest producer of rice, wheat and cotton, the fourth largest coarse grains producer and the fifth largest oilseeds producing country.

Despite being a major producer and consumer, inward-looking trade policies have meant that India has played a relatively minor role in world agricultural trade for many products. But this is changing as India is increasingly becoming more integrated into world markets. Some experts in India expect agricultural exports to increase at about a 9% per annum while imports are expected to grow at double this rate, averaging 18% per annum to 2007. Some estimates also indicate that in the absence of significant technological changes in the agricultural sector, growing income and population would lead to an import demand of 115 to 142 million tonnes of food grains.

India has entered a path of policy reform in the 1990s and has taken a more active role in the international arena through active participation in the G-20 group. As its agricultural sector becomes more outward-looking, structural issues, such as limits on farm size, infrastructure inadequacies, access to inputs, and the adequacy of investments in less favoured areas may emerge. Another policy issue that might need to be considered is a well targeted safety net to ensure that the rural poor are not left worse off in the transition from high support and protection to a more open and market oriented agricultural environment.

A major policy issue

Policies influence prices

A repeated theme in the *Outlook* is that both domestic and trade policies are important factors, as they influence markets directly by controlling the availability of produce through supply management tools such as production quotas or border measures such as tariff rate quotas, or indirectly through their influence on prices and the transmission of these prices to market agents. Government policies matter in influencing the degree of integration and variability of domestic and world prices

Some policies reduce downside price risks for producers

The section "Sensitivity Analysis: Impact of Stochastic Macroeconomic Variables on the Outlook for World Commodity Prices", at the end of the Economic and Policy Assumptions chapter, illustrates how one provision of the US's FSRI Act, the marketing loan benefits, and the EU's intervention price mechanism, both prevent market signals from reaching producers when prices are low. For example, the soyabean producer price in the US under random variations of income and exchange rates can fall as low as USD 167/t or rise as high as USD 239/t when marketing loan benefits are excluded. However, the marketing loan benefits shift the lower prices higher, and in this case the incentive price is truncated at about USD 196/t. In almost half of the observations, the US supply of soyabeans does not respond to the world price. US producers will continue to receive about the same total returns per unit regardless of how far prices fall.

Similarly, the EU's intervention price reduces the downward price risk for its wheat and coarse grain producers. Japan's deficiency payments on rice, beef and oilseeds also serve similar functions as do barriers to imports, particularly tariff rate quotas and special safeguard measures that increase tariffs, stifle the incentive to import and maintain higher domestic prices thus also reducing downward price risks. All of these types of policies that automatically restrict market signals may have destabilising effects on world markets as domestic farmers and consumers respond only to incentives set by policy rather than prices.

The URAA attempted for the first time to discipline trade and domestic policies and thus reduce their distorting effects. Almost ten years later significant price gaps between domestic and world prices persist and in some cases these gaps are projected to increase. But it is not only policies that create these price gaps. For homogeneous products, transportation costs and transaction costs can also help explain such gaps. However, in the projection period, transportation costs have similar effects on all importing countries and should not diverge. If the pattern of price gaps differs for the same commodity across different countries, than one can assume that changes in transport costs are not the primary reason for the divergence, and that other factors, such as exchange rates, tariffs and tariff rate quotas, are large contributors.

Resulting in domestic prices much above world levels...

Figure 3 to 6 for selected commodities and selected countries (those with border and domestic policy measures in place) illustrate the evolution of the price gaps since 1997, the third year of the URAA implementation period. The figures represent the tariff equivalent of all the measures that result in domestic prices to be higher than world prices in the historic and projection period. The prices used are those reported in the detailed commodity tables.

In order to reduce the influence of changes in exchange rates, the calculations are based on a three year average exchange rate for each country based on 1999 to 2001 annual exchange rates. The reader is reminded that under the URAA, border measures were scheduled for reduction between 1995 and 2000 for developed countries. Assuming everything else constant, one would expect the price gaps to narrow as a result of these reductions.

... for butter

Figure 3 shows the percentage difference between the world butter price and domestic prices in Japan, Canada, the EU and the US, commonly referred to as the Quad countries. This figure shows that in 1997, the butter price in Japan was a little less than 4 times the world price. Subsequently, the gap expanded substantially, and even though it is expected to decline somewhat during the projection period it is still more than four times the world price. Similarly, the gap is expected to increase in Canada to about twice the world price but to decline somewhat in the case of the EU and the United States to a level of about 75% greater that the world price at the end of the period.

Figure 3. **Price gaps – butter**

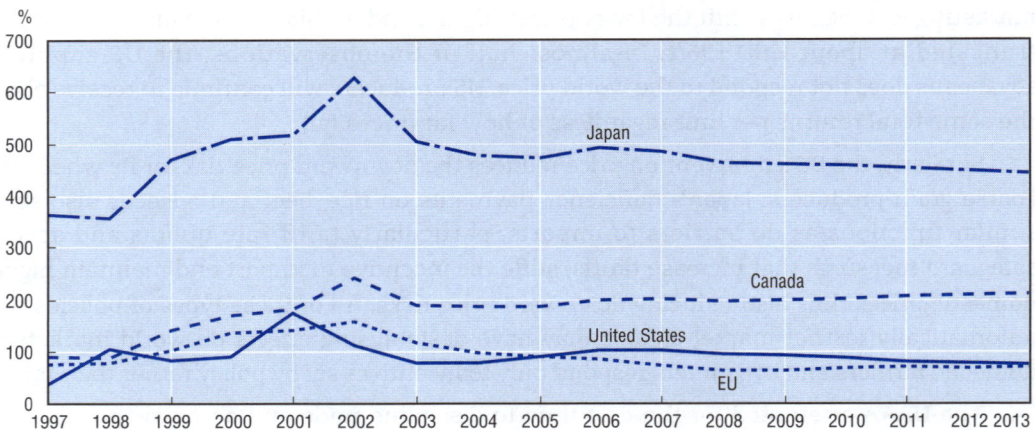

Source: OECD Secretariat.

... for skimmed milk powder

Figure 4 shows the evolution of the price gap for skim milk powder in the Quad countries. Relative to butter, the gap is not as large, suggesting skim milk producers are protected relatively less than butter producers or alternatively, skim milk consumers are taxed relatively less. In the case of skim milk powder, as in the butter case, the price gap in Canada is expected to increase over time. A larger price difference at the beginning of the period prevailed in Japan, but the difference is projected to slowly narrow during the projection period. For the US and the EU, the gap is expected to remain relatively flat and at a substantially lower level.

Figure 4. **Price gaps – Skimmed milk powder**

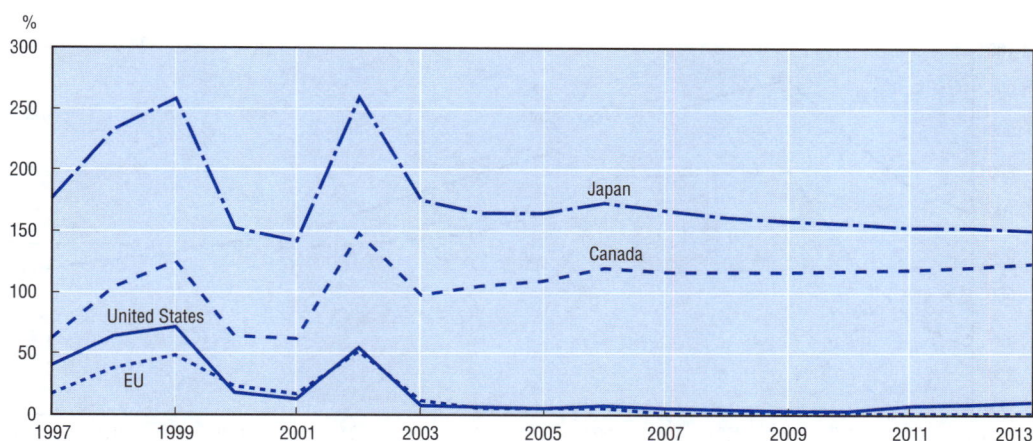

Source: OECD Secretariat.

... and for sugar and rice

In sugar, the case is illustrated for Japan, the EU and the US in Figure 5. The domestic price gap in each of these countries is larger at the end of the projection period compared to the 1997 level. Although fluctuating somewhat, for Japan the gap is more than 650% in 2013. The gap for the US and the EU also fluctuates over time and although lower than that for Japan, it is still more than 150% for each country. Finally, the price gap in the Japanese and Korean rice markets is shown in Figure 6. While the gap is expected to narrow over time, it will remain at more than 6 times the world price in Korea's case and somewhat lower for Japan.

Figure 5. **Price gaps – Sugar**

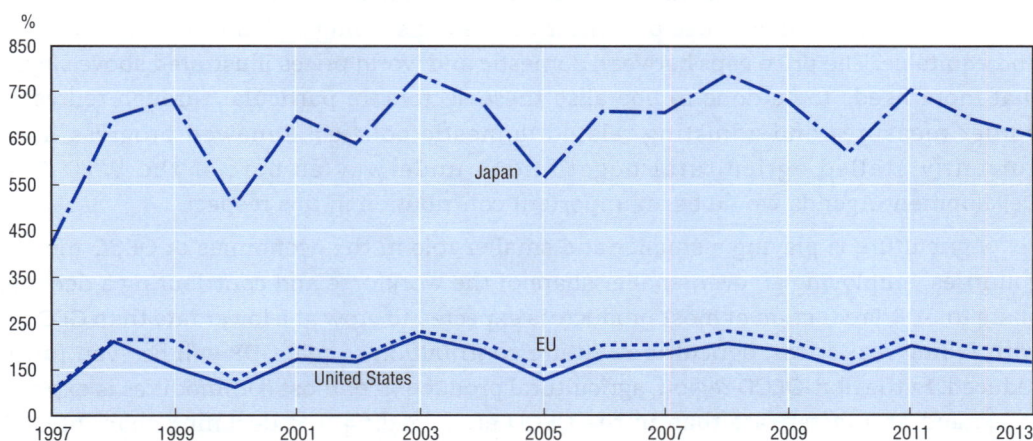

Source: OECD Secretariat.

Figure 6. **Price gaps – Rice**

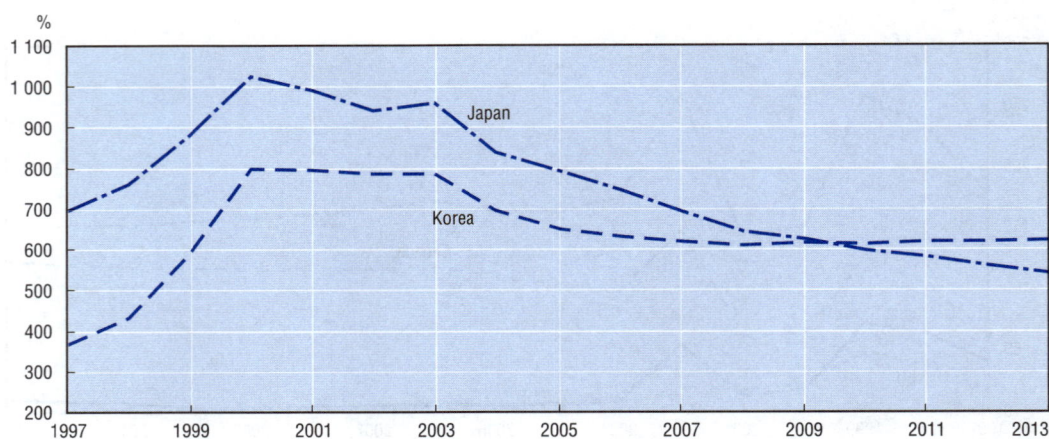

Source: OECD Secretariat.

The price gaps during the projection period are a result of changing prices as policy is assumed constant. A reason that price gaps may increase in the historical period even as tariffs are reduced, is the changing world price and the prevalence of specific tariffs by these countries for these commodities. As world prices change, the *ad valorem* equivalent of specific tariffs also changes.

The case for reform

These figures illustrate the potentially high costs some policies impose on domestic consumers and the lack of responsiveness of domestic markets to changing scarcities on international markets. OECD analysis indicates that market price support, maintained behind border protection and export subsidies, is the most production and trade distorting policy instrument in place today. It is also among the most inefficient in transferring income to producers. OECD analysis also shows that decoupled support, targeted to explicit objectives and intended beneficiaries is less distorting and more effective, efficient and equitable. The price gaps between domestic and world prices illustrated above suggest that more needs to be done to liberalise these sectors, in particular through redressing border measures and adjusting related domestic policies. Renewed progress in the currently stalled agricultural negotiations underway as part of the WTO Doha Development Agenda would be an important contribution in this respect.

Agriculture is playing a smaller and smaller role in the economies of OECD member countries, employing an ever smaller share of the workforce and contributing a declining share to GDP. Production of most products is expected to grow at a lower rate than GDP, and with falling real prices, agriculture's future contribution to real GDP will be even further reduced. In the non-OECD region, agricultural production of most commodities is expected to expand at a faster rate than in the OECD area, leading to a declining share of OECD member countries in world production and trade. However, higher population growth rates and high and broad based income growth in the non-OECD region could provide opportunities for expanded trade by OECD member countries, especially in higher value-added products such as dairy and meats. But, domestic and trade policies should not hinder market forces so that producers and consumers can take advantage of trading opportunities as they arise.

ISBN 92-64-02008-X
OECD Agricultural Outlook: 2004-2013
© OECD 2004

Chapter 1

Economic and Policy Assumptions

Key economic assumptions

- *Restrained optimism prevails for a recovery in the world economy as growth momentum takes hold in North America, Asia and the United Kingdom. Continental Europe has begun to rebound, but has shown a more hesitant recovery. Major Non-member economies such as Brazil and China continue to enjoy substantial economic activity and growth, as does Russia in the near term.*

- *Short term interest rates are at historical lows across the OECD area and there is a concerted resolve to maintain targeted rates low for the near future. Although fears of deflation have subsided for all countries other than Japan, it is expected that low core inflation conditions will endure for most of the OECD area.*

- *The US dollar has depreciated against most other OECD currencies in 2003, in particular, the euro, the Japanese Yen and the Australian, Canadian and New Zealand dollars. While this adjustment should increase United States export competitiveness, it should reduce that of other competing exporters, and in particular limit export related growth in the Euro area. In Japan, the stronger Yen and increased economic activity have improved the affordability of US dollar denominated imports.*

Key policy assumptions

- *The 2004 Agricultural Outlook assumes a continuation of existing agricultural and trade policies or announced policies within well defined programmes. Of particular importance in this Outlook are the continuation of the 2002 FSRI Act in the United States and the implementation of the 2003 EU CAP reform.*

- *For the first time the Agricultural Outlook includes the expanded European Union of twenty five countries. The EU's Common Agricultural Policy (CAP) will be phased in during the implementation period in the new EU countries. The expected impacts of the 2003 EU CAP reform have been taken into consideration.*

- *In light of the lack of progress on multilateral trade negotiations to date, trade policies are assumed to be governed by the Uruguay Round Agreement on Agriculture (URAA). No new bilateral, regional or multilateral trade agreements have been accounted for in the projections. All existing trade commitments are assumed to be met.*

Macroeconomic assumptions

Strong economic activity in North America and Asia but weak economic recovery in Europe

The persistent pessimism and hesitant upswing of early 2003 has given way to solid, but unbalanced OECD-wide growth towards the end of the year. The momentum is clearly established; after languishing below potential for the last three years, real GDP growth has risen above 3% (annualised rate) for the second half of 2003. It is anticipated to remain at 3.1% for 2004-05. International factors, such as relatively steady oil prices and strengthened global confidence levels, have contributed to spur a rebound of economic activity in North America, Asia and the United Kingdom. However, continental Europe is more hesitant in moving along the path to recovery, with both domestic concerns such as labour market conditions in individual countries and the depressing effect on exports of an appreciating euro, retarding growth. Other OECD exporters whose currencies have appreciated recently against the US dollar are expected to suffer a decline in their export price competitiveness with the United States, reducing export related growth in the near future.

Weaker US dollar relative to most other OECD currencies

The US dollar's depreciation *vis-à-vis* other OECD currencies through 2003 has modified trade prospects, in particular reducing export growth expectations in competing agricultural commodity exporters like Australia, New Zealand, Canada, and the euro area. Japan, a leading OECD importer of US dollar denominated agricultural products, finds itself in an enviable situation with accelerating growth and a relatively stronger yen. Imports should increase, at least in the short term, because of both income and price effects. The dollar's slide also reflects domestic and world-wide concerns about the sustainability of fiscal and monetary policy measures in the United States. Within the OECD, the underlying economic environment is one of historically low short term interest rates, in nominal terms, and correspondingly low long term rates that have in many countries bottomed out in mid-2003. Strong consumption expenditure has been a persistent driving factor behind the upturn, in particular, expenditure on durable goods and residential investment. Given this healthy demand environment, it is expected that prospective income gains shall flow through to increased consumption of agricultural commodities. Typically, this increased demand concerns products that are highly responsive to income: increased demand for livestock products in general; a shift in consumption from less to more expensive meats; and a similar shift in consumption from bulk dairy commodities to more value-added products, such as cheeses. Indirect demand for cereals and oilseeds for use in animal rations is also expected to strengthen as consumption of livestock products expands.

Nevertheless, it is hoped that economic growth and stable OECD-wide economic conditions such as contained inflation and low interest rates will translate into more investment expenditure. In particular, capital intensive production could benefit from the momentum transferred to investment. However, increased investment in this environment has wide effects throughout agriculture. For example, farmers are more easily able to adjust production in response to relative and absolute price changes in crops and livestock. But also, investment in capital intensive livestock production in turn flows through to higher feed crop – cereals and oilseeds – production because of increased feed demand.

Sustainability of expansionary policies is a concern

The **United States**, as leader out of the recession, has shown a firm recovery in 2003. Latest evidence confirms renewed growth and instils confidence for a maintained momentum through 2004-05. There has been an aggressive reduction in short term interest rates accompanied by actively increased public spending and the implementation of significant tax cuts. This monetary easing and fiscal stimulus is expected to continue for some time; the current expansionary course has been held firm since 2001, resulting in a generally buoyant economy and manifest in high consumer and business confidence levels. Nevertheless, two predominant concerns resound. First, nominal short term interest rates are close to the zero lower bound, with the Federal Reserve having cut the targeted federal funds rate to an historic 45 year low of 1%. The second concern is the aforementioned sustainability of fiscal and current account deficits that have put downward pressure on the dollar. At 5% of GDP, the current account deficit is among the highest on record. These two unprecedented situations constitute a perhaps overly pessimistic downside risk for the otherwise strong recovery. Moreover, both have contributed to expectations that the US dollar should depreciate further against most OECD currencies in 2004. The projection of renewed and robust business investment, maintained consumption levels, and strong exports, mean that real GDP growth in the United States will exceed 4% in 2004 and then slow slightly to remain at a sustainable, non-inflationary rate of roughly 3% from 2006 until the end of the *Outlook* period (Figure 1.1).

Robust recovery expected for Canada and Mexico

Prospects remain bright for **Canada**, despite the series of adverse shocks experienced throughout 2003. The strengthening economy was hit in the spring with an epidemic of Severe Acute Respiratory Syndrome (SARS) which affected tourism. Moreover, the sharp appreciation (of 19% over the 12 months to November) of the Canadian dollar against its US counterpart has more generally weakened export growth. The discovery in June of a case of Bovine Spongiform Encephalopathy (BSE) has further weakened large livestock exports through the imposition of import bans by Canada's trading partners. Reactive monetary policy has helped limit the downside effects, and persistently strong consumer demand has provided some degree of resilience. Once the effects of these adverse shocks have been worked through, it is expected that a robust recovery will resume, with GDP growth to once again exceed 3% from 2005 onwards. A rebound in export growth is anticipated from 2004, with the negative influence of a stronger currency being offset by more vigorous US demand.

In a similar spirit, the sluggishness of **Mexico's** recovery in 2003 is expected to be short-lived. Economic activity is expected to increase in 2004, as the hesitant US export demand eventually firms up, pulling up Mexico's GDP growth because of the direct links between the two economies under NAFTA, and providing some latitude for monetary and fiscal policy measures that have recently remained cautiously restrained. The Mexican currency is one of the few OECD currencies that have depreciated against the US dollar,

Figure 1.1. **Real GDP growth for selected countries**

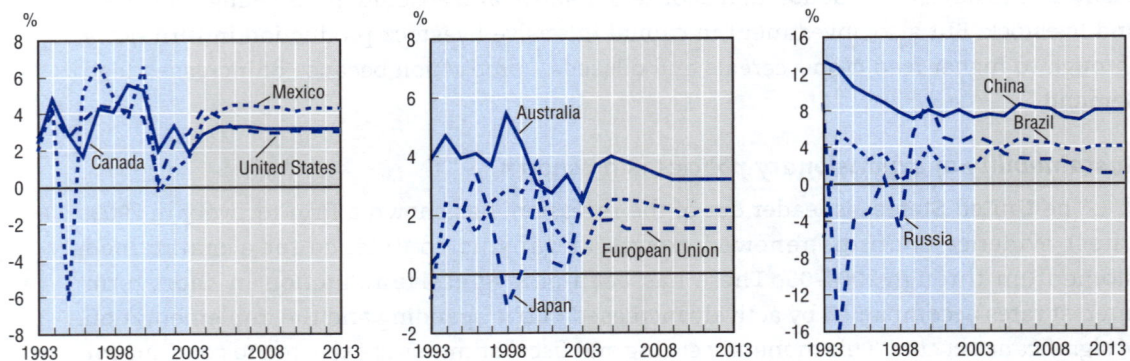

Source: OECD Secretariat.

which also adds to the attractiveness of Mexican agricultural exports to the United States. Prospects are strongly optimistic from 2004 onwards, in an underlying economic environment of contained inflation, and low interest rates. Moreover, in 2004 Mexico finished the negotiations to reach an Economic Partnership Agreement with Japan, which opens opportunities of entry to a new market.

Optimistic outlook in Japan despite continuing short-term deflationary pressures

In **Japan**, the past year has shown an acceleration of the economic upturn that started in 2002 after 10 years of recession. There has been strong growth in business investment, and a steady increase in private consumption. Business confidence indicators have rebounded, with higher profits stemming from past restructuring in the corporate sector. Export growth has been a key factor in the turnaround, and despite the slight weakening of exports recently, expectations are for renewed acceleration through 2004. Fiscal policy continues to be expansionary, in part due to significant recent tax cuts, and monetary easing has kept long term interest rates low. The resulting real disposable income gains have driven the aforementioned rise in private consumption. The combination of these factors provides a basis for a prevailing optimism throughout the *Outlook* period. Nevertheless, on the downside, moderate deflation persists and unemployment remains at a record high of 5%. The Bank of Japan has stated its objective of policy stability in the aim of achieving zero or positive inflation; we assume that moderate deflation will persist for the short term, followed by price stability from 2007.

Slower economic growth to continue in Europe over the near term

Europe continues to sputter forward, with economic recovery lacking the resolve found in North America and Asia. Growth in 2003 has been uneven, with the **United Kingdom** showing a solid upturn and strong demand, in contrast to its continental partners. In the first half of 2003, only negligible growth was recorded in the **euro Zone**, due in part to the appreciating euro. More recently, there have been welcome signs of a modest rebound in the making. However, several factors dampen the prospects. Unemployment remains persistently high and the strong euro hampers export growth. Active stimulus is burdened with policy concerns regarding the validity of the 2% inflation target and the prospective easing of the Maastricht Treaty fiscal stability limits of 3% of GDP. Short term prospects are nevertheless positive. The euro is anticipated to remain strong and inflation should fall slightly to 1½%. Real GDP growth is expected to rebound from an estimated ½% in 2003 to attain 2½% by 2005. By that time recovery in Europe should be well established, with stable GDP growth rates slightly exceeding 2% during the period 2005-13.

Strong economic growth in Australia and New Zealand

In **Australia**, economic growth is expected to accelerate in the short term and to remain strong throughout the *Outlook* period. The effects of last year's drought have contributed to negative export growth (along with SARS, weak global demand and the appreciation of the Australian dollar), which has offset the recent strength in total domestic demand. Drought effects continue in the south, and may affect the *Outlook* into 2005, in particular for dairy products and rice because of lower production and reduced incomes throughout the rebuilding phase. Nevertheless, several positive factors set the scene for renewed growth: business and consumer confidence levels are high; unemployment levels, at 5.8%, are close to the estimated structural rate; and inflation remains low and is expected to stay within the Reserve Bank's targeted range of 2% to 3%. Australia, as opposed to **Europe** and **Japan**, still has the luxury of being able to set accommodating monetary policy, as witnessed in the recent tightening to counterbalance the dollar's appreciation. In the short term, this gradual tightening is expected to continue, with higher interest rates helping to cool residential investment, slow private consumption and shifting emphasis to maintained growth in business investment and revived exports. GDP growth rates are projected to peak at nearly 4% in 2005, tapering back to 3.2% by 2009 and remaining at this steady state for the remainder of the *Outlook*.

With similarly strong domestic demand, housing sector growth, low unemployment, and cautious fiscal policy, **New Zealand** has shown exceptional economic vigour in the last several years. Output is currently above potential, but it seems likely that the pace of activity may slow to OECD-wide average GDP growth rates of 3.1%, particularly in light of past exchange rate appreciation and its effect on the export sector. Nevertheless, this growth rate is anticipated to be maintained throughout the *Outlook* period to 2013, with accompanying healthy demand.

Asian economies to gain further momentum

The GDP growth prospects remain buoyant for **Korea**, despite several adverse shocks experienced in 2003 (namely SARS, labour unrest and a serious accounting scandal). Exports, in particular to **China**, are anticipated to spur recovery and perhaps strengthen sagging business and consumer confidence.

The restrained expectations with regards to the OECD economies contrast with the vigorous economic activity and bright outlook for certain **Non-member economies (NMEs)**. A rapid rebound following the containment of SARS in **China** should continue, with bank lending inspiring strong investment and consumption spending, helping to maintain GDP growth rates in the neighbourhood of 8% throughout the *Outlook* period. Other Asian countries such as **Taiwan, Thailand and Singapore**, have also rebounded quickly and will gain further momentum as their exports to China continue to develop.

Export opportunities are the key to GDP growth in South America

Export demand is also a key factor for continued GDP growth in the main South American economies. After weakness in the first half of 2003, both **Brazil** and **Argentina** are picking up strongly. Inflationary pressures have decreased markedly in Brazil, and should continue to do so, and a similar situation is anticipated for Argentina in the short to medium term. Both economies will benefit from reform efforts to pursue a more stable macroeconomic environment. Strong domestic demand and new export opportunities have led to acceleration in GDP growth in **Chile**.

Slowing in rate of economic growth foreseen in Russia

The real GDP growth rate in **Russia** is estimated to recover substantially to around 6% in 2003 from a rather healthy 4.3% in 2002. However the anticipated trend is for a decline in this growth rate, which should remain in the neighbourhood of 2½% for 2006-09, before stabilising at roughly 1% from 2011 onward. Economic activity should maintain its solid foundations of the past few years, with broad based growth and stronger domestic demand. On the one hand, fiscal stimulus is continuing, while on the other, the stance for monetary policy is consciously balanced between the competing aims of reducing upward pressure on the exchange rate and trying to further reduce inflation below double digit levels. This latter task is further complicated by active trade growth in the oil industry; indeed, low global oil stocks and high oil prices have meant that this sector has contributed significantly to recent GDP growth in Russia. In the longer term, with more balanced economy-wide growth, GDP growth rate will decrease to more conservative levels.

Higher incomes will encourage increased demand

The high economic growth rates in these Non-member economies (NMEs) are the main driving force for consumption growth and increased trade for agricultural commodities over the medium term. The price and income elasticities of demand for most agricultural products are higher in the NMEs than in the mature markets of OECD countries. Higher incomes will therefore translate more directly into stronger demand, in particular for livestock products and other high value-added meat and dairy products.

Table 1.1. **Where population and income is projected to grow**

Average annual percentage increase over 10 year period

	1994-2003	2004-2013	1994-2003	2004-2013
	Population %		Income %	
World	1.32	1.05	2.66	3.12
Africa	2.41	2.05	3.18	3.99
America	1.34	1.02	3.06	3.05
Europe	0.15	0.08	2.11	2.50
Asia	1.36	1.05	2.73	3.78
Oceania	1.30	0.87	3.66	3.34

Note: Income is at 1995 USD market prices.

Source: World Bank, October 2003.

There are significant differences in projected population and income growth rates among world regions. (Table 1.1). Globally, over the 2004-13 *Outlook* period, average annual population growth should just exceed 1%, slightly less than the average rate over 1994-2003. Population growth rates decrease in all regions. Of particular note is that population growth in Africa is nevertheless expected to exceed 2% per year, but is projected to be negligible in Europe over the next ten years. Real GDP growth however is anticipated to increase for the *Outlook* period to a world average of 3.1% per year. Areas of strongest growth in GDP are Africa and Asia, with average growth rates nearing 4% per year.

Several conclusions can be drawn from these projections. The combination of strong population and income growth is a characteristic of developing countries. In Africa and Asia, rising populations and higher total income mean increased demand from new emerging markets. Once again, because of higher price and income elasticities than in the other three regions (which contain a higher proportion of more developed countries), more of this growth translates directly into higher demand for agricultural commodities. Developing country markets are expected to play an ever increasing role in world trade in agricultural commodities, not only as exporters of primary products, but as consumers of domestically produced and imported goods. Moreover, as per capita income increases in these countries there is a shift in consumption towards more value added commodities. The ageing populations and mature markets of Europe and Oceania have different underlying dynamics with growth in agricultural markets driven by improvements in productive technology, cooperative trade links and stable economic conditions.

Prices of primary commodities are important international factors

The price of oil is an important **international factor** to be considered, because of its direct influence on the macroeconomic variables of interest. The *OECD Economic Outlook* (December 2003) adopts an underlying hypothesis of stability for the nominal Brent price for crude oil of USD 27 per barrel. This is lower than price expectations in the period leading up to the Iraq war, but higher than both the ten-year average, and the mid-point of OPEC's target range. OPEC has cut oil production back starting in November 2003. Low global oil stocks following the Iraq crisis, upward pressure due to recovery in the US and Japan and nascent demand from China have driven prices higher. Natural gas prices have also risen since 2002. Increased activity in North America and Asia is also a key factor behind the recent rise in non-energy commodity prices, such as industrial raw materials, and in particular base metals. This is a short term trend tied to the stage of the economic cycle, and is not expected to persist, once primary production catches up with the increased demand. Thus, the low core inflation setting that prevails in most OECD member countries should not be affected.

Policy developments and assumptions

Agricultural policies affect markets

Agriculture and trade policies are important to both the domestic and international markets because they affect prices, production and lead to market distortions. Domestic agricultural policies are fundamentally linked with international trade. Trade and protection policies are normally an integral part of domestic support programmes based on market price support, as they limit market access for imports or help to dispose of unwanted surpluses on world markets in order to maintain internal prices above world levels. Export competition policies include export subsidies, export credits or other measures that can affect exports, such as food aid under certain conditions. Through their impacts on domestic production and trade, agriculture policies of OECD member countries have a direct effect on international agricultural markets and are therefore important elements to consider when assessing future market developments and prospects.

Agricultural policy assumptions used in the *Outlook* are based on measures in place or announced within well-defined programmes. Therefore the projections assume a continuation of current agricultural and trade policies except when new policies have been announced for implementation within the *Outlook* period. The following section provides a brief overview of the major policy developments in OECD countries which affect the *Outlook*. (For more detailed policy developments, refer to *Agricultural Policies in OECD Countries: At a Glance, 2004*.)

US Farm Act provides more market insulation to producers

The various provisions and programmes of the **United States'** Farm Security and Rural Investment Act of 2002 (often called the FSRI Act or the 2002 Farm Act) are assumed to continue over the *Outlook* horizon, even though the legislation will only apply until 2007. While the 2002 Farm Act re-authorises the Export Enhancement Program, only limited use has been made of it in recent years. This situation is assumed to continue over the *Outlook* period.

Under the current US farm legislation, farmers are now eligible for fixed direct payments for crops and counter-cyclical payments (CCP). CCPs are made when the higher of the loan rate or the season-average farm price of grains, oilseeds or upland cotton is below the target price less the direct payment rate for that crop. Direct payments are based on pre-determined rates and past production. The 2002 Farm Act continues the marketing assistance loan programme for cereals and oilseeds. Milk and dairy products in the United States are supported by minimum prices with government purchases of butter, skim milk powder, cheese and a payment per tonne of milk marketed as well as by tariffs, tariff-rate quotas and export subsidies. For more information on the 2002 Farm Act provisions the reader is referred to the *OECD Agricultural Outlook, 2003*.

Market dislocation due to BSE outbreak expected to be temporary

Following the discovery of a single BSE affected cow in the State of Washington in 2003, more than fifty countries have implemented full or partial bans on US beef imports. In addition to a ban on feeding meat and bone meal from ruminants to ruminants in 1997 in Canada and the US, new regulations have been put into operation in the United States to prevent the potential spread of BSE. Beef destined for consumption, dietary supplements and cosmetics must no longer contain:

- Downer cattle material.
- Material from cattle that die before reaching the slaughter plant.
- Specified risk materials, such as the brain, skull, eyes or spinal cord of cattle 30 months or older or a portion of the small intestine and tonsils of all cattle.
- Meat mechanically removed by means of advanced meat recovery system.

Additionally, a ban on the use of air-injection stunners has been implemented and the establishment of a nationwide cattle identification programme is scheduled to be operational in 2006. Also, the US government has decided to introduce an enhanced BSE testing programme which should test between 201 000 and 268 000 cattle considered to be most at risk over the next 12-18 months. The market disruption caused by the BSE incident, and resulting trade bans, have been assumed to be short-lived in the *Outlook* projections. In early 2004, it is assumed that Canada, Mexico and the United States would choose to resume their beef trade with each other, although meat would only come from animals under thirty months of age. For the purpose of the 2004 *Outlook*, it was assumed that the Asian market will remain closed until January 2005. Oceania, where producers may benefit from the lack of supply of North American beef to the Asian market, may see a temporary increase in exports for that period. The extent to which this may happen depends on the ability to substitute grain-fed beef from North America with grass-fed beef from Oceania. Finally, it is also assumed that some small quantities of beef will be supplied to the Pacific market by non-traditional exporters.

CAP reforms implemented in the EU

For the **European Union**, the main elements of the Common Agricultural Policy (CAP) 2003 Reform have been included in the 2004 *Outlook*. One of the most significant changes is the new Single Farm Payment (SFP) which will enter into force in 2005 at the earliest. However, member states can delay implementation up to the year 2007, at the latest. SFPs will now be linked much less directly to production than the payments that they replace, such as the former crop payments based on fixed reference yields and the beef direct payments tied to animal numbers. Farmers will be allotted payment entitlements based on historical reference amounts received during the 2000-02 period, with payments established at the farm level.

For this *Outlook* it is assumed that member states opt for maximum decoupling. Hence, the largest part of crop and beef payments is assumed to enter the Single Farm Payment scheme by 2005, the earliest possible point in time. The levels of the SFP then increase in 2007 when dairy payments are included as well. The new member states will benefit from payments with an adjustment between 2004 and 2013. They are, however, assumed to top up payment rates to the maximum degree. Consequently, payments in the EU accession countries are assumed to be 55% of the full EU rate in 2005, and to gradually increase to 100% by 2010.

The European Commission will implement the majority of the agreed CAP changes to crops, beef and dairy products in 2004. The expected market impacts of the CAP reforms over the projection period are explored in fuller detail in a special section in this *Outlook*. For the first time, the *Agricultural Outlook* covers the expanded European Union of 25, rather than 15 members. Agricultural production in the 10 accession countries has been assumed to be subject to the provisions of the EU CAP. Benefits include the phasing-in of a single farm payment to the new member states under CAP. Export subsidies are another part of the CAP that had an important effect on EU exports. Pending a final agreement, the WTO export subsidy limits for the projection period of the expanded Union of 25 members are assumed to be equal to the sum of the export subsidy limits of the current fifteen members and the limits (if any) of each of the 10 new member states. Finally, no assumptions have been made about any further enlargement of the European Union during the ten year projection period to 2013.

NAFTA encourages trade opportunities

Canadian agricultural policies administered by the federal and provincial governments are assumed to remain unchanged from existing legislation. Supply management and price support measures continue to be applied with supporting trade measures to the more heavily regulated milk and poultry sectors. The discovery of a single incident of bovine spongiform encephalopathy (BSE) in a cow in Alberta in May 2003 resulted in an embargo on Canadian beef exports by several countries. However since then, the United

States and Mexico have re-opened their markets to certain beef cuts. The assumptions with respect to renewal of trade with other countries are as outlined for the US.

Mexican agricultural markets have become more open with less trade distorting measures than previously, due to significant recent reforms in agricultural policy. Agricultural policies consist mainly of market price support maintained by border measures, supplemented by direct payments under the PROCAMPO programme based on historical entitlements and payments based on input use. The use and adjustment of TRQs in response to domestic pressures is a feature of trade policies. Mexico's border protection with Canada and the United States is being reduced within the NAFTA framework. Almost all tariffs with the US were eliminated in 2003; the remaining five will be eliminated in 2008.

Mexico has implemented a number of measures to reduce the penetration of high fructose corn syrup (HFCS), mainly imported from the United States, in its domestic sweeteners market. These include a countervailing duty on US exports of HFCS, a 20% tax on the production and distribution of beverages that use any sweetener other than cane sugar and a tariff quota on HFCS imports from the United States that is equal to the raw sugar import quota granted to Mexico by the United States under NAFTA for the period to 2008. Unrestricted trade in sugar after 2008 combined with the restructuring process initiated by the government is expected to create opportunities for the Mexican sugar industry.

Japanese and Korean market orientation remains limited

Japanese agriculture is characterised by continued high support levels and only limited market orientation for rice, which is the major crop. Support to agriculture is primarily provided through border measures, with direct payments, deficiency payments and administered prices playing a role in some commodity markets. Supply management regimes persist for several commodities including continued reliance on state trading enterprises to regulate imports and domestic markets. Nevertheless, since the mid-1990s, due to changes in domestic policies and tariff reductions, markets for livestock products, oilseeds and cereals have become more integrated with world markets. Some efforts have also been made towards improving market orientation for rice since the Food Control Law was repealed in 1995, allowing prices within Japan to move relatively freely. A scheduled rice policy reform has the potential to improve market orientation and reduce the high costs of agricultural protection, but this has not been reflected in the present *Outlook*.

Korea's agricultural policies consist mainly of market price support implemented through restrictive trade measures and domestic price stabilisation based on government purchases and public stockholding. Whilst the government has been lowering the level of support over the past decade it is still high compared to most other OECD member countries. Progress with reform has been made for certain commodities such as beef, while other commodity markets, including rice, the main staple crop, remain largely isolated from international market signals creating high consumer costs.

New Zealand and Australia most market orientated

New Zealand and **Australian** agricultural support continues to be extremely low, with New Zealand reporting the lowest level of support in the OECD. The agriculture sector in these two countries is more market orientated compared to the OECD average, with producer incentives being determined by world markets. Support to the Australian agricultural sector is mainly provided through budget financed programmes, regulatory arrangements and general tax concessions. In the case of New Zealand, support is provided through general budget outlays for basic research and for the control of diseases and pests. Low border measures in the form of tariffs apply to some imported products such as pig meat and poultry.

NME trade affected by policy changes

Policies in the NMEs are assumed to remain as defined under current policy settings, except when policy changes have been announced for implementation. The tariff quotas that were introduced for beef, pork and poultry in **Russia** in 2003 are assumed to continue for the entire projection period. On 1 January 2004, Russia introduced a floating duty tied to average market prices for raw sugar along with a fixed tariff on white sugar; these trade barriers are also assumed to remain in place throughout the *Outlook* period. **China**'s accession to the WTO in 2001 has opened up trade possibilities to and from China. As part of the accession arrangements to the WTO, China has established tariff quotas for a number of agricultural products, with increases in quotas to be phased in over coming years. However, these market access provisions remain non-binding on China. **Brazil** and **Argentina** are two NMEs which play a significant role in the world market, mainly as low cost competitors with OECD exporting countries. The export performance of both countries has been assisted by substantial currency devaluations relative to the US dollar. In addition, agricultural policies have been extensively reformed in recent years with producer prices essentially aligned with world markets.

Another NME where agriculture is critical to the economy is **India**. The National Agricultural Policy announced in 2000 aims to improve agricultural production through, amongst other measures, increased efficiency, improved infrastructure, rural electrification and increased private sector participation. More detailed information on the Indian agricultural sector and its agricultural policies can be found in a special focus section of this *Outlook*.

International trade agreements shape policies

National commitments under **international trade agreements** remain a major influence on shaping agricultural and trade policies of OECD member countries and many NMEs alike. The Doha round WTO multilateral trade negotiations launched in November 2001 is currently stalled following the collapse of discussions in Cancún. Since no new global trade agreement has yet been reached, this *Outlook* assumes that the provisions of the 1994 Uruguay Round Agreement on Agriculture (URAA) hold for the entire projection period to 2013. In addition, the trade provisions of a number of regional agreements such as the North American Free Trade Agreement (NAFTA) between Mexico, the United States and Canada, the MERCOSUR agreement between Argentina, Brazil, Uruguay and Paraguay, the Everything But Arms (EBA) Initiative of the European Union with the least developed countries and its Contonu convention with the African, Caribbean and Pacific (ACP) countries, are assumed to apply to trade for particular commodities over the projection period. Finally, as noted earlier, the enlargement of the EU with the accession of the ten new member states from May 2004 has been taken into account as well. Although a number of other FTAs have recently between negotiated, or are under negotiation, among OECD member countries or between OECD member countries and NMEs, these are not taken into account in this *Outlook*.

High support and protection to continue in many OECD countries

Despite a trend towards policy reform in OECD countries to improve market orientation, high support and protection to agricultural sectors will continue to be a significant feature of the domestic markets of many of these countries over the projection period. Although some changes have occurred in the composition of support towards less production and trade distorting forms, the dominant form of support continues to be market price support in many OECD member countries. Furthermore, the situation of limited market access for agricultural products due to high border protection and use of export competition policies to dispose of unwanted surpluses on world markets, will continue to distort production, trade and world prices for many agricultural products over the period to 2013. These policies will restrain the growth in agricultural trade, reduce export returns for other exporters, including those in developing countries, and lead to lower welfare of many consumers.

Sensitivity Analysis: Impact of Stochastic Macroeconomic Variables on the Outlook for World Commodity Prices

Stochastic macroeconomic input

The projections presented in this report rest on the assumption of a steady, favourable macroeconomic environment. This setting serves as a useful starting point for scenario analysis, but may be criticised as unrealistic. Historic patterns do not show many decades where the macroeconomic conditions of all countries are uniformly well described as either steady or favourable. The purpose of this section is to exploit recent work on partial stochastics as it relates to the macroeconomic assumptions of the Outlook in order to highlight the sensitivity of the market projections with respect to economic growth, general price levels and exchange rates.

Partial stochastics represent a simulation method that may serve as a useful alternative to the point estimates of the *Outlook*.[1] In the *Outlook*, one particular constellation of macroeconomic variables (as described already and as represented at the very beginning of the tables) serves as an input into the model. Then, these data will be used as inputs into a partial equilibrium model of temperate zone commodity markets: assumed exchange rates will determine some part of the price links between countries; these GDP data will serve as an explanatory factor in the demands for different commodities; and these general prices indices will be used to represent the prices of other inputs in production and of other goods in consumer demand. The alternative offered by partial stochastics would be to replace this particular set of macroeconomic variables with a distribution, or range of possible values, which could be sampled repeatedly to generate a sequence of different inputs into the model simulation. If fifty different observations were drawn from the random distribution, then the model would be solved fifty times to calculate all the different market outcomes that correspond to these various settings.

This description hints at two problems with partial stochastics: the difficulty of defining an appropriate distribution of macroeconomic variables and the no less significant difficulty of communicating the output. As regards the first of these problems, the distribution of macroeconomic variables is defined by historic data with a view to account for certain influences across countries and across variables, as well as some of the delayed effects. For example, a distribution that treated a country's inflation as though independent of that country's exchange rate could give implausible results. Similarly, distributions of GDP in each country are likely to depend on the evolution of the world economy. At the same time, however, the macroeconomic database offers only short time series for many countries, making a complicated analysis of all possible effects impossible – as well as a far more ambitious project than is strictly necessary for the purposes at hand.

In order to acquire distributions for the macroeconomic data that are required for the *Outlook*, a simple model has been constructed to capture as many of the cross effects as possible. In the most elaborate case, the model is specified as follows.

- $LN(GDP_{i,t}) = a_{i,0} + a_{i,1} * LN(GDP_{i,t-1}) + a_{i,2} * LN(GDP_{i,t-2}) + a_{i,3} * LN(GDP_{US,t}) + a_{i,4} * LN(trend_t) + GDPError_t$

- $LN(GDPD_{i,t}) = b_{i,0} + b_{i,1} * LN(GDPD_{i,t-1}) + b_{i,2} * LN(GDP_{i,t}) + b_{i,3} * LN(GDPD_{US,t}) + b_{i,4} * LN(XR_{i,t}) + b_{i,5} * LN(trend_t) + GDPDError_t$

- $LN(XR_{i,t}) = c_{i,0} + c_{i,1} * LN(XR_{i,t-1}) + c_{i,2} * LN(GDP_{i,t}/GDP_{US,t}) + c_{i,3} * LN(GDP_{i,t-1}/GDP_{US,t-1}) + c_{i,4} * LN(GDPD_{i,t}/GDPD_{US,t}) + c_{i,5} * LN(trend_t) + XRError_t$

In these equations, *GDP* is, obviously, Gross Domestic Product, *GDPD* is the general price deflator associated with the Gross Domestic Product and *XR* is the country's exchange rate *vis-à-vis* the US dollar. The final term in each equation represents the corresponding error term. As regards subscripts, i denotes the country in question, with United States represented by US, and t is the time indicator (so t – 1 is the previous year). In this specification, developments in the United States are taken as indicators for the world economy and many of the possible links across variables are represented by parameters that can be estimated.[2] In an ideal case, then, this representation would capture some part of the cross-country, cross variable and delayed effects of a shock in one variable as they exist in historic data, without embarking on an experiment in complicated macroeconomic modelling. Moreover, the error terms of each estimated equation are exactly the information required to construct the distribution of future shocks which, when fed back into the model over the projection period, give a new set of macroeconomic variables for the *Outlook*. This process is demonstrated Figure 1.2.

Figure 1.2. **Identifying variability of macroeconomic data**

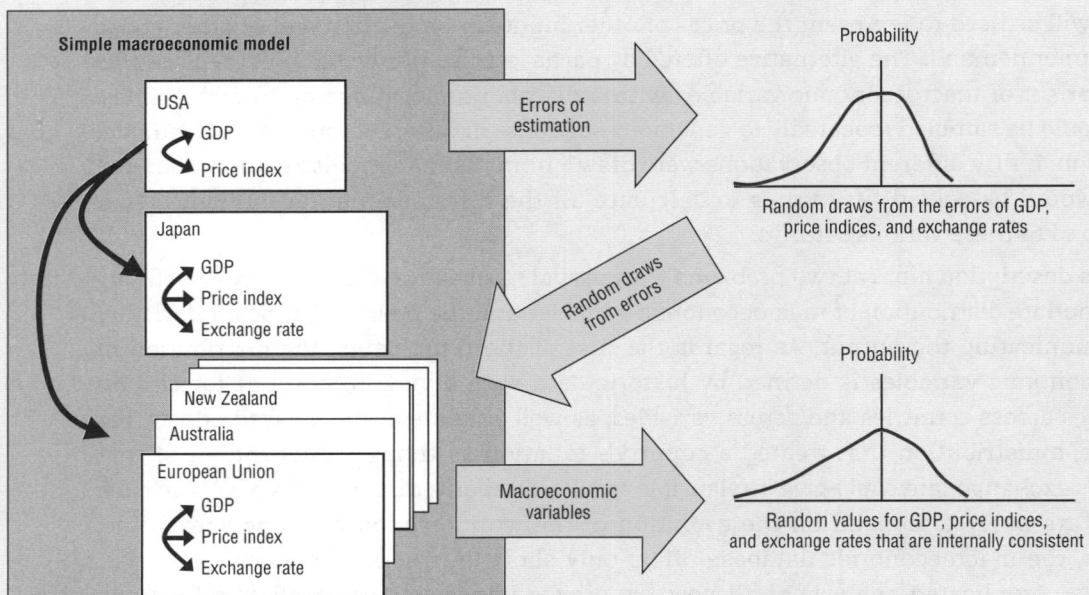

Source: OECD Secretariat.

The data do not, however, allow for an ideal representation. While the example above is relevant for some countries, in most cases there is not enough information to explore so many possible links and, in any case, the full data set is too limited to allow more advanced experiments (such as time-series analysis or system estimation). Moreover, equations are not tested individually for the presence of various complicating factors, such as structural breaks, that would further complicate this representation. The interrelationships – across indicators, across countries and across time periods – may nevertheless be captured, but limits to the exercise should also be noted: the historic data do allow an approximate, but not completely satisfactory development of random distributions.

Given this process for determining random distributions of the macroeconomic data for the *Outlook* and then simulating the model repeatedly on the basis of many samples from this distribution, the second problem of stochastic simulation becomes relevant: how can the outcomes of such a large number of simulations be usefully communicated? In this section, we will address this issue from two perspectives: first, what do varying macroeconomic data imply for the market projections and, second, what do they indicate about OECD countries' agricultural policies and their impact on commodity markets?

Stochastic macroeconomic output: world markets

The results of fifty simulations with macroeconomic data randomly determined from 2004 to 2013 are first introduced by summarising some of these macroeconomic data.[3] The ranges of a few variables, namely the GDPs for the United States and Argentina, are illustrated in the first figures of this section. These ranges are bound below by the mean value of the 50 random draws minus their standard error and extend up to the mean plus one standard error. In 2004, the standard error of the United States' GDP is about 1.4% of the mean value, implying a fairly narrow range of results. On the other hand, the standard error of Argentina's

Figure 1.3. **Mean GDP indices +/– one standard error**

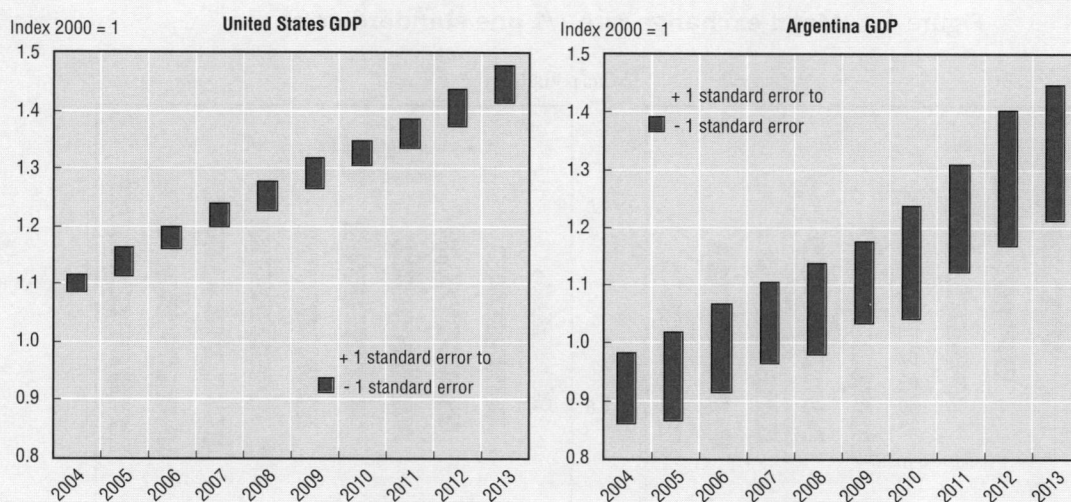

Note: The values of GDP indices are presented here, so these data are only indirectly comparable to the growth rates reported in the table providing the economic assumptions underlying this year's *Outlook*.

Source: OECD Secretariat.

GDP is substantially higher relative to the mean, about 7%, implying a much larger range of variation. Moreover, the extreme values will lie outside the ranges described in these figures: the maximum random draw for Argentina's GDP is 10% above the mean value in 2004 and the minimum is 18% lower. This last point serves to highlight one point not apparent in the figures below: the distribution of these random draws is not necessarily symmetric as historic evidence may point to a greater incidence of certain types of shocks (such as occasional large negatives in the case of Argentina's GDP).

Figure 1.4 illustrates some of the results for exchange rates. Whereas the *Outlook* macroeconomic assumptions taken from other sources often depend upon flat – or nearly flat – nominal exchange rates, the partial stochastics generates a wider range of possible values. The cases of the yen and the euro are used to illustrate this point in the figure here. By construction, the mean of the former currency follows the path set out in the baseline assumptions. However, the variations around that path are substantial owing to a standard error of JPY/USD 11 to 16, or 9%-15% of the mean. In 2004, for example, the mean value of about JPY/USD 118 is quite close to the level assumed in the *Outlook*, but the maximum random value is 136 and the minimum is 94. The standard deviation for the euro defined by historic data is hardly less significant at 8% to 15% relative to the mean, with the more extreme values representing a decrease of 25% on the low side and an increase of 30% on the upper side in 2013.

The implications for a single, domestic market of higher or lower income and a weaker or stronger exchange rate may be readily understood, at least in terms of direction, but it is more difficult to quantify the results when the global economic environment is subject to random shocks *a priori*. For example, a shock to incomes may affect oilseed meal markets through two indirect paths. First, higher or lower demand for vegetable oil will affect the rest of the oilseed complex and, hence, oilseed meal supplies. Second, fluctuations in income may impact significantly on demand for livestock products as these goods are often relatively more sensitive to income, implying changing incentives to livestock production which, of

Figure 1.4. **Mean exchange rate +/1 one standard error**

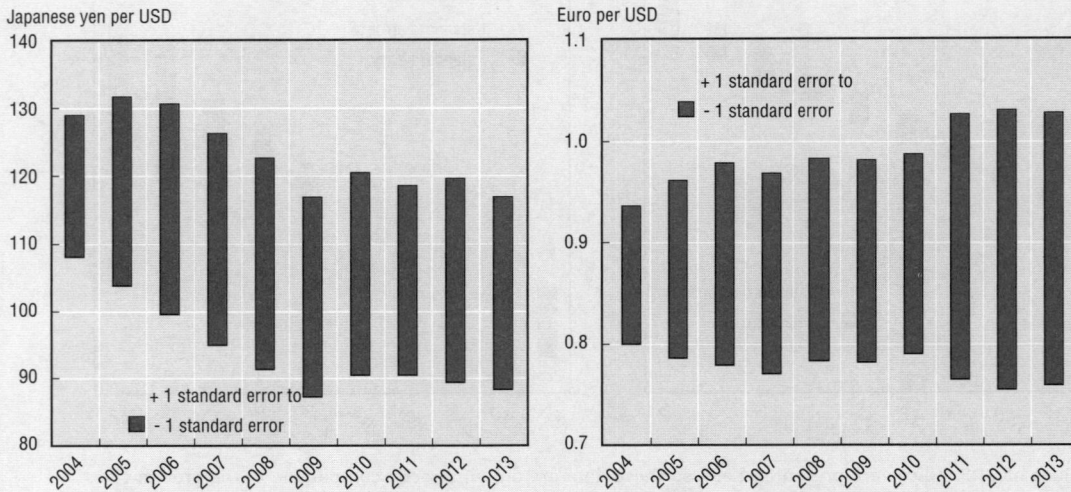

course, implies indirect, delayed changes in feed demand for oilseed meal. The consequences of differing constellations of exchange rates is no less obtuse and, in either case, policies that entail intervention in the transmission of price signals or in decision making may further complicate the link from macroeconomic environment to market outcome. In some cases, these policies will have implications for the variability of domestic prices and producer returns, as well as direct or indirect effects on the variability of world prices. The OECD's *Aglink* model was used in connecting the random macroeconomic variables to their consequences on world markets. Figure 1.5 demonstrates the ranges of some world market prices that serve as useful indicators of the wider situation.

Figure 1.5. **Mean world prices +/1 one standard error**

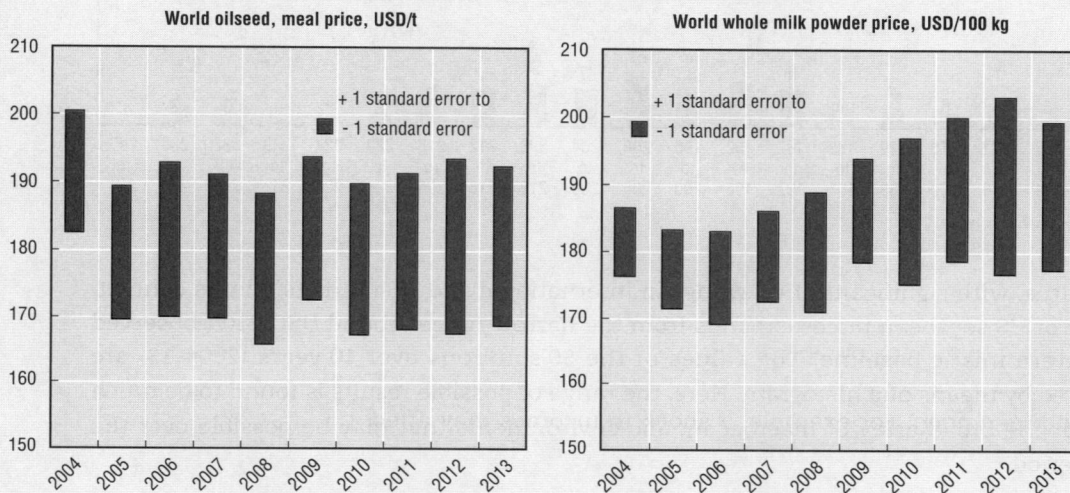

Source: OECD Secretariat.

Stochastic macroeconomic output: OECD policies

Many of the agricultural policies of OECD countries not only affect the market by, for example, expanding production or reducing imports, but also are in a certain sense determined by market outcomes. Without speculating as to whether decision makers would opt for discretionary policies in the event of low prices, several countries have automatic mechanisms that lead to increasing market intervention when prices are low. As regards OECD crop production, chief among this category of policy in terms of impact may be the marketing loan program of the United States and the intervention price of the EU, although other countries, including Japan, also conduct support, such as deficiency payments, that may have similar effects. Or there may be more complicated interactions between policies and markets, such as safeguard clauses that are triggered by strong increases in imports and consequently depend on world and domestic prices only indirectly.

The US marketing loan program for soyabeans offers a particularly interesting example. Marketing loan benefits are expected to be triggered in only one year of the projection period after having been triggered frequently, if not routinely, in recent years. In the *Outlook*, the nominal price lies just high enough to prevent government support under this program. When subject to varying macroeconomic conditions, however, the world oilseed market is

Figure 1.6. **US soyabean price in 50 simulations over 10 years, USD/t**

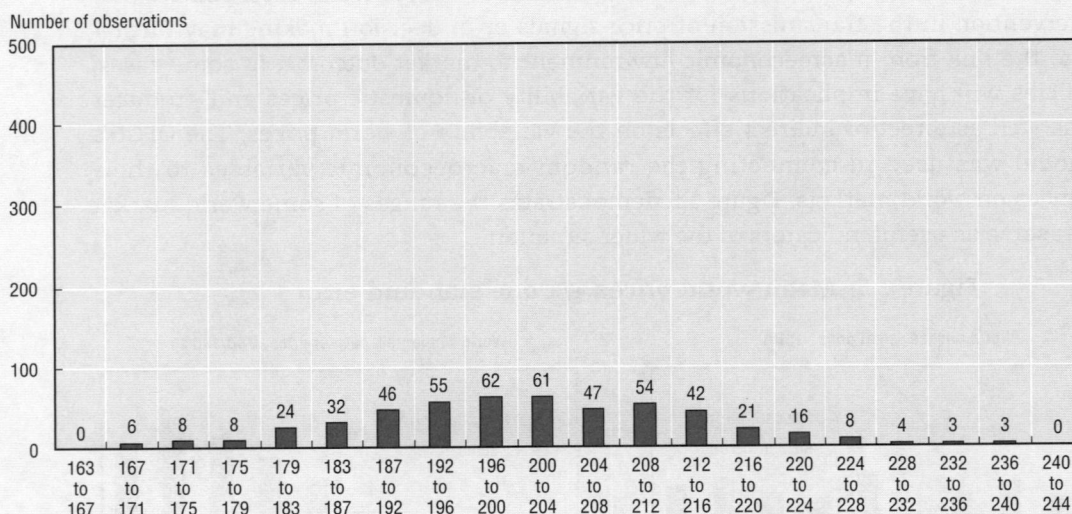

Number of observations

Range	Observations
163 to 167	0
167 to 171	6
171 to 175	8
175 to 179	8
179 to 183	24
183 to 187	32
187 to 192	46
192 to 196	55
196 to 200	62
200 to 204	61
204 to 208	47
208 to 212	54
212 to 216	42
216 to 220	21
220 to 224	16
224 to 228	8
228 to 232	4
232 to 236	3
236 to 240	3
240 to 244	0

Source: OECD Secretariat.

more volatile, with significant price swings in international and US markets. In this context, the range of US soyabean prices expands from the narrow range around USD 210/t observed in the deterministic baseline. The values of the 50 solutions over 10 years (2004-13) are represented by means of a histogram. Here, the range of possible results is found to be much wider over the broad array of macroeconomic environments that may be possible over the 10 year period.[4]

Adding the marketing loan benefits accruing to soyabean producers to the price, offers a more accurate measure of their returns per unit sold. Producers need not respond to prices below the marketing loan rate, as the program will offset such low prices through a combination of deficiency payments and marketing loan gains. In other words, all observations of prices that fall below a certain point will be associated with an offsetting amount of marketing loan benefits; the sum of price plus marketing loan benefits is not allowed to fall below a certain level, unlike the price alone. Thus, whereas the observations of the soyabean price portrayed in Figure 1.6 comprised a fairly symmetric distribution with about as many above the mean as below, the total returns per unit (the sum of soyabean price and marketing loan benefits) will not be symmetric as low price observations will be shifted higher by the marketing loan benefits. As shown in Figure 1.7, the sum of market returns plus marketing loan benefits are truncated at about USD 196/t.[5] Out of 500 total observations, 241 fall at about this level of returns per unit. In all these cases, the US supply of soyabeans will no longer depend on world prices; US producers will continue to receive about the same total returns per unit regardless of how far prices fall.

The EU offers some of its crop producers a similar, one-sided bet. The intervention price reduces the downward price risk for wheat and coarse grain producers. Rather than direct payments to farmers along a formula, as in the US case, the EU system provides for levers with which to push the domestic market prices higher in the event that they fall relative to a given support price. Thus, import barriers, export subsidies and intervention stocks are all harnessed to maintain internal prices at or above support price levels. In simulation, this

Figure 1.7. **US soyabean price plus marketing loan benefits in 50 simulations over 10 years, USD/t**

Number of observations

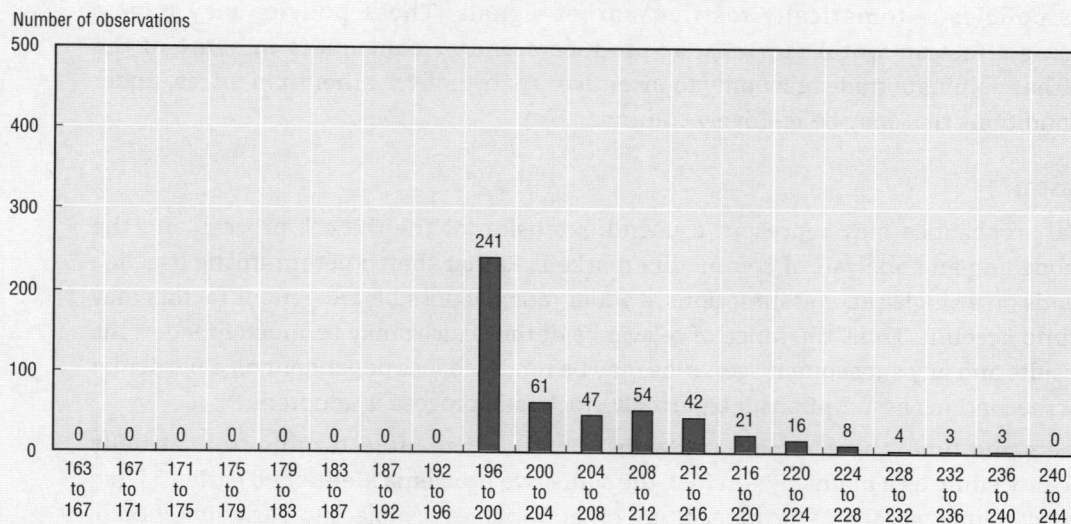

	163 to 167	167 to 171	171 to 175	175 to 179	179 to 183	183 to 187	187 to 192	192 to 196	196 to 200	200 to 204	204 to 208	208 to 212	212 to 216	216 to 220	220 to 224	224 to 228	228 to 232	232 to 236	236 to 240	240 to 244
	0	0	0	0	0	0	0	0	241	61	47	54	42	21	16	8	4	3	3	0

Source: OECD Secretariat.

policy regime is reproduced: falling domestic prices relative to the predetermined intervention price will stimulate greater subsidised exports and higher public stocks. As a consequence, as shown in the histogram, producers face less risk of low prices and, conversely, EU consumers have little prospect of finding a bargain even in the event of plentiful stocks or a strong euro that would otherwise make purchases from the world market cheaper.

These and similar policies in OECD countries encourage supply even at times of low commodity prices. Japan's deficiency payments and barriers to imports, particularly

Figure 1.8. **EU wheat price in 50 simulations over 10 years, EUR/t**

Number of observations

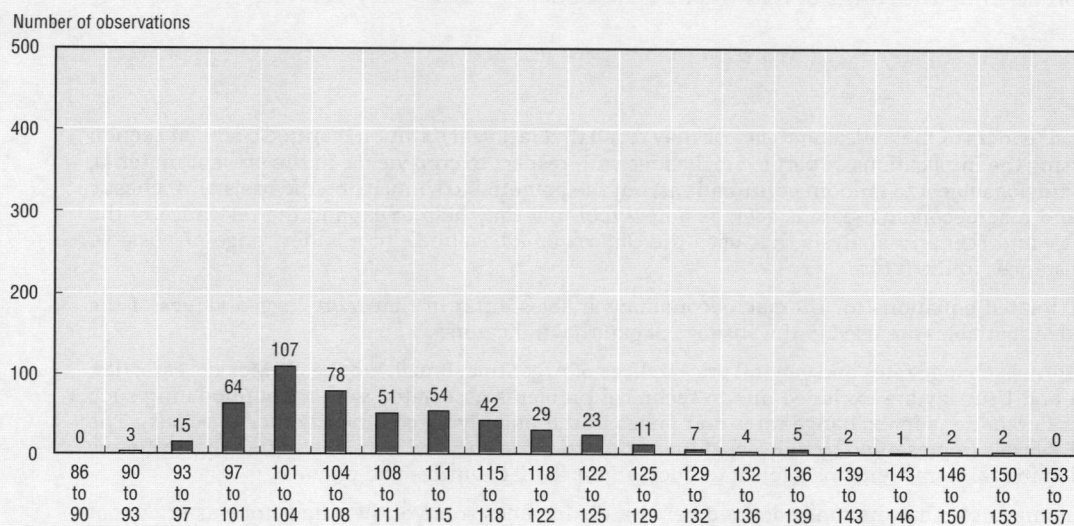

	86 to 90	90 to 93	93 to 97	97 to 101	101 to 104	104 to 108	108 to 111	111 to 115	115 to 118	118 to 122	122 to 125	125 to 129	129 to 132	132 to 136	136 to 139	139 to 143	143 to 146	146 to 150	150 to 153	153 to 157
	0	3	15	64	107	78	51	54	42	29	23	11	7	4	5	2	1	2	2	0

Source: OECD Secretariat.

safeguards that rise to offset the incentives to import when market prices most favour doing so, have the same implication: response by agents in Japan to changing world prices are muted as policies automatically restrict market signals. These policies may have a destabilising effect on world markets, as producers and/or consumers in some of the countries that dominate trade react only to incentives set by policy, rather than prices, under certain, conditions that may be relatively common.

Conclusion

Partial stochastics may represent a useful extension of the *Outlook* process and the related modelling and analysis of commodity markets. Rather than a deterministic baseline that depends on a single set of assumptions, a wider range of possible exogenous factors may be taken into account. Thus, the range of relevance of the *Outlook* may be much broader; the *Outlook* results and any subsequent scenarios may be rendered less dependent on a particular set of macroeconomic assumptions if a partially stochastic process is adopted.

At the same time, there remain obstacles to overcome before turning to a partially stochastic procedure as a mainstay – in fact, the same two problems mentioned earlier.[6] First, the macroeconomic variables – or indeed any other variables – subject to variation should, ideally, be all-inclusive. Those few variables that have not been randomly determined to date (*i.e.* prices and exchange rates in a few countries) should be replaced with distributions as well, if possible. Second, methods of communicating results must be further developed. Multiple simulations over a ten-year period do not lend themselves to easy representation in figures or in tables. Ranges around the mean bound by plus/minus one standard error have many applications in other contexts and may be useful here as well, but do not hint at some of the asymmetric responses frequently generated by agricultural policies. In fact, problems of communication will only multiply when turning to matters of policy analysis, when full sets of multiple simulations will be conducted at least twice, first for "baseline" results and second for the "scenario" results, and comparisons of these sets of multiple simulations would ensue. Nevertheless, the Secretariat recognises the potential for these techniques, which are being used by researchers elsewhere in a similar context, to be applied in future reports, broadening the range of relevance of the *Outlook* and of policy analysis.

Notes

1. Habitual readers of the *Agricultural Outlook* may recall that last year's edition featured a similar section discussing the implications of partial stochastics with respect to crop yields. In the present material, yields are not subject to random perturbations, but the potential to simulate on the basis of stochastic yield and macroeconomic data is seen as a new tool that may help to expand the relevance of the *Outlook* – and scenario analysis that uses the *Outlook* as a baseline – to a wider range of possible circumstances in the future.

2. The estimated equations for US macroeconomic variables differ in that more lagged values of the dependent variables are used in the absence of external indicators.

3. At present, exchange rates and general price indices of Argentina, Brazil, Indonesia, Mexico, Paraguay, Russia and Uruguay are excluded due to technical problems associated with accommodating such serious fluctuations in exchange rates and prices in the underlying commodity model (*Aglink*). The real GDP levels (represented by indices) of these countries are shocked nevertheless, so the economic situation in real terms remains subject to fluctuations in all countries and regions.

4. Readers may note that the range defined here hardly includes some of the values for this price that have been seen in the past – including preliminary estimates for 2003 – thereby inviting two

important reminders. First, the procedure described here remains *partially* stochastic as only macroeconomic factors, rather than the complete array of random elements imaginable, are drawn subject to variability. Second, this process is neither capable nor intended for producing probabilities with respect to future prices. The commodity model that supports this work, Aglink, is intended for projections and policy analysis, not prediction.

5. The actual loan rate for soyabeans is lower, at about USD 184/t. However, past experience has shown that producers are able to receive an amount of support through the marketing program that is greater than the amount implied by a simple comparison of national average prices and loan rates. The difference may be due to the difference between national level and the relevant local data or due to timing as producers are able to claim a loan deficiency payment when prices are perceived to be low but retain ownership nonetheless, speculating on future prices. Regardless of the reasons, the relevant point is that past evidence suggests an effective lower bound on price plus marketing loan benefits closer to USD 196/t.

6. Another hurdle, namely marrying the process described in this section to earlier work based on variability in yields, has been successfully undertaken in preliminary testing.

ISBN 92-64-02008-X
OECD Agricultural Outlook: 2004-2013
© OECD 2004

Chapter 2

Cereals*

Main projections – outlook in brief

- *World wheat and coarse grain nominal prices are expected to increase to 2005 and thereafter to show little change until 2013, but to continue their long-term decline in real terms. Rice prices to increase both in nominal and real terms over the same period.*

- *Global grain production to reach a total of more than 2.1 billion tonnes in 2013 – 17% higher than in 2003. Wheat and coarse grains production should grow more rapidly than for rice.*

- *World grain consumption, currently larger than total supplies, to increase less strongly, with most of the growth in feed use, particularly in the NMEs.*

- *Slow down in the decline in global stocks, followed by a moderate increase from 2007, except in China. Global stock-to-use ratios to fall to 18% across all cereals.*

- *OECD net exports of wheat and coarse grains are expected to rise as import demand of NMEs increases with lower stocks.*

- *Levels and changes in stocks, in particular in China, represent the main uncertainty to the cereal projections, and could increase world grain market variability in the future.*

---

* All crop data provided in this chapter are on a crop year basis unless otherwise specified.

World market trends and prospects

International grain prices expected to increase in near term

After several years of lower grain prices, droughts in Australia and North America, as well as the economic turbulence in Latin America, have increased world wheat and coarse grain prices significantly in 2002. This, however, was not yet the start of a long-term price rise. With production recovering in 2003, prices for wheat and maize declined again, even though levels were still well above those observed in the recent past.

A moderate change in medium term price trends is expected with reduced global stock levels and with smaller supplies coming from stock rundowns. World wheat prices are projected to increase by 4% in 2005, as the rundown in Chinese wheat stocks slows and imports increase. Nominal world prices to remain relatively flat thereafter, reaching USD 154 per tonne by 2013. Similarly, as the decline in world coarse grains stocks – and mostly in China – slows down, world maize prices are expected to increase by 5% in 2005, and then continue a more moderate increase to USD 114 per tonne in 2013. Rice prices, which fell to low levels in 2000, and have since started to recover, are expected to jump by 14% and 9% in 2004 and 2005, respectively, and thereafter, to rise to USD 316 per tonne in 2013. This will bring the price ratio of rice to wheat back to the level of around 2 where it was during most of the 1990s and earlier periods.

Global cereal production on the rise again

The market years 2002 and 2003 have seen relatively low harvests of wheat and coarse grains in a number of major producing regions. Among the largest producers, yields in 2002 were significantly below trend in the harvests of the US, India (coarse grains), Canada, Australia, and Argentina (wheat), partly offset by above trend yields in Brazil (coarse

Figure 2.1. **Nominal world prices to remain largely flat after 2005, except for rice**

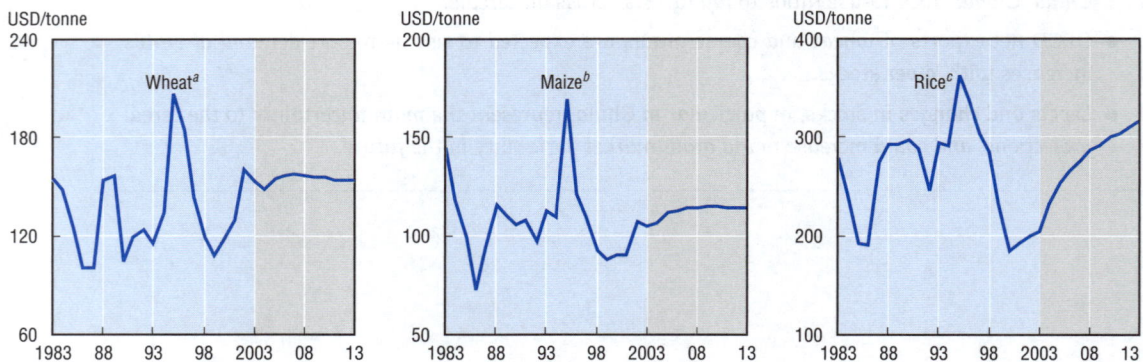

a) No. 2 hard red winter, ordinary protein, wheat, USA, f.o.b. Gulf Ports.
b) No. 2 yellow corn, USA, f.o.b. Gulf Ports.
c) Milled, grade b rice, f.o.b. Thailand.
Source: OECD Secretariat.

grains), Russia and Ukraine (for the second consecutive year). In 2003, very dry conditions and/or winterkill resulted in particularly bad harvests in the European Union, India (wheat), Russia (wheat), Argentina (coarse grains) and Ukraine, while above trend yields were experienced in the US (wheat) and Australia. Globally, average wheat and coarse grain yields in 2003 were found to be about 2% lower than what the recent trends would have suggested. In response to rising cereal prices, however, and assuming average weather conditions, production of wheat, coarse grains and rice is projected to increase. In 2013, wheat and coarse grain production should be about 19% and 17% larger, respectively, than in 2003, while the increase is less pronounced for rice, at 14%, over the same period. The growth rates are generally higher in earlier years when the price increases are more significant, but production should continue to rise throughout the projection period.

... while grain area is mostly unchanged

By far the largest part of the additional production is expected to come from higher yields. With the exception of some Non-member economies (NMEs), most notably Argentina and Brazil, the harvested grain area is largely unchanged in most countries, or even reduced. Yields are expected to continue their increasing trend. Assuming average weather conditions, global wheat and coarse grain yields should increase by about 1.3% per annum within the 10-year projection period (just over 1% p.a. on average after 2004), while rice yields are expected to grow more moderately. Stronger yield growths in 2004 are expected for some regions, mainly for the EU and Russia, which were seriously affected by drought conditions in 2003.

With rising livestock production, NME feed use to increase rapidly

With rising prices, global consumption of cereals is expected to grow less rapidly than production. Total use is projected to increase by 1.2% p.a., with slightly higher rates for wheat and coarse grains than for rice. Strong growth is expected particularly for feed use in developing countries. As meat production is expanding rapidly (see the Meat chapter for details), feed use of wheat and coarse grains in NMEs is projected to increase by more than 2% p.a. over the 2003-13 period, respectively, while growth in food use, which rarely exceeds population growth, is above 1.2% p.a. in a few countries and is negative in some

Figure 2.2. **Real prices to decrease for wheat and coarse grains, but to increase for rice**

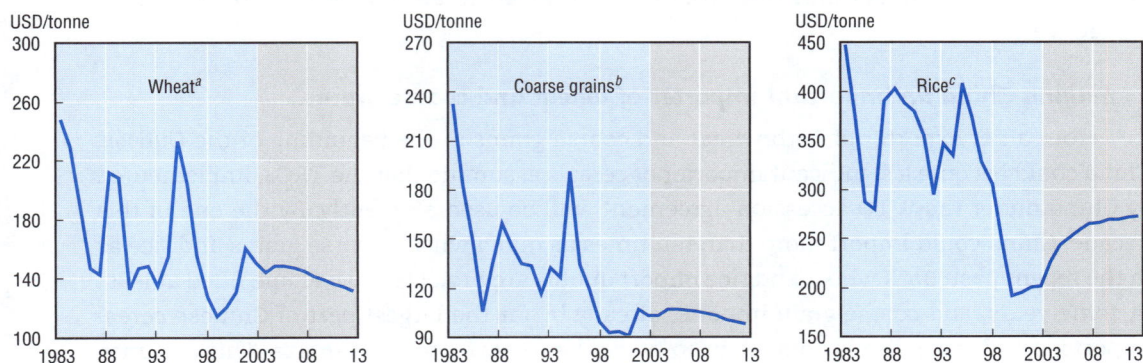

a) No. 2 hard red winter, ordinary protein, wheat, USA, f.o.b. Gulf Ports.
b) No. 2 yellow corn, USA, f.o.b., Gulf Ports.
c) Milled, grade b rice, f.o.b. Thailand.
Source: OECD Secretariat.

Figure 2.3. **Feed use of cereals to increase in NMEs**

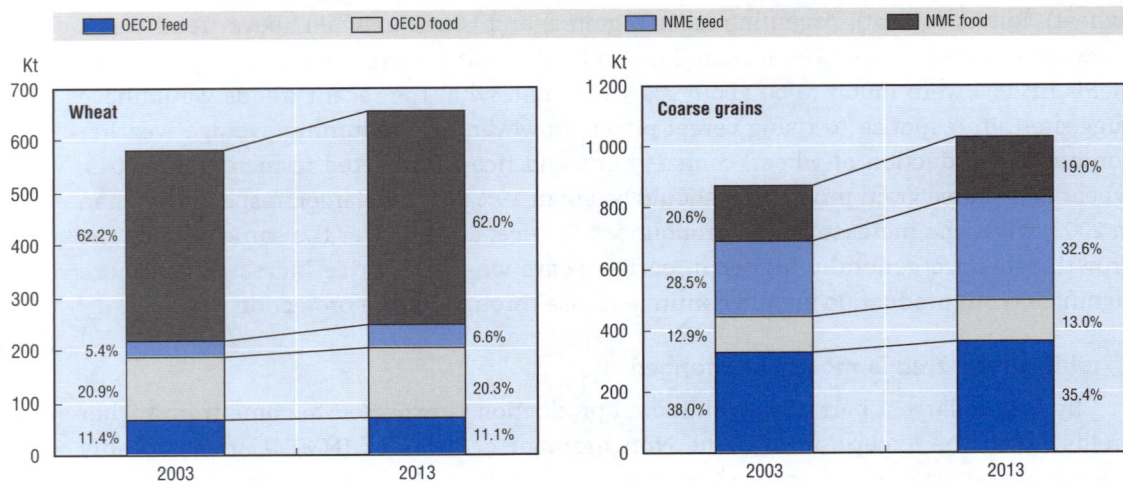

| ■ OECD feed | □ OECD food | ■ NME feed | ■ NME food |

Wheat (Kt)

2003: 62.2%, 5.4%, 20.9%, 11.4%
2013: 62.0%, 6.6%, 20.3%, 11.1%

Coarse grains (Kt)

2003: 20.6%, 28.5%, 12.9%, 38.0%
2013: 19.0%, 32.6%, 13.0%, 35.4%

Source: OECD Secretariat.

others. As a consequence, the feed share in total grain consumption (wheat, coarse grains and rice) should increase further and is likely to reach 40% by 2013. Obviously, feed use of cereals will expand the most in countries with large increases in meat production: China and Latin America, but also Russia.

Chinese stock run-downs to end soon

This baseline assumes a significant reduction over future years in the amount of supplies coming from the rundown of world cereal stocks, which has been a feature of the last couple of years. The decline in global cereal stocks, which has supplied more than 70 million tonnes in 2003, is expected to bottom out in 2006, with remaining stocks at barely more than 60% of the levels seen in the year 2000. In contrast to recent years, total grain production is expected to keep pace with global consumption. The main contributor to this stock adjustment is China, where wheat, coarse grain and rice stocks are expected to be reduced by almost 70% in total by 2008, after which they should stabilise. This reduction is particularly pronounced for coarse grains' stocks which are projected to fall by nearly 90%, while wheat and rice stocks are expected to be reduced by 66% and 60%, respectively.

... making China an important importer of wheat and coarse grain...

From a net exporter of both wheat and coarse grains at the beginning of the *Outlook*, China could become a significant importer of cereals assuming that the TRQs, implemented by China under the WTO accession agreement, will be used efficiently. By the end of this decade, China could import more than ten times as much wheat, coarse grains and rice as in the recent past. Both wheat and rice import quotas are projected to become filled at least in some years, and coarse grain imports, already by far the largest part of Chinese cereal imports, could reach levels equivalent to twice the import quota for maize. This assumes that other grains like feed barley and sorghum could be added to the feed mix in the increasingly modern livestock industry, which currently is dominated by maize.

Figure 2.4. **Significant stock declines in China come to an end**

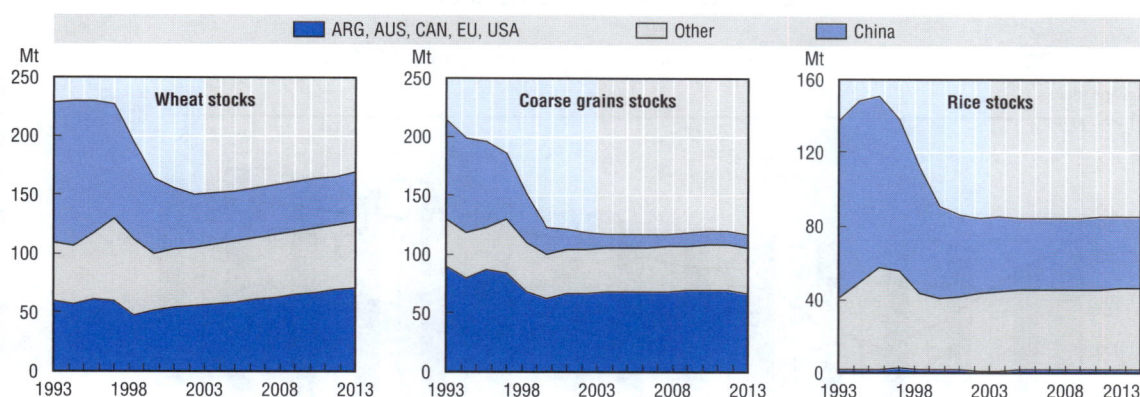

Source: OECD Secretariat.

... and with global stock-to-use ratios historically low

Bearing in mind the uncertainty related to estimates of actual stock levels (see below), global stock-to-use ratios for cereals are projected to fall to levels that have not been seen in the recent past. Declining grain stocks in China and, in the early years of the projection period the EU, and only moderately increasing stocks in other regions with continued consumption growth, result in declining global stock-to-use ratios for wheat, coarse grains and rice to touch a low point by the end of this decade. Ratios for wheat and rice, predominantly used for human consumption and hence considered crucial for assuring food security, remain higher at 25% and 19%, respectively, whereas the stocks-to-use ratio for coarse grains could fall below 12%. In contrast to the NMEs, where stock-to-use ratios are well below those in the OECD, cereal stocks within the OECD are not projected to decline. Indeed, OECD stock-to-use ratios for wheat and rice are expected to increase despite rising grain consumption, while the ratio for coarse grains is projected to change only little over the *Outlook* period.

OECD grain exporters increasingly compete with non-member suppliers

OECD net exports of wheat and coarse grains are expected to increase markedly. The largest increase in wheat exports is projected for the EU. These suffered significantly from the drought-reduced harvest in 2003 and the rising value of the euro, but are expected to reach close to 25 million tonnes in 2013. Despite the assumption of a continued relatively strong euro, the WTO limit on subsidised exports should not be an issue for wheat during the projection period. In contrast to rising EU wheat exports, those for coarse grains are not expected to reach the high levels seen in the recent past in the foreseeable future. With less coupled support due to the 2003 CAP Reform, coarse grains production largely stagnates after recovering from the poor harvest in 2003. As a slow increase is projected for consumption, EU coarse grain exports are expected to change only little after 2004, and to reach 12.7 million tonnes in 2013. Significant increases of coarse grain exports are expected for the United States instead, where international sales could reach 73 million tonnes by 2013, some 34% higher than in 2003.

Figure 2.5. **Increased net imports of cereals by NMEs as demand outstrips production**

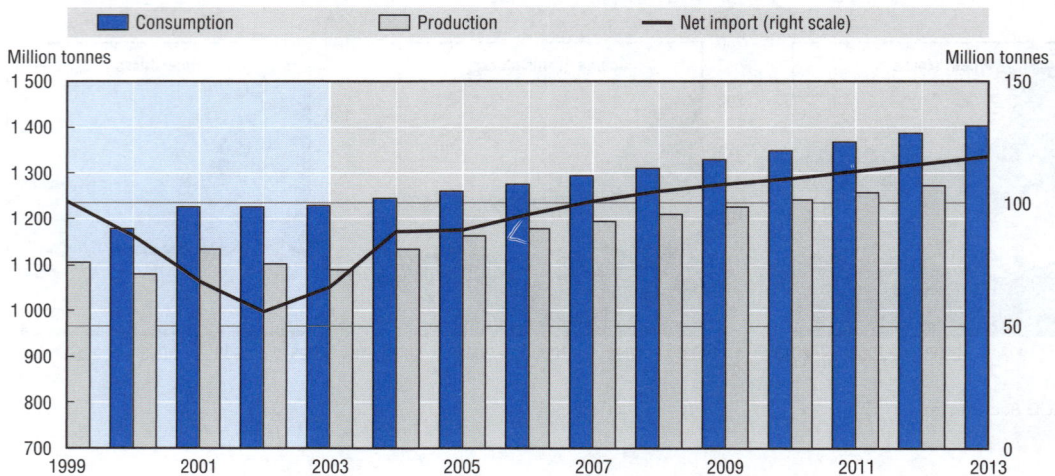

Source: OECD Secretariat.

Despite these increases in OECD exports, however, a few NME exporters are projected to play an increasing role in international cereal trade. In particular, Argentinean wheat and coarse grain exports are expected to grow by 2.7% p.a., respectively, resulting in slightly increased shares for both commodities in total world trade. Similarly, some of the CIS Republics, most notably the Ukraine, could establish significant grain exports.

In total, global trade of wheat and coarse grains is expected to grow moderately. World wheat and coarse grain trade should expand by 2.7% and 2.4% p.a., respectively, stimulated by the large imports by China. This is somewhat faster than global production growth. Global rice exports are projected to grow more substantially at almost 4% p.a. In particular, India should continue its trend towards larger export supplies, and Thailand as well as some smaller rice exporters, such as Argentina, are also likely to expand their rice trade. In contrast, the rice trade of most OECD member countries is not expected to change much. The EU is an exception here. The reform of the EU rice market organisation, in particular the 50% cut in the intervention price, the shift towards less coupled support and the "Everything-But-Arms" agreement that allows exports from least developed countries to enter the EU duty-free, should result in a significant increase of rice imports.

Issues and uncertainties

Policy to become somewhat less distorting in some OECD countries...

Not only in the case of rice, but also for other crops, the Common Agricultural Policy (CAP) of the EU becomes less distorting due to the 2003 reforms. With the cut of monthly increments of the cereal intervention price, the abolition of rye intervention, and the shift of area payments for arable crops towards a largely decoupled Single Farm Payment, the impact of agricultural policies on markets is becoming less significant (see Chapter 7 entitled Medium term market impacts of the 2003 EU Common Agricultural Policy reform). At the same time, as grain prices are projected to increase, marketing loan benefits as well as counter-cyclical payments in the United States are expected to play only a minor role in production decisions over the *Outlook* period. Obviously, in the event of above-average harvests and hence lower-than-expected market prices, both support measures might become more important again (see the Analysis of the 2002 US Farm Security and Rural Investment Act in the 2003 edition of *Agricultural Policies in OECD countries: Monitoring and Evaluation*. Similarly, the lower EU intervention prices may have an impact in the event that world prices, when denominated in euros, are below these support levels, as in the cases of coarse grains during the *Outlook* period.

... while those in other countries remain important for market developments

On the other hand, continuing high support to grain producers has a more distorting impact in several countries. High market price support and restrictions on international trade is maintained for rice in both Japan and Korea, and the same commodity remains significantly supported in the EU (despite a major cut in support prices) and the United States as well. A number of other OECD member countries with smaller markets, also have policies that raise incentives for producers to supply grains (as well as other commodities).

Obviously, this *Outlook* report does not make any presumption on the outcome of the Doha Round negotiations on a further multilateral trade liberalisation currently ongoing in the World Trade Organization (WTO). Any further steps towards more liberalised markets and trade in agriculture could have implications for future developments in production, use, trade and prices of agricultural commodities.

Uncertainty of actual Chinese grain trade developments

The radical change of China from a net exporter to a large net importer of cereals during the Outlook period is clearly uncertain. Two elements in the market projections are particularly uncertain. First, the level of existing grain stocks, as well as the actual rates of stock decline, is imperfectly known at best. Wheat, coarse grain and rice stocks are suspected to be significantly higher than assumed by current market statistics, and the estimates of supply and demand may need to be revised. Second, Chinese imports might remain well below the *Outlook* projections if the tariff rate quotas cannot be filled. An earlier OECD report Proceedings, *China's Agriculture in the International Trading System* (2001, pp. 21 ff.) has found that even with the TRQs which China has agreed to when it joined the WTO, imports might well remain limited for several reasons, including infrastructural problems and other obstacles to trade, and the large shares of the trade quotas remaining under control of Chinese state trading enterprises (if only for the first three quarters of a calendar year). A special section of this report analyses the market implications of these important unknowns.

... while the low stock levels might increase global price variability

With global stock-to-use ratios to decrease to levels not observed in the recent past, there is a risk of increased world price variability. Even though stocks in the main exporting regions are projected to moderately increase (Figure 2.4), they remain low relative to past decades. In years of exceptionally high import demand in major importing regions, the reduced short-term export supplies could induce significant price increases that could threaten food security for poorer parts of the importing countries' populations. A similar situation would occur with a poor harvest in major exporting countries as this would aggravate the low stock situation and further reduce total available export supplies.

Increasing freight rates might well impact market developments

This *Outlook* does not take into account any longer-term increase in transportation costs. However, freight rates have increased significantly in the recent past. Given the large share of wheat and coarse grains that is traded internationally, persistently high freight rates could have a dampening effect on trade quantities, and to some degree could separate markets from each other.

Chinese Grain Import Surge – Market Implications of Alternative Stock Estimates and Trade Policies

Introduction

Chinese grain market developments are of particular importance for global agricultural market projections for at least two reasons: First, the sheer size of the Chinese market makes world markets sensitive to changes in the Chinese trade regime induced by even small relative adjustments in domestic supply and demand. Second, market analysts largely agree that some of the current estimates for domestic market variables are likely to be imprecise (*i.e.* potentially subject to large error).

In the recent past, the Chinese grain market was found to be characterized by a significant deficit in domestic production when compared to consumption. Despite production failing to meet consumption by 34 million tonnes per year, on average, between 2000/01 and 2003/04, China remained an important net exporter of grains, exporting 10 million tonnes of wheat, coarse grains and rice per year, on average, in the same period, according to the *Aglink* database. Obviously, the deficit is filled from a rundown in large cereal stocks, which were built up particularly during the 1990s, when domestic production out-performed consumption by significant amounts.

Clearly, with consumption remaining larger than production in China, its status as a net exporter cannot be maintained forever. Once stocks are reduced to levels considered to be minimum reserves by the Chinese authorities, the deficit will have to be filled by imports from abroad. Recent developments indicate that Chinese grain prices have increased significantly over several past months, indicating some first signs of scarcity in domestic markets. Between February 2003 and February 2004, maize and wheat prices increased by 12% and 36%, respectively, while rice prices rose by between 40% and 60%. Furthermore, policy changes on the rural programme, announced at the beginning of 2004, indicate that the Chinese authorities are concerned about the large gap between supply and demand as well. Among other elements, these policies aim to increase farmers' net incomes and the profitability of grains production by means of direct subsidies, better access to rural credit and reduced tax burdens faced by farmers. In addition, recent changes in the Chinese constitution enforced property rights, which could help to stimulate agricultural investment.

If the domestic supply-demand ratios remain as in the past four years, the drop in stock supplies would mean a switch from 7 million tonnes net exports to 34 million tonnes net imports. This change in the net trade position by some 40 million tonnes per year compares to current global trade quantities of around 230 million tonnes of wheat, coarse grains and rice – a ratio that indicates the relative importance of the issue.

When China joined the WTO, it agreed to open its domestic market through the introduction of TRQs for wheat, maize and rice, totalling to some 22 million tonnes per year, with a certain and, in the case of maize, increasing share of import quotas allocated to private traders other than the official Chinese state trading enterprise. In addition, any STE quota shares unused by the third quarter of a given year were to be opened to non-STE traders.

Indeed, the baseline reported in this year's *Agricultural Outlook* projects a significant rise in cereal imports. With vanishing supplies from domestic stocks, this year's baseline expects both wheat and rice import quotas to become filled, and coarse grain imports (comprising both maize and other coarse grains) to far exceed the TRQ for maize. There are several unanswered questions, though. First, the exact magnitude of current Chinese grain stocks is unknown (and is indeed regarded as a state secret by the Chinese authorities). Furthermore, the quality of the remaining stocks, which may to a large part date back over several years now, is subject to uncertainty as well and may have deteriorated over time. Obviously, any change in the total stock figures will result in a change in the pressure for cereal imports. Second, opening up TRQs may not be sufficient to allow for the necessary grain imports; domestic policy reforms, *e.g.* the reduction of trade control power of the current Chinese STEs – as well as infrastructural improvements are equally necessary to allow the import quotas to be actually filled.[1]

This section aims to discuss the impacts on China and the international cereal market of changes in two factors driving Chinese grain imports. First, hypothetical alternative developments of Chinese grain stocks are considered based on recent findings. Second, an analysis is provided of the market implications that would arise from a significant under fill of import quotas for technical or political reasons, for example, because the Chinese state traders could prevent imports within the TRQ shares that remain under their control, or from an overfill if Chinese authorities decided to expand the import quotas.

Market implications of two alternative scenarios for Chinese cereal stocks

Using new information derived from recent FAO work on historical Chinese grain market data and stocks, hypothetical market balances for the period 1994-2002 were generated that allow for significantly higher stock levels (Figure 2.6). These aggregate grain balances were then broken down to wheat, coarse grains and rice components following two simple approaches which result in different adjustments to the respective market data (see the annex for a detailed description of the data adjustments).

Corresponding to the two simple approaches to adjust historical market balances outlined in the annex, the market implications of two scenarios are analysed: First ("Scenario 1"), stocks for wheat, coarse grains and rice are assumed to be proportionally larger, while the patterns of supply and demand remain unchanged.[2] In effect, this scenario simply assumes a slower pace of stock draw-downs by one or two years.

A second scenario ("Scenario 2") assumes disproportionate change in stocks of wheat, coarse grains and rice and additionally takes into account the changes in production shortages. Stock reductions are therefore assumed to slow down only in 2006, 2007 and 2013 for coarse grains, rice and wheat, respectively (as opposed to 2004 for all cereals in the baseline).

Figure 2.6. Chinese cereal market developments 1990 to 2002, adjusted to recent FAO estimates

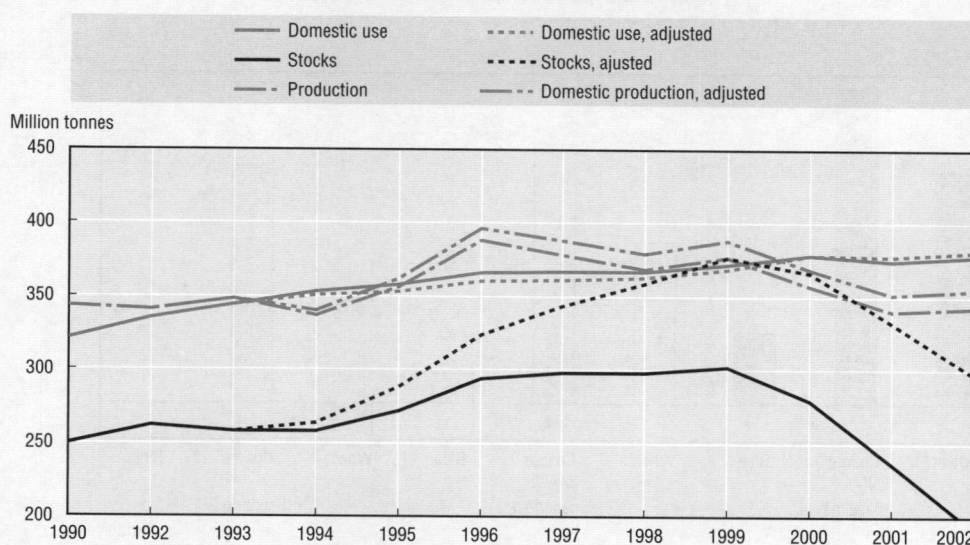

Source: OECD Secretariat based on Aglink Database and FAO estimates.

The following table shows the stock variations in the baseline as well as those assumed in the two scenarios:

Table 2.1. Stock variation of wheat, coarse grains and rice in baseline and scenarios, 2003-2013

Million tonnes

Scenario	Grain	2003	2004	2005	2006	2007	2008	2009	2010	2011	2012	2013
Baseline	Wheat	−17.6	−10.9	−6.5	−3.0	−0.8	−0.7	0.0	−0.2	−0.3	−0.1	0.3
	C. Grain	−18.2	−5.9	−3.6	−2.3	−0.6	−0.1	0.1	0.1	0.2	0.2	0.1
	Rice	−18.2	−6.2	−3.3	−0.9	−0.4	−0.2	0.2	−0.1	−0.2	−0.1	−0.1
Scen. 1	Wheat	−17.6	−17.6	−17.6	−17.6	−9.4	−3.0	−0.8	−0.7	−0.3	−0.1	0.0
	C. Grain	−18.2	−18.2	−14.0	−2.3	−0.6	−0.1	0.0	0.0	0.0	0.0	0.0
	Rice	−18.2	−18.2	−18.2	−10.8	−0.9	−0.4	−0.2	0.0	0.0	0.0	0.0
Scen. 2	Wheat	−17.6	−17.6	−17.6	−17.6	−17.6	−17.6	−17.6	−17.6	−17.6	−17.6	−6.5
	C. Grain	−18.2	−18.2	−18.2	−5.9	−3.6	−2.3	−0.6	−0.1	0.0	0.0	0.0
	Rice	−18.2	−18.2	−18.2	−18.2	−6.2	−3.3	−0.9	−0.4	−0.2	0.0	0.0

Source: OECD Secretariat.

As Table 2.1 shows, the two scenarios generally assume significantly higher grain supplies from Chinese stocks. Consequently, import requirements fall relative to baseline levels, as shown in Figure 2.7.

This has two direct implications. First, with lower net imports (and indeed continued net exports instead for some years), world prices are significantly lower than projected by the baseline. On average, for the 2004-2013 simulation period, international prices would be between 2% and 4% lower than under baseline assumptions. Second, domestic prices, which depend on world prices as well as on actual import levels through TRQ tariffs, are reduced even more strongly (Figure 2.8).

Figure 2.7. **Impact of alternative stock assumptions on Chinese net grain imports, average 2004-2013**

Million tonnes

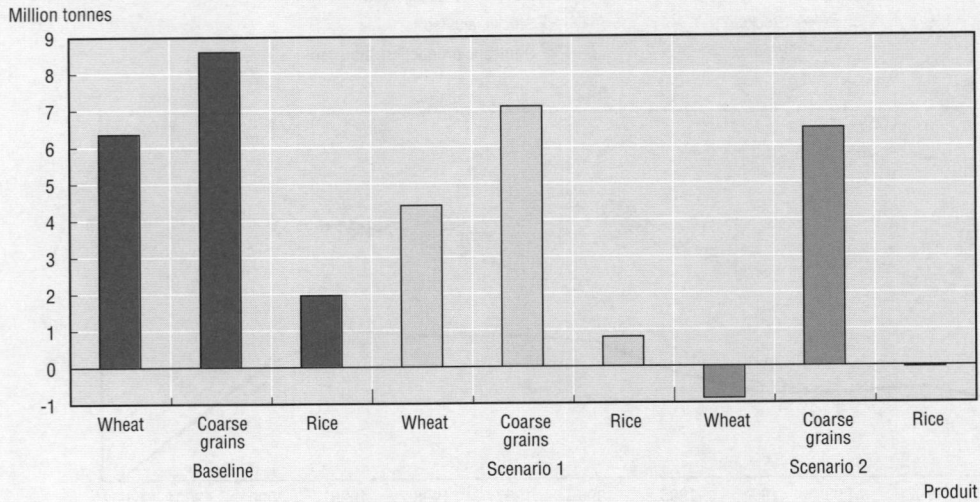

Source: OECD Secretariat – Aglink simulation results.

Figure 2.8. **Impact of alternative stock assumptions on Chinese and world grain market prices, average 2004-2013**

Change relative to baseline (%)

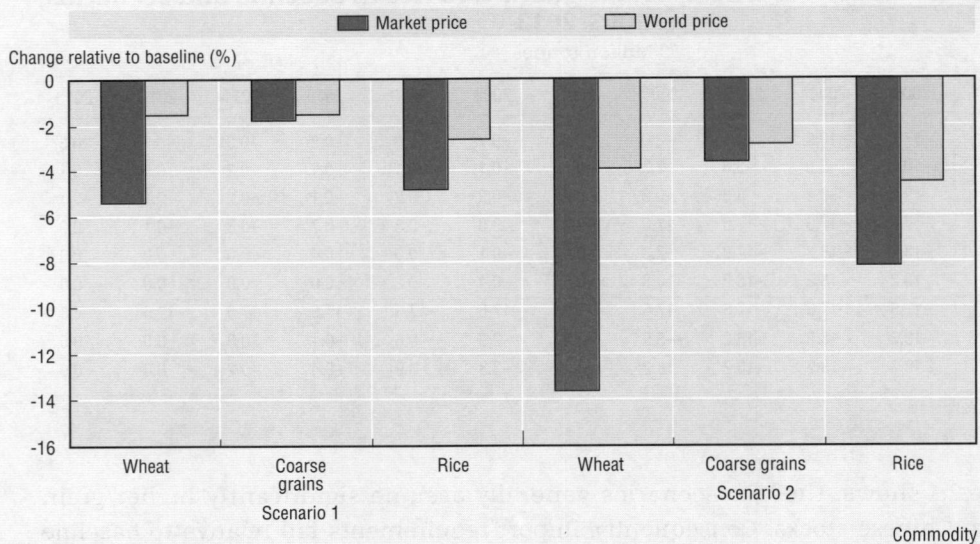

Source: OECD Secretariat – Aglink simulation results.

It is interesting to note that, despite the smaller changes in rice trade, price changes on domestic and world rice markets are relatively large when compared to wheat and coarse grains. The main reason for this is the relatively thin world market for rice as compared to those for wheat and coarse grains, which tends to increase price response to shocks in supply and/or demand.

Obviously, the reduced domestic prices for cereals cause production and consumption to adjust. On average, total area for wheat, coarse grains and rice is reduced by 0.5% and 0.9% under the two alternative stock assumptions, and the average reduction of total grain production would reach 0.8% and 1.6%, respectively. Wheat and rice are reduced more strongly than coarse grains in the first scenario, and most of the reduction in the second one would come from wheat, corresponding to the stronger price changes. At the same time, food use, particularly of wheat, tends to be higher with the lower prices. While in the first scenario average food use of rice and coarse grains is hardly changed, the significantly lower wheat prices in the second scenario cause some shift from rice (–1%) to wheat (+4%) for food consumption. Lower grain prices would trigger higher feed use as well – with livestock production almost 1% higher on average. Feed use of coarse grains, the largest contributor to overall feed, would on average be 0.2% and 0.5% higher in the two scenarios than under baseline assumptions, respectively. Feed use of wheat, although small in absolute quantity, would increase by 5% and 13% under the two alternative stock assumptions due to the stronger drop in wheat prices.

Market implications of restricted access to TRQ import licenses

As mentioned above, the baseline assumptions of completely filled TRQs for cereals may not be entirely realistic, if the introduction of the TRQs is not supported by domestic reforms and necessary infrastructural development. In particular, it was said that the persistent system of STEs controlling large shares of the total TRQs might represent a *de facto* reduction of the import quotas. On the other hand, in a severe deficit situation the Chinese authorities may opt to expand imports at in-quota tariffs beyond the agreed TRQ levels.

Obviously, there is no reason to assume that the share of TRQs that is left under STE control would remain entirely unfilled even if domestic market conditions would ask for the respective imports. In fact, given the possibility of non-STE traders to import within the unfilled STE quota shares after the third quarter of any given year, it seems very unlikely that the STE shares would remain completely unfilled in a Chinese deficit situation. Nor does this analysis aim to suggest that the non-STE shares would become completely filled under all circumstances. For illustrative purposes, however, the third scenario assumes effective TRQs for wheat, maize and rice to be reduced to the non-STE shares as given in the last column of Table A.2.1. In contrast, a fourth scenario assumes an expansion of import quotas to a degree that would make the TRQs unbinding.

With effective TRQs, and hence imports significantly reduced, domestic market prices for grains would be markedly higher than under baseline conditions. While the TRQs for wheat and rice would remain binding, prices for these food grains would increase by 20% and 9% on average, respectively (Figure 2.9), resulting in market prices moving closer to, though still significantly below, the price levels that would be defined by the over-quota tariffs. On the other hand, the decline in imports by China reduces world cereal prices by 3% to 4% on average for the simulation period.

These significantly higher prices create larger incentives for domestic producers. Chinese cereal production would on average be about 2% higher than in the baseline, with stronger effects on wheat output than for coarse grains and rice. On the other hand, however, there is a strong negative effect on domestic use of grains, as shown in Figure 2.10. In particular, with reduced livestock production, feed use of cereals would be reduced by 2% on

Figure 2.9. **Impact of restricted and extended import quota access on Chinese and world grain prices, average 2004-2013**

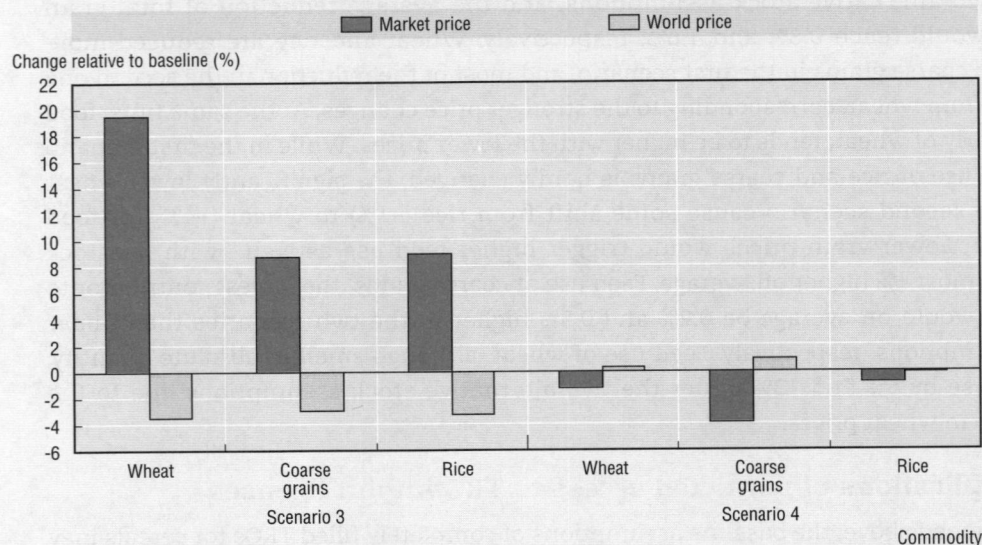

average. Food consumption would decline as well, with a significant shift from wheat to rice as the staple food.

Allowing for demand-driven imports at in-quota tariffs, on the other hand, would result in imports larger than current TRQ commitments. Wheat imports could be up to 11% higher than with binding TRQs, whereas the increase in rice imports could reach 5%. Coarse grain

Figure 2.10. **Impact of restricted and extended import quota access on Chinese domestic grain use, average 2004-2013**

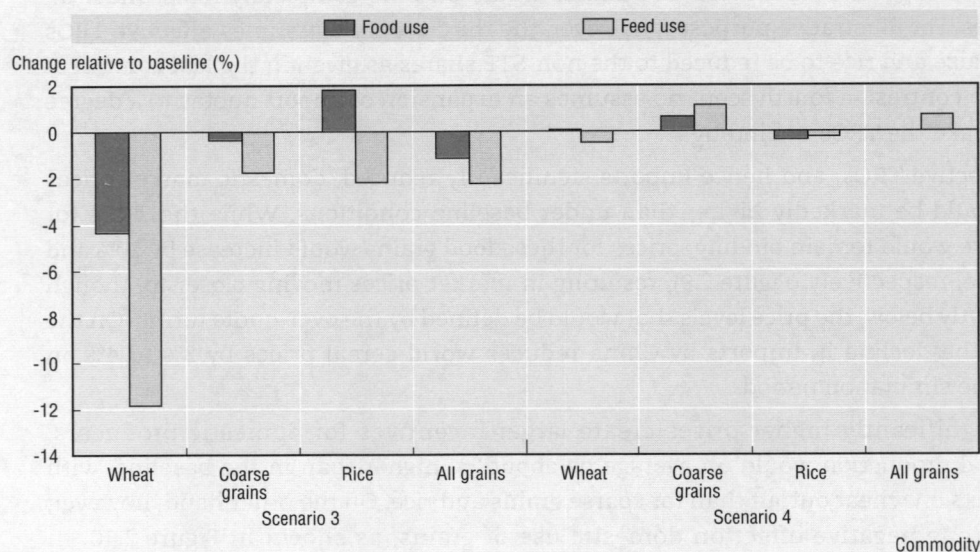

Source: OECD Secretariat – Aglink simulation results.

imports, which are only partly determined by the TRQ for maize, would increase by up to 38% if the in-quota tariff were applied to all maize imports. Consequently, domestic wheat and rice prices, now determined by the respective in-quota tariffs, would be 1% lower, on average, than under baseline assumptions, while the average coarse grain price would be reduced by 4%. Domestic consumption, and particularly feed use of coarse grains, would benefit.

Conclusions

With baseline projections being influenced by the marked increase of Chinese cereal imports after available supplies from grain stocks run out, the focus of this section is on two main factors that are likely to impact these trade quantities. Two scenarios on levels and depletion rates of Chinese grain stocks have shown that both domestic and world cereal prices would be significantly lower than expected if the existing stocks were to provide supplies for a longer period than anticipated. Obviously, this should benefit grain consumers and have a positive effect on meat production in China.

The other two scenarios have focussed on the levels of available cereal import quotas. For several reasons, actual imports could be more limited than suggested by the TRQ quantities even if domestic markets would ask for larger supplies from exporting countries. One of them is the remaining large shares of TRQs that are controlled by STEs and that might maintain an import-limiting strategy. On the other hand, Chinese authorities may opt to allow for larger than agreed imports at in-quota tariffs if the domestic market shows significant grain deficits. The scenario results show that a restrictive import policy would have severe implications on domestic markets, with livestock producers as well as consumers suffering from significantly higher grain prices. Clearly, such implications cannot be in the interest of China which repeatedly has made clear the high priority it accords to food security issues. In contrast, allowing for larger imports at low tariffs would benefit grain users through lower prices.

Notes

1. See OECD (2001) China's Agriculture in the International Trading System, pp. 21ff. and OECD (2002) Agricultural Policies in China after WTO Accession, pp. 115, for discussions on possible implications of the introduction of TRQs in China.

2. As discussed above, the higher stocks could not have developed without adjustments in supply and demand. As the gap between production and consumption would hardly be changed, and for the sake of simplicity, we ignore this here without reducing the meaningfulness of the analysis.

Annex

Hypothetical revision of Chinese grain market balances

The FAO has undertaken a serious effort in analysing other sources of data on Chinese grain consumption for food and feed, as well as on Chinese grain stocks,[1] than official statistics. The general findings are that, while food consumption of cereals could be significantly overestimated, feed use appears to be higher than stated in the recent statistics. Finally, total grain stocks could be underestimated by as much as 25% for the peak year 1999 according to adjusted data from the National Bureau of Statistics, NBS.[2]

It is commonly argued that, while production and trade quantities tend to be relatively well documented, the basis for consumption and stock quantities is weak. Trade is generally channelled through a limited number of ports and can therefore be considered as rather accurately reported. For production, this is true only to a limited extent given the relevance of on-farm use in rural China, but it could nonetheless hold for marketed production quantities.

For this analysis, FAO findings are utilised to adjust food and feed consumption data, as well as data on grain stocks and to a lesser extent production, to obtain another hypothetical set of market balances for wheat, coarse grains and rice as the starting point for our scenarios. In particular, total grain stocks are assumed to be higher by 25% in 1999, and food and feed consumption in 2001 are adjusted by –7% and +17%, respectively, according to the mentioned report. Furthermore, assuming that most of the under-reporting of production would be feed used directly on farm, 2001 production is assumed to be 4% (or half the level change of feed use) higher than what is given by the database.[3] Through minimising the sum of the squared changes in absolute growth figures of domestic use for feed and food, and of production, and allowing for adjustments to start in 1994/95 in order to maintain a smooth development of these data, a hypothetically adjusted time series is obtained for the market balance of total cereals as given in Figure 2.6. Under the given assumptions and with the rather mechanical adjustments made, total grain stocks would have reached their peak in 1999/00 at the level of 377 million tons, hence significantly higher than what is found in the current *Aglink* database. Significant stock declines would not have started before 2001/02, with rates slightly lower than given by the current data base.

Clearly, this alternative data set would suggest that stocks could be sufficient to help balancing the markets for a long period even in the case of significant reductions in grain production, and that the Chinese target of high self-sufficiency would be threatened much later than foreseen by current projections. This view seems to be supported by the continued efforts of the Chinese authorities to export large amounts of grains, particularly maize.

Adjustments on the market balances for wheat, coarse grains and rice

The above adjustments need to be broken down to the individual grains represented in *Aglink*, i.e. wheat, coarse grains and rice. The same rates of adjustments for feed and food use are applied to the three crops, which results in quite different changes in total domestic use reflecting difference in the relative shares of feed and food. Thus, domestic use of coarse grains is adjusted upwards by more than 9% in 2001, while total wheat use is reduced by 7% and that for rice decreased by 2%.

In the absence of more detailed information, two simple adjustments to stocks and production would give consistent time series of grain market balances. First, it could further be assumed that stock adjustments are proportionally equal across cereal types. In consequence, stocks in 1999 are assumed to be 25% higher than given in the database for wheat, coarse grains and rice. In consequence, adjustments for grain production would largely correspond to those for total domestic use. Therefore, the rates of stock declines in 2002 would be roughly equal to what is given by the database for all cereals.

Second, it could be assumed that the full adjustment to grain production (equal to half the adjustment of feed use) would be assigned to coarse grains, the most important type of feed grains. Adjustments in grain stocks would then become residual, with the largest ones for coarse grains (+36% in 1999) and the lowest adjustment for rice (+17%). In this case, rates of 2002 stock declines would be 7% higher than in the database for coarse grains, but 16% and 49% lower for rice and wheat, respectively.

Chinese WTO commitments on grain import quotas

Table A.2.1 shows the development of the Chinese import quotas for wheat, maize and rice, as well as the respective shares that remain under the control of Chinese state traders.

Table A.2.1. **Development of Chinese grain tariff rate quotas after WTO accession**

Commodity	Year	Total import quota 1000 t	Share reserved to STEs %	Remaining non-STE quota 1000 t
Wheat	2002	8 468	90	847
	2003	9 052	90	905
	From 2004	9 636	90	964
Maize	2002	5 850	68	1 872
	2003	6 525	64	2 349
	From 2004	7 200	60	2 880
Rice	2002	3 990	50	1 995
	2003	4 655	50	2 328
	From 2004	5 320	50	2 660

Source: Schedule of the People's Republic of China.

Notes

1. FAO (2004): Critical Review of China's Cereal Supply and Demand and Implications for World Markets. Paper presented at the Joint Session of the Intergovernmental Group on Grains (30th Session) and the Intergovernmental Group on Rice (41st Session). Rome: Food and Agriculture Organization of the United Nations.

2. The paper notes that this estimate on cereal stocks would need "a more in-depth review of all the other main variables in China's cereal balances" and proposes to change stock statistics only

marginally. In this analysis, we still use the higher stock estimate in order to give an idea on the magnitudes of the possible impacts.

3. Two reasons could support the assumption of production to be under reported in the late 1990s: First, as mentioned above, a large proportion of grain production could have been used for feed on farm, a strategy particularly interesting to farmers in the second half of the 1990s when domestic prices and hence incentives to sell commodities on the markets were low (OECD 1999) Agricultural Policies in Emerging and Transition Economies, p. 125; (OECD 2002) China in the World Economy p. 66. Second, the existing grain quotas were supplemented by so-called "protective" price that was intended to represent a floor price to farmers. However, even though this protective price was continuously reduced after 1997, the market price fell below this protective price in the year 2000. While the protective price legally was not subject to quantity restrictions, the financial burden could have stimulated local administrations to buy and report less than the actual production.

ISBN 92-64-02008-X
OECD Agricultural Outlook: 2004-2013
© OECD 2004

Chapter 3

Oilseeds*

Main projections – outlook in brief

- *After a sharp initial drop, oilseed prices are projected to remain flat in nominal terms to 2013, with lower oilseed meal prices, higher vegetable oil prices and with all real prices continuing to decline.*

- *Rate of growth in global oilseed production expected to slow to 2% per year, on average, as the growth in oilseed area expansion moderates in South America.*

- *World oilseed meal use to grow at a marginally faster pace. Demand to increase faster than average in countries with rapidly expanding livestock sectors.*

- *Growth in world vegetable oil use to outpace that of oilseed meal use. Production from oilseed crush is supplemented by palm oil production, which grows at some 4% per year.*

- *Key uncertainties are policy impacts on producer and consumer incentives, area expansion in South America and the potential for animal diseases to disrupt oilseed meal markets.*

---

* All crop data provided in this chapter are on a crop year basis unless otherwise specified.

World market trends and prospects

Matched supply and demand at flat nominal prices...

The projection period assumptions of normal weather and a supportive macroeconomic environment favour a stable price path in oilseed markets. Prices of oilseeds and oilseed meal – and, to a lesser extent, those of vegetable oils – are expected to be lower in the early years of the projection period than the adverse weather-induced higher levels of the recent past. Over the longer term, additional demand created by economic and population growth is matched by trend yield growth and some expansion in oilseed area, leading to fairly steady nominal prices for oilseeds. At the same time, a somewhat greater demand growth for vegetable oils tends to lead to some upward pressure on these prices, even as palm oil production continues its fairly rapid expansion. With global oilseed meal supplies outpacing the growth in demand, world oilseed meal prices are expected to decline slightly in nominal terms over the *Outlook*.

... but only after some recovery in production at the beginning of the Outlook

Production is expected to rebound in 2004 because of two factors. First, assumed normal weather results in better yields. In the United States, a key producer that accounted for 34% of world oilseed production in the 1998-2002 base period, yields in 2004 are expected to rise by about 19%. The second factor is an increase in area planted to oilseeds as producers respond to the rising prices of 2003. Area is expected to expand by at least half a million hectares each in Brazil (1.2 mha more), Canada (0.9 mha), Australia (0.7 mha) and China (0.6 mha) in 2004 relative to 2003. These two factors help to explain the increase in world oilseed production by more than 10% in 2004. The jump in production drives 2004 prices lower, with oilseed prices down by 20% and oilseed meal

Figure 3.1. **Relatively flat nominal world prices for oilseeds and oilseed products**

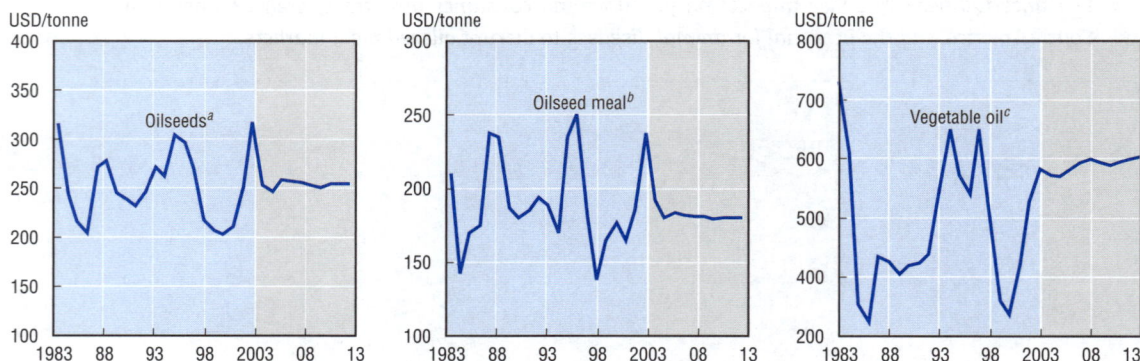

a) Weighted average oilseed price, Europe.
b) Weighted average oilseed meal price, Europe.
c) Weighted average price of oilseed oils and palm oil, Europe.
Source: OECD Secretariat.

prices off by almost as much. The smaller decrease in 2004 vegetable oil prices follows from the less significant price increase in the previous year as well as a smaller relative increase in production.

Brazil and Argentina production continues to boom as others stabilise

Beyond the near-term adjustments, growth in production is expected to be concentrated in many of the countries that have accounted for much of the expansion in recent years. Whereas yield growth raises production in most countries, further increases in harvested areas in Brazil and Argentina also lift the shares of these two key producing countries. Argentina's share of world production rises from under 14% in the base period to over 17% by 2013, representing an increase in production of 3.6% per year over the projection period, with about two-thirds of that growth attributable to greater area harvested. The increase in Brazilian oilseed area is expected to be even greater, more than two-thirds higher relative to the base period. In fact, although a substantial share of the increase over the 1998-2002 base period average takes place by 2003, the addition of more than 5.7 million hectares of oilseed area over the period corresponds to an average increase of over 2% per year – more than double the rate of yield growth in Brazil. The projected expansion of oilseed area represents an uncertainty about future oilseed markets that is explored in a special section, entitled "The impact of further area expansion for oilseeds in Brazil". Nevertheless, in the baseline Brazil's share of world oilseed production is expected to rise from 15% in the base period to over 22% at the end of the projections.

Within the OECD area, oilseed production is expected to increase only modestly, largely owing to expanded yields in the absence of strong incentives to plant more area to oilseeds. It is true that some area expansion occurs in 2004, as producers respond to higher 2003 prices, and this transitory effect causes some part of the projected 21% increase in OECD oilseed production in that year. Even so, the OECD share of world oilseed production in 2004 is still well below the 46% share of the 1998-2002 base period and this share falls further to 37% by the end of the projection period. At the same time, the United States' share of world production falls from over a third to just over a quarter. Rising yields in the United States are partly offset by marginal decreases in oilseed area over the period

Figure 3.2. **Falling real world oilseed and oilseed product prices**

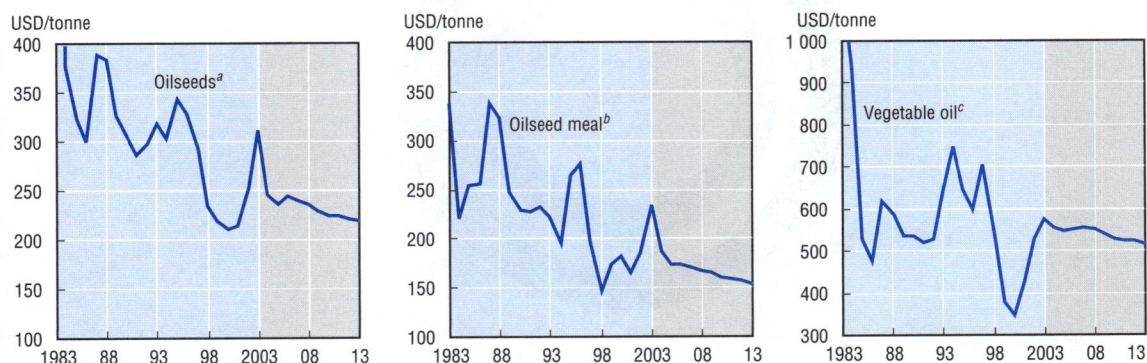

a) Weighted average oilseed price, Europe, deflated by USA GDP deflator 2002 = 1.
b) Weighted average oilseed meal price, Europe, deflated by USA GDP deflator 2002 = 1.
c) Weighted average price of oilseed oils and palm oil, Europe, deflated by USA GDP deflator 2002 = 1.

Source: OECD Secretariat.

and the net effect is a much slower pace of increase than is the case in South American competitors. The consequences of these relative paths in future oilseed production are represented in Figure 3.3.

Figure 3.3. **Oilseed area to increase in Brazil and Argentina, but not in the United States**

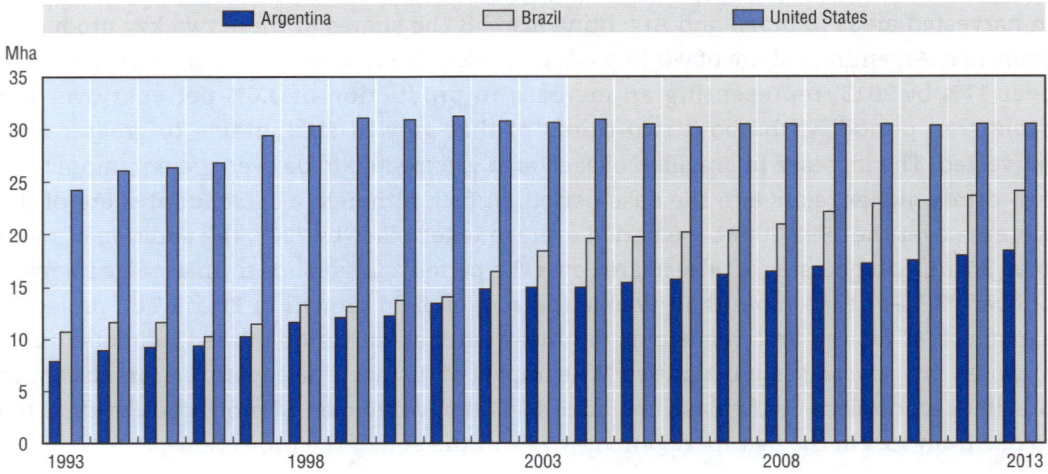

Source: OECD Secretariat.

Figure 3.4. **Changing shares of world oilseed production, average 1998-2002 compared to 2013**

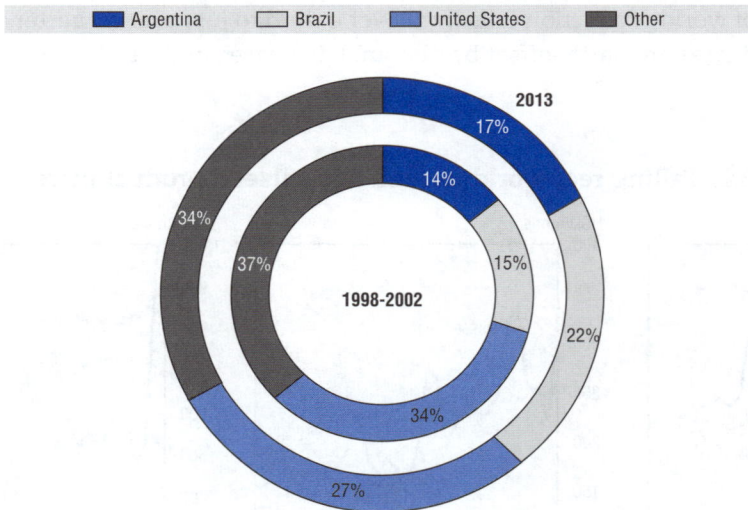

Source: OECD Secretariat.

OECD AGRICULTURAL OUTLOOK: 2004-2013 – ISBN 92-64-02008-X – © OECD 2004

Gradual shift in trends in crushing capacity

Oilseed use is, as usual, dominated by the crushing industry which is not always centred on the same regions that are responsible for most of the oilseed production. Many countries or regions, such as Japan, Korea and Mexico, but especially the EU and China, crush a far greater amount of oilseeds than they grow. In the projection period, it is expected that the shares in most of these countries, except China, will decline as their quantities of oilseeds crushed grow at a slower pace than the world total.

In the key oilseed producing countries, oilseed crush tends to follow a similar pattern as oilseed output, as exporters seek to capture the extra value added, though moderated by the adjustment costs associated with changing capacity and also by competition from importers who crush their own seeds. The United States' share of world crush slides from 23% in the base period to 20% at the end of the projections. Brazil and Argentina will account for greater shares of world crush, with shares rising from under 12% for each in the 1998-2002 base period to 15% and 13%, respectively by 2013.

Oilseed stocks recover part of the way

Falling oilseed prices will encourage greater stock-holding, so stocks that had previously been drawn down recover substantially at the start of the projection period when production rises. Relative to consumption, however, they are not expected to return to the same average level, 9.2%, as observed in 1998-2002. Nevertheless, the stocks-to-use ratio of about 8.6% over the projection period does provide more of a buffer against shocks to the market than the estimated 6.6% ratio in 2003. Geographically, the stock distribution is expected to shift away from the OECD region as Brazil, in particular, is expected to increase its stocks with climbing production, both in absolute terms and also relative to domestic consumption.

Oilseed meal demand expanding with incomes

Due to the role of oilseed meal in animal feed, livestock production drives the demand for oilseed meal. Given the baseline assumptions of higher income growth and the expectation that consumer demand for livestock products will remain sensitive to income, it is unsurprising that global oilseed meal consumption increases at over 2% per year. Nor is it surprising that oilseed meal use is concentrated in the same regions as expanding livestock inventories.

Much of the growth in oilseed meal consumption occurs outside the OECD area. China has been and is expected to continue to be the fastest growing of the major NME markets as consumption ends the projection period at more than double the base period volume. During the projection period, China's oilseed meal consumption is expected to rise from 22 mt in the 1998-2002 base period (already at 30 mt in 2003) to 43 mt by 2013. At this level, China's use will be close to that of the European Union, the largest single oilseed meal using region – an unsurprising possibility given that China's population is much larger and per capita income is assumed to continue to increase strongly. On the other hand, China is expected to be relatively less dependent on imports for its supplies: even after rising by 4 mt over the period imports would only account for 10% of domestic consumption. In contrast, oilseed meal imports by the EU account for more than half of annual consumption.

Brazil and Argentina are also expected to post increases in oilseed meal use, particularly in relative terms. In the baseline, both countries realise some part of their potential for greater production of livestock products, thus lifting their oilseed meal demand. However, base period use in these countries is quite low relative to oilseed meal production and is certainly quite low relative to oilseed production. Thus, while notable in relative terms, doubling in Argentina and increasing by almost as much in Brazil over 1998-2002 base period levels, the absolute levels remain small.

Oilseed meal use also grows throughout the OECD region with substantial increases in certain countries. Growth in consumption in Mexico and Korea is particularly pronounced, rising from the base period by 76% and 41%, respectively, but feed use of oilseed meal in Australia, United States, Japan, Canada and the EU rises as well. The increase in oilseed meal use reflects increased production of meat in these countries or regions, whether for domestic consumption or for export.

Vegetable oil demand rises a bit more quickly, outpacing supply in the OECD

As with oilseed meal demand, vegetable oil demand is expected to expand with rising incomes and growing populations, particularly in the NMEs. Worldwide, vegetable oil consumption is expected to grow by just under 3% per year during the period to 2013. China, by 2003 already the world's largest vegetable oil consuming country, looks set to end the projection period at double the level of consumption of the 1998-2002 base period. China is expected to rely on imports for a growing share of domestic use, with the ratio of imports to use rising from about a fifth in the base period to a third by the end of the baseline.

The majority of vegetable oil consumption and production, about two-thirds in either case, takes place outside the OECD area and the role of the OECD in these markets is set to decrease further. The 5% reduction in its share of world vegetable oil consumption (from 37% to 32%) is attributed to the slower rate of growth as OECD consumption increases by a quarter over the period, which amounts to less than half the change in consumption that takes place outside the OECD. Vegetable oil derived from oilseed crush is expected to continue to account for a large part of supplies, but palm oil production is expected to grow more quickly. Relative to the base period, palm oil production rises by over 60% – equivalent to an annual rate of almost 4%, but 1% lower than in the past 10 years owing to falling real prices – while oilseed oil grows by about 36%. As the bulk of palm oil production takes place outside the OECD area, as well as a growing portion of oilseed crush, the OECD's share of vegetable oil production is expected to fall rapidly, ending the projection period at only a quarter of the world total.

Uncertainty due to animal diseases

As discussed in the meat chapter, there are several questions about how different animal diseases' outbreaks that have struck in certain regions will impact on markets and trade. Feed demand is centred where the animals are raised. Thus, demand may shift across regions in the event that a disease causes a shift in the location of meat production. If, for example, the instances of BSE in Canada and the United States lead either to a prolonged absence of their beef from the Pacific market or to price rises in other markets that producers see as accurate indicators of future prices, and therefore lead to future herd expansion, then feed demand and its composition may be shifted from one region to another. Bans on the feeding of animal-based meals would cause a shift to alternative feed inputs, such as oilseed meal. On the other hand, if consumer demand for certain meats is reduced temporarily owing to concerns about consumer safety, then the negative implications for livestock producers in curtailing their production will, in turn, feed back to oilseed meal markets. The recent outbreak of an avian virus struck too late to be assessed in this year's *Outlook*, but may be of even greater importance for oilseed meal markets due to the large portion of this protein source in poultry feed.

Uncertainty due to policy

Governments continue to intervene in oilseed and oilseed product markets. Setting aside indirect implications of policies that affect other commodity markets, such as for other crops and livestock products, many countries retain tariffs that protect domestic oilseed producers or crushers at the expense of consumers and foreign suppliers of oilseeds, oilseed meal and/or vegetable oil. In addition, many OECD member governments continue to provide direct subsidisation of oilseed farmers. Some part of these subsidies have been shifted into forms that are not directly linked to contemporaneous production, such as the fixed direct payments of the United States' FSRI Act and the single farm payment of the European Union's recently announced CAP Reform. However, the amount of support provided through these less coupled subsidies remains substantial. In addition, the United States' marketing loan program for oilseeds, which effectively sets a minimum return per unit for crop producers, similar in operation to the Japanese deficiency payment system, remains in place. At the level of nominal prices of this year's *Outlook*, US farmers are projected to receive marketing loan benefits in only one year, yet this policy could have serious implications for the market were prices lower as US producers would be completely insulated from market signals at taxpayers' expense. In these market circumstances, the likelihood is that US producers would maintain rather than cut back production, so the largest oilseed producing country would not adjust supply during a period of low world prices.

Existing policies described above will be changed and new measures implemented during the *Outlook* period. Agricultural policies such as those listed above are subject to routine review and amendment. In the event that the on-going WTO talks result in an agreement by members to impose further restraints on the degree to which they disrupt the market, domestic and trade policies would have to be adjusted with direct impacts on oilseed and oilseed product markets. Indirect impacts could also be significant if other crop markets, such as cereals and sugar, and downstream markets, in particular dairy, were subject to a substantial policy reform. Apart from policies specific to the agriculture sector, government support for fuels derived from biomass in some countries or regions, such as Brazil, the EU and the United States, may have consequences for oilseed markets, as discussed in the *OECD Agricultural Outlook of 2002*. The point of the *Outlook* exercise is, as ever, to project a plausible baseline against which these sorts of policy changes or initiatives can be evaluated, not to predict how policies will evolve.

The Impact of Further Area Expansion for Oilseeds in Brazil

Planting oilseeds in Brazil

Brazil has experienced an impressive expansion in oilseed area that the *Outlook* envisions continuing for the next decade. This assessment is largely based on an extrapolation of producers' decision-making and the development of infrastructure observed in the past, but these factors may prove to be unreliable indicators of future events. Certainly, total area in Brazil is finite and so area expansion must someday stop. Prior to having every available space filled with oilseeds, other factors may impinge upon oilseed producers not to expand further. For instance, environmental concerns relating to deforestation may conceivably lead to the creation of some disincentives to area expansion; economic signals may change if prices to alternative uses of land or costs of other inputs shift to favour other uses, whether related to agriculture or not; or government policies that indirectly favour greater area, such as development of the infrastructure to facilitate market penetration into even the heart of the country, may be minimised. Alternatively, a more significant continuation of past trends, including development of infrastructure, could have the opposite implications for the pace of area expansion in the future. Given the importance of Brazil's oilseed production in world markets and the uncertainty of how much more land can be brought into production, this section explores the consequences of more or less expansion in oilseed area in Brazil.

How important is Brazil in world markets and what does the *Outlook* project for the future? Brazil, accounting for a fifth of world production in 2003, is second only to the US in oilseed production. Moreover, whereas US production is expected to rise by less than 1% per year, on average, after the 2004 recovery in yields, Brazilian oilseed production growth is projected to average well over 3% each year. Greater area accounts for about two-thirds of the increase in production; in Brazil, the increase in area each year is expected to exceed yield growth by 1%. But area growth is assumed to slow, and in fact, the 2.4% annual growth in oilseed area is well below the average of almost 6% observed during the preceding ten years.

What if producers did not increase oilseed area so quickly?

If producers allocate land to alternative uses or some other limiting factor develops, then Brazil would necessarily produce less and export fewer oilseeds and oilseed products, leading to higher world prices for these goods. In this scenario, Brazilian oilseed area supply is reduced by 3% in 2004, 6% in 2005 and so on to a 30% reduction in 2013. These reductions are relative to the baseline and before price effects, so some part of the reduction in land supply is offset as producers respond to rising prices.

The results represented in Table 3.1 show the per cent changes relative to the *Outlook* in the first year of the scenario, 2004, and the last year, 2013, as well as the average. After price

Table 3.1. **Brazilian oilseed area growth rate impacts on world markets**

	2004		2013		Average, 2004-13	
	Level	Change	Level	Change	Level	Change
Brazil oilseed area, mha						
Outlook	19.6		24.2		21.7	
Low area	19.1	−2.7%	18.9	−21.6%	19.0	−12.5%
Higher area	20.1	2.7%	28.7	19.0%	24.2	11.5%
Brazil oilseed production, mt						
Outlook	52.0		70.4		60.6	
Low area	50.7	−2.4%	57.0	−19.0%	53.9	−11.1%
Higher area	53.2	2.4%	81.5	15.8%	66.6	9.8%
Brazil oilseed exports, mt						
Outlook	20.7		28.8		25.2	
Low area	19.4	−6.1%	15.5	−46.2%	18.6	−26.4%
Higher area	21.9	6.1%	39.8	38.5%	31.1	23.4%
World oilseed production, mt						
Outlook	269.2		318.1		291.2	
Low area	268.2	−0.4%	313.9	−1.3%	288.8	−0.8%
Higher area	270.3	0.4%	322.1	1.2%	293.3	0.7%
World oilseed price, USD/t						
Outlook	253		254		253	
Low area	258	2.2%	276	8.7%	266	5.1%
Higher area	247	−2.2%	236	−7.1%	242	−4.3%
World oilseed meal price, USD/t						
Outlook	192		179		182	
Low area	195	1.7%	195	8.8%	190	4.8%
Higher area	189	−1.7%	167	−6.8%	174	−3.9%
World vegetable oil price, USD/t						
Outlook	570		603		589	
Low area	574	0.7%	615	2.0%	597	1.4%
Higher area	567	−0.7%	593	−1.7%	582	−1.2%

Source: OECD Secretariat.

effects, the reduction in area is 22% in 2013, or about two-thirds of the initial shift, and the final impact on production is moderated further, to an 19% reduction, owing to the modest, but offsetting price effect on yields. Exports of oilseeds from Brazil are halved at this lower production in 2013. In fact, even world production is noticeably lower by the end of the *Outlook* at 1.3% lower.

World oilseed prices rise by 8.7%, in 2013, in this scenario, with oilseed meal prices 8.8% higher and vegetable oil prices up by 2.0%. The greater impact on oilseed meal markets reflects that Brazilian soyabeans are of higher meal content relative to rapeseed or sunflower seed. Conversely, soyabeans have a lower oil content and so are less important in determining world vegetable oil markets, particularly when the additional supplies of palm oil are added. The stronger effect on oilseed prices relative to oilseed product prices implies a narrowing of the crush margin. As world supplies are reduced, particularly in a country that exports a substantial share of its oilseeds, competition among users would be expected to increase, driving somewhat lower the margin between output and input prices on which crushers depend.

What if producers increased oilseed area even more quickly?

On the other hand, given past trends, it may be no less reasonable to ask what would happen if Brazilian oilseed area were to grow at an even faster pace. Thus, a second scenario is undertaken with a 3% shift in oilseed area supply in 2004, 6% in 2005, and so on to plus 30% in 2013. The results are also represented in the table, showing the more modest 19% 2013 increase in area planted after allowing for price effects. Here, again, the world production change is significant by 2013, leading to substantial price changes. The world price falls by 7.1%, which is a sufficient change to trigger US marketing loan benefits for soyabean producers in the last four years of the scenario. Once again, the world oilseed meal price change, a reduction of 6.8%, is stronger than that for vegetable oils. In this case, as oilseed supplies rise the crush margin is expected to widen, encouraging more crushing capacity to come on line.

What are the wider implications?

The implications for other oilseed producers and consumers, where they sell or buy at prices that are determined by world market prices are clear. With everything else unchanged, in the event that Brazilian oilseed area expansion is constrained, whether by economic forces or natural limits, then prices will tend to rise faster, implying better revenues for producers but higher costs for consumers. On the other hand, should Brazilian oilseed area expand even more quickly than in the baseline, then lower world prices will follow and be transmitted to other producers and consumers throughout the world, save in those cases, as in the US, where lower prices may be offset by greater government support.

ISBN 92-64-02008-X
OECD Agricultural Outlook: 2004-2013
© OECD 2004

Chapter 4

Sugar

Main projections – outlook in brief

- The fundamentals of the world sugar market are expected to remain bearish over the Outlook period, with world raw sugar prices to remain in a band of USD 7-9 cents/lb (USD160-200/t) over the period to 2013/14.

- World sugar production to increase by nearly 28 million tonnes to reach about 174 million tonnes in 2013, or at an annual rate of growth of 1.7%. Brazil, the lowest cost producer, shows the largest absolute increase of 9.5 million tonnes, with half the cane crop continuing to be used for ethanol production.

- Global consumption of sugar to continue to grow at around 1.9% per annum, with the fastest growth taking place in developing countries, particularly in Asia, due to faster population and income growth.

- World production exceeds global consumption in most years of the Outlook, allowing only a small reduction in the global stocks-to-use ratio and limited scope for higher sugar prices.

- OECD exports of sugar are projected to reach 12.6 million tonnes in 2013.

- Pressures are emerging for reform of some national sugar regimes. However, substantive reform of the heavily distorted world sugar market will likely depend on the details of an eventual agreement in the Doha round of WTO multilateral trade negotiations.

World market trends and prospects

World prices to recover slightly from depressed levels

World sugar production is expected to continue to expand faster than global consumption in most years to the *Outlook* horizon. The resulting mismatch leads to some accumulation of global stocks that overhang the market and a continuing, although smaller, structural surplus in sugar. In the final analysis, the fundamentals of the world sugar market are expected to remain bearish. Although the global stocks to use ratio is expected to fall by 2013, this will remain too high to have much impact on prices. With ample sugar supplies available, and Brazil's reserve capacity to increase production, world sugar prices are projected to remain low over the period to 2013/14 (see Figure 4.1). The world prices of raw and white sugar are projected to stay within a band of USD 7-9 cents/lb (USD 160-200/t) and 8.5-10.5 cents/lb (USD 180-230/t), respectively, up to 2013/14. This level of nominal world prices implies a continuing decline in real sugar prices (Figure 4.2).

World production to expand further

Despite a continuation of low world sugar prices, the harvested area of sugar cane and sugar beets is projected to increase by 4% and 1%, respectively between 2003/4 and 2013/14. Most of the projected growth in crop output, however, will be accounted for by increasing yields of sugarcane and sugar beets. Sugarcane production which, in general, has lower production costs than for sugar beet, continues to be the dominant global source of sugar production. A steady improvement in yields and sugar extraction rates results in world sugar production expanding by nearly 28 million tonnes or at an annual average rate of around 1.7% per year, to reach about 174 million tonnes in 2013/14. OECD production is projected to increase to nearly 45 million tonnes over the same period.

Figure 4.1. **World sugar prices to remain under pressure**

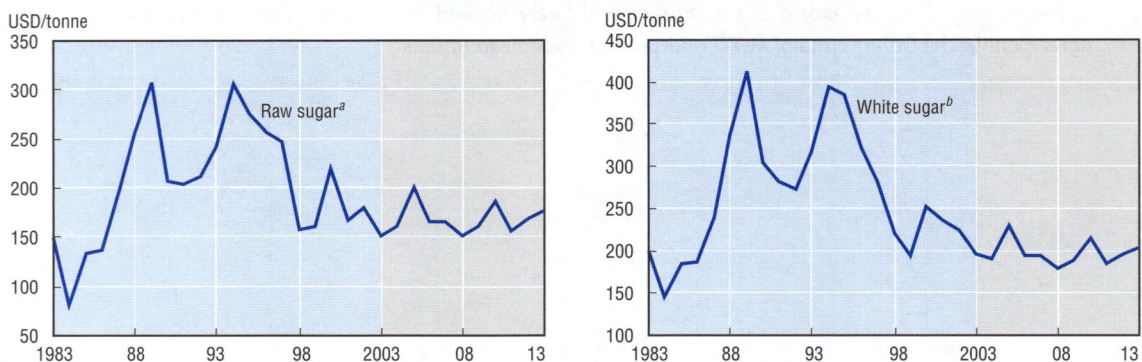

a) Raw sugar world price, New York No. 11, f.o.b., bulk spot price, September/August.
b) Refined sugar price, London No. 5, f.o.b. Europe, spot price September/August.
Source: OECD Secretariat.

Figure 4.2. **Real sugar prices to continue to decline**

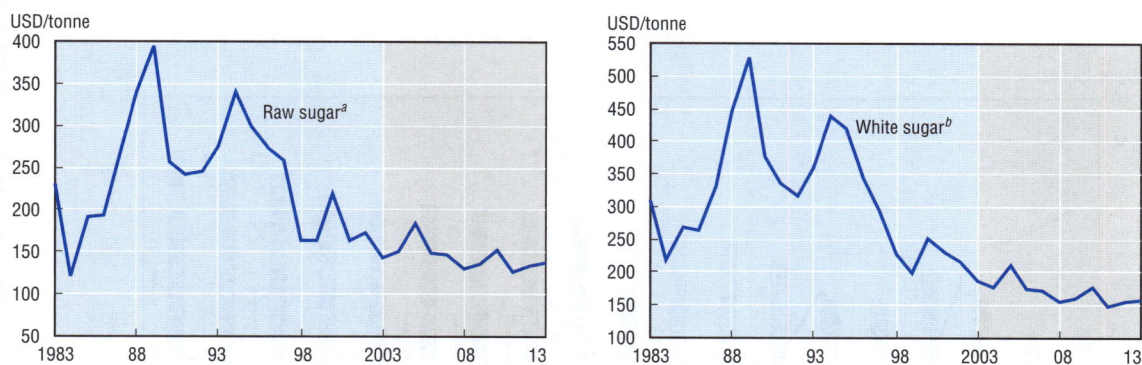

a) Raw sugar world price, New York No. 11, f.o.b., bulk spot price, September/August.
b) Refined sugar price, London No. 5, f.o.b. Europe, spot price, September/August.
Source: OECD Secretariat.

... lead by Brazil and other low cost producers

Brazil, as the world's lowest cost producer leads the pack of producing countries in expanding sugar production year on year. The size of the Brazilian crop and, by extension, the change in the global sugar surplus has become the weathervane for international market prospects and world prices. The main factor in this regard is how much of its sugarcane crop Brazil will convert into alcohol. Rising demand for ethanol over the projection period for use in the hydrous form or as anhydrous ethanol for blending with gasoline, as well as some increase in exports are expected to result in around 50% of sugarcane production continuing to be utilised for the production of alcohol. This competing demand for sugarcane will continue to provide indirect support to the Brazilian sugar industry and represents a large reserve capacity that can be tapped to increase the growth in sugar production for domestic use and exports. Sugar production in Brazil is projected to rise by around 9.5 million tonnes to reach nearly 34 million tonnes in 2013/14,

Figure 4.3. **Global stocks to use ratio set to decline**

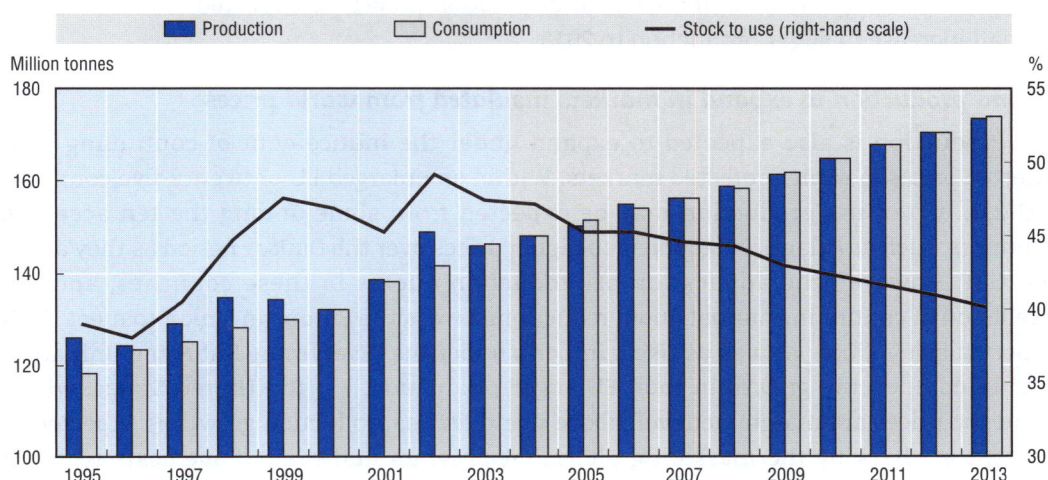

Note: Data are in raw sugar equivalent.
Source: OECD Secretariat.

Figure 4.4. **Brazilian sugar production and exports to expand**

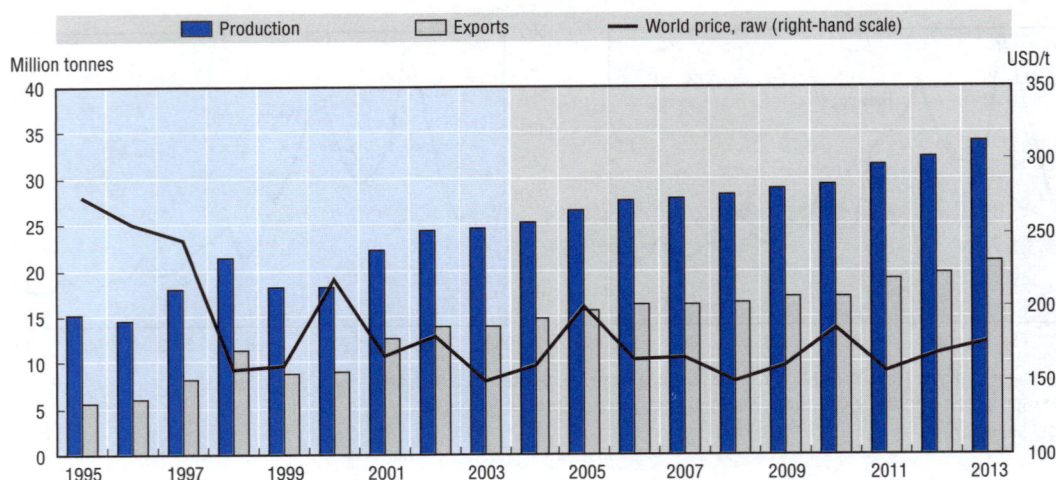

Note: Data are in raw sugar equivalent.
Source: OECD Secretariat.

equivalent to 34% of the projected global increase. The further weakening of the Brazilian real over the period to 2013, following the massive devaluation at the start of the current decade, can be expected to both help sustain industry expansion, despite low world prices, and the price competitiveness of Brazilian sugar on world markets. However, it will also provide a stimulus to increased ethanol production and use to replace fuel oil imports with higher world prices for fossil fuels. Of the other low cost producers open to world markets, such as Australia, Thailand and Cuba, continuing low world prices can be expected to result in some slowdown in production growth in these countries during the course of the projection period, unless necessary structural adjustments occur to improve industry productivity to sustain growth. Measures to improve industry structures and lower costs are assumed to be implemented in most of these countries during the course of the projection period. As a consequence, sugar production is projected to expand by 1.7 million tonnes in Thailand and 0.6 million tonnes in Australia by 2013, when compared to 2003 levels. For Cuba, a drawn out industry rationalisation process is expected to result in only a small increase in sugar production to 2013.

... and production to expand in markets insulated from world prices

Production is also expected to expand under the inducement of continuing high internal prices in some protected markets. Within an enlarged EU of twenty-five countries (EU-25), increasing production may be expected from some among the ten acceding countries, such as Poland, in response to higher prices over the *Outlook* period as they adopt the CAP. However, the application of production quotas in these countries, and the likelihood of continuing annual quota reductions across the European Union to meet WTO export subsidy limits, as well as rising imports under the "Everything But Arms" Initiative, should dampen the growth in total EU sugar production. For the United States, where domestic prices are maintained well above world levels, production growth is expected to be kept in check by the marketing allotment program and the storage costs to sugar producers of holding any surplus sugar off the market. Japanese beet production is expected to increase by around 12% from a low base, but with sugar cane output remaining relatively stable. Mexican production is projected to increase by over 20% to 2013, partly in response to the incentive of increasing domestic demand with a tax on HFCS use in the manufacture of soft drinks and greater access to the higher priced US market as import duties are gradually eliminated under NAFTA by 2008. Indian sugar production, which expanded rapidly in the early years of the current decade due to the maintenance of relatively high administered prices for sugarcane, is expected to be better aligned with growth in domestic consumption. Russian sugar beet production is projected to grow by around 9% to 2013, in response to rising domestic prices reinforced by higher and less porous trade barriers. Tighter government controls on the use of saccharine and other artificial sweeteners in China is expected to increase demand for sugar to be met mainly by higher imports rather than larger domestic sugar production.

Consumption growth remains a major driver, especially in Asia

Growth in sugar consumption remains a fundamental driver of the world sugar economy, and is a major determinant of trade and world prices. Although the rate of sugar consumption growth has been on a downward spiral since the early 1950s, it has settled at around 2% per annum in recent years. An annual rate of global sugar disappearance of 1.9% p.a. is projected over the *Outlook* period to 2013. Sugar consumption patterns are largely determined by growth in population and per capita incomes as well as changes in dietary preferences and lifestyles, with increased reliance on processed and convenience foods that contain sugar. In the advanced OECD member countries, sugar consumption in the form of processed products represents more than 70% of annual demand, with direct consumption being the residual. While sugar remains the preferred sweetener worldwide, its share of the global sweetener market has fallen steadily in recent decades due to inroads made by other caloric and artificial sweeteners.

... With population and income growth underpinning demand

As is the case for other commodities, changing prices and per capita incomes can have an impact on sugar demand. Low and declining world prices in real terms to the *Outlook* horizon would normally be expected to stimulate global sugar demand. This will likely be the case for those developing countries open to world markets. These countries as a group, and especially those in Asia that consume the most, currently account for more than 60% of world sugar consumption and are expected to remain at the centre of any further demand growth over the *Outlook* period as well. The problem is that only a small proportion of world sugar supply and demand is exposed to changes in world market prices. Many consumers remain isolated from world market prices due to a predisposition of many governments for sugar market intervention reinforced by high trade barriers. Historically, governments have had a tendency to fix domestic sugar prices above world levels to ensure the survival of domestic industries and little has changed in this respect. Moreover, the price elasticity of demand for sugar is often low, such that a change in price has only a small impact on demand, particularly in the mature markets of developed countries where consumption has reached near saturation levels. In other cases, such as, for example, the new member states of the EU, consumers are likely to face higher, rather than lower, sugar prices under the CAP, leading to a slowdown in growth of sugar demand.

An additional factor in the consumption equation is the prospect for per capita income growth. The world economy remains in a precarious state at the beginning of the *Outlook*, although economic growth rates are projected to accelerate as recovery proceeds and gathers strength. The projected upward path in economic growth and associated rise in per capita incomes will contribute to higher sugar consumption, particularly in developing countries, over the period to 2013. This is less the case for a number of the advanced industrial countries where the income elasticity of demand is often low at high levels of consumption, thus limiting the positive impact of rising per capita incomes on sugar consumption. Furthermore, the appearance of alternative sweeteners and mixtures of artificial sweeteners with sugar in industrial applications to achieve cost savings impinge directly on sugar consumption in the advanced economies. In addition to these economic factors, population growth remains a primary determinant of sugar consumption. In those OECD countries which have reached a saturation point in consumption, slow or negative population growth has become the main factor explaining static or declining growth in sugar consumption. As virtually all of the increase in the world population will take place in developing countries over the projection period, this provides another reason why these countries are expected to remain at the centre of any future growth in global sugar use.

Sugar trade to increase

Brazil holds the key to the world sugar industry's fortunes

Further growth in sugar trade is projected over the period to 2013, with OECD exports of raw and white sugar projected to total 12.6 million tonnes in 2013/14. Brazil holds the key to the global sugar industry's prospects to 2013 with its considerable reserve capacity to expand production. Accordingly, it is expected to maintain the dominant position as the leading sugar exporter. Brazil's exports are projected to increase by 50% to 2013, along with its share of raw and white sugar trade. For the other major exporters, the projected growth in exports will be affected, in some instances, by a slowdown in production growth and/or by the appreciation of their currencies relative to the US dollar and the Brazilian real. As a result, exports of sugar are projected to increase by 15% for Australia and to stabilise for Cuba to 2013. Thailand's sugar exports are projected to grow by over 20% over the projection period. Tightening export subsidy limits in the case of an enlarged EU of twenty

Figure 4.5. **Brazil leads the pack of sugar exporters**

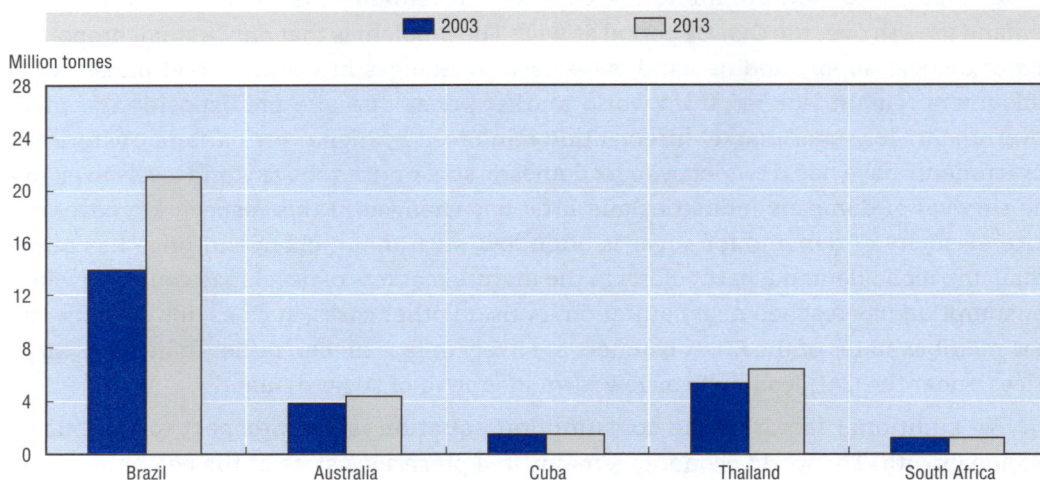

Source: OECD Secretariat.

five countries, effectively restrains the growth in total EU exports to C sugar production with the result that its international market share declines to 2013. Mexico is projected to emerge as a consistent and larger sugar exporter, primarily to the higher priced US market from 2007 onwards. Exports from South Africa are projected to show little overall growth in response to a low world price environment. Indian exports are expected to remain intermittent, and to respond to export opportunities that may arise from time to time over the projection period to dispose of surplus stocks, as domestic consumption takes a larger share of rising production.

Economic recovery and low world prices to boost imports

The driving force on the import side of the trade equation is essentially consumption growth. The projected recovery in the world economy is expected to underpin rising consumption and higher import demand in those countries where additional trade is not prevented by prohibitive import barriers. In addition, the assumption of continuing US dollar weakness will make sugar even more affordable to importers whose currencies are not pegged to the US dollar. World sugar imports are less geographically concentrated than is the case of exports. Many smaller importers exist, including a number of toll refiners who import raw sugar for refining and re-export as white sugar. Russia is expected to remain the world's largest (raw) sugar importer in absolute terms, but rising domestic sugar beet production is projected to lead to some import replacement by 2013. Chinese imports are projected to increase but to remain below the TRQ quota volume it established on accession to the WTO. Apart from China, a number of other countries in Asia, including Korea and Indonesia, and a host of small importers represented in the Rest of the World region, are expected to account for the bulk of additional imports of sugar to 2013. In the case of the United States, which is assumed to maintain prohibitive tariffs on over quota imports under its general WTO sugar tariff quota regime, some additional imports are expected to originate from Mexico as NAFTA duties decline and are eliminated by 2008.

... but structural surplus to continue, limiting price rises

Overproduction in previous years has left a legacy of high sugar stocks and low prices. The accumulation of surplus stocks over consecutive crop years has given rise to a structural surplus and this has become a major problem plaguing the world sugar market. These surplus stocks have arisen from a number of sources including the effects of extensive government intervention in many national markets to shield domestic industries from world market developments. As a consequence, world production has been slow to adjust to low world prices, prolonging the imbalance. With the maintenance of existing policy settings, the structural surplus is expected to persist as production continues to exceed consumption in most years of the Outlook to 2013. From an average of around 63 million tonnes in the 1998-2002 period, global sugar stocks are projected to rise to over 70 million tonnes by 2008, representing over 5 months of consumption and a stock to use ratio of 44%. While the stocks to use ratio in 2013 will be reduced to 40% from the peak of over 47% at the beginning of the Outlook period, this is not expected to permit much strengthening of world prices.

Issues and uncertainties

The sugar projections discussed in this chapter are conditioned by a number of key assumptions. These include the absence of weather shocks to disrupt production in key producing regions (historically the cause of a number of earlier world price spikes), and a favourable global economic environment with high and sustained economic growth. Another crucial assumption is that of an unchanged policy environment to the end of the *Outlook*. However, there are a number of pressure points emerging for national sugar policies from a range of sources including from internal developments, the WTO multilateral negotiations and regional trade agreements. Examples include the outcome of the current review of the Common Market Organisation (CMO) for sugar in the European Union to determine what will replace the existing regime in 2006. Another source is the Doha round of multilateral trade negotiations in the WTO. Considerable efforts are being made by the leading OECD trading countries to restart these negotiations, which stalled in Cancún, with a new negotiation framework based on more substantive offers of trade concessions. At the same time, existing regional trade agreements such as NAFTA and the EBA initiative imply mounting pressure from rising imports on the sugar regimes of the United States, with free trade with Mexico from 2008, and for the European Union, with unrestricted imports from LDC countries scheduled from 2009, respectively. Negotiations on other regional trade agreements are also underway with countries that are significant sugar producers and these may also involve additional sugar trade. Granting open access to the US and EU sugar markets, through existing or prospective regional trade agreements, could result in cutbacks in domestic production or lower institutional prices, or some combination of the two, in the absence of policy reform and with limits on public sugar stock accumulation.

EU reform options announced

Within the group of OECD member countries with high sugar support and protection to their sugar industries, the European Union is the only trading bloc where there are currently signs of possible sugar policy reform. After much discussion of the need for reform of the CAP regime for sugar and the decision taken by EU Heads of Government to delay any changes until 30 June 2006, the European Commission has announced reform options for discussion in September 2003. The intention is to make the EU sugar sector more market orientated and economically, environmentally and socially sustainable. The three options put forward by the Commission include a "*status quo*" scenario as a reference for the two alternative proposals. The *status quo* scenario would involve maintaining the existing regime while allowing the domestic market to be open to import quantities under current and future commitments. The second option included a reduction in the EU internal price, custom duties and a phase out of production quotas, including for sugar substitute products: isoglucose (high fructose corn syrup) and inulin. The level of imports and production would be allowed to stabilise to the lower internal price and then production quotas would be phased out. The extent of competition from isoglucose supplies would be a consideration in this respect. The third option was a complete liberalisation of the whole market with the removal of the guaranteed price, production quotas, import barriers and export subsidies, including the production controls on isoglucose and inulin. Depending on which option is finally agreed, the implied policy changes can be expected to have a significant impact on EU production and trade, as well as the sugar sectors of those developing countries benefiting from preferential access to the high priced EU market. To the extent that significant reforms are eventually agreed and implemented by the EU, they may, in turn, create pressure for reform in the other countries with high support and protection to their sugar sectors such as Japan and the United States.

Doha Round holds the key to more fundamental sugar reform

The URAA was an important watershed in the reform of domestic and trade policies for a number of agricultural commodities, but did not have much impact on sugar support measures and trade. This lack of reform, in part, explains the parlous state of the world sugar market in terms of slow adjustment in production, excessive stocks and unremunerative world prices for most producers. Prospects for substantive reform of the sector will likely depend on when the Doha Round of multilateral trade negotiations in the WTO will be restarted and brought to a successful and timely conclusion. A multilateral agreement that results in wide-ranging sugar policy reform and further trade liberalisation in those countries with high support and protection to their sugar industries, will promote a wider and deeper world market for sugar, with less distortions and potentially more remunerative and less volatile prices for efficient producers, than is contemplated in this set of sugar market projections. In the absence of progress in the multilateral negotiations, some countries, such as the United States, have been giving attention to the negotiation of bilateral or regional trade agreements. To the extent that these agreements include market opening arrangements and reform pressures for domestic sugar industries, and not all do, they are usually required to be phased in over a long period of time and offer little immediate relief for depressed world markets. Pre-existing trade agreements such as the URAA, NAFTA and the EBA Initiative, with the latter two involving various degrees of market opening for participating countries over the next several years, are taken into account in developing the projections for the participating countries to 2013.

Efficiency gains will be a major motivation

In an environment where prices are projected to remain low, achieving further efficiency gains through reform of industry structures to lower costs will be critical for the survival of many sugar producers. This will be necessary for producers in some high cost countries due to inevitable pressures for reform, but also for those in low cost countries who, in the meantime, seek to etch out a livelihood in a highly distorted world market where depressed prices do not reflect costs of production of many of the efficient producers. Sugar producers have been able to achieve substantial cost reductions by improving sugar yields in the field and extraction rates in the factory, through the effective utilisation of by-products, because of technological, labour reducing advances and by vertical integration to eliminate middlemen and maximise efficiencies. Future cost reductions will hinge on further structural adjustment and continued investment for technological gains that, in turn, will depend for their justification on eventual market reforms leading to increased market opportunities and returns.

ISBN 92-64-02008-X
OECD Agricultural Outlook: 2004-2013
© OECD 2004

Chapter 5

Meat

Main projections – outlook in brief

- Reflecting import bans put in place since the occurrence of several cases of BSE in North America, the Pacific market price for beef is expected to rise by 33% while the US domestic price will fall by 14% in 2004.

- If no further BSE cases are discovered, the impact on North American beef markets may be modest and short-lived. Otherwise, beef demand could drop with lower prices in the future.

- Low priced supplies from new member states increases competition within EU meat markets, particularly for poultry meat and these affect trade opportunities for other external low cost producers such as Brazil and Thailand.

- The global market prospects for pig meat and, to a lesser extent, the poultry meat sector over the Outlook period is one of lower profitability with steady feed costs in a weak product price setting. The recent outbreaks of avian virus have not been considered in this Outlook.

- OECD net exports of pig meat and poultry meat should remain stable throughout the Outlook period mostly due to increased competition from Non-member economies, such as Brazil and Thailand.

- OECD sheep meat net trade is expected to decline over the Outlook period due to lower supplies from both Australia and New Zealand. Consequently, lamb prices are expected to remain strong until 2013.

World market trends and prospects

Drought in Europe and Oceania, a weaker US dollar, along with disease outbreaks in North America, Europe and Asia resulted in stronger meat prices in 2003. But the disruption in beef markets due to BSE in North America is expected to be short-lived. Following a sharp price drop in 2004 with increased internal supplies, US beef price developments should return to their historical cyclical pattern, reaching a peak in 2006 at USD 336/100 kg carcass weight (cw). Following this peak, beef prices should gradually decline until 2012, to USD 283/100 kg cw, before rising again. For the OECD as a whole, supply of beef and veal will continue to grow at a faster pace than consumption. Markets for pig meat and, to a lesser extent, poultry meat are expected to weaken with falling product prices in real terms over the *Outlook* period. Coupled with stable feed costs, this should result in lower profitability over the *Outlook* period. Poultry meat consumption in OECD countries increases by 19% in 2013, when compared to 2003; this continues to be the strongest growth rate of all types of meat. Finally, with declining production of sheep meat, with a projected 5% decline by 2013, when compared to 2003, and a global increase in OECD demand of 1% for the same period, nominal lamb prices should remain strong over the entire *Outlook* period (see Figures 5.1and 5.2).

Multiple diseases cripple the meat industry...

The year 2003 brought its share of animal diseases in OECD member countries and NMEs. After a recovery from BSE in the EU and Japan, North America was affected by the discovery of several cases. Following the identification, in May 2003, of a BSE-affected cow in Canada, import bans were immediately imposed by Canada's main beef trading

Figure 5.1. **Cyclical movement in nominal world prices for meat**

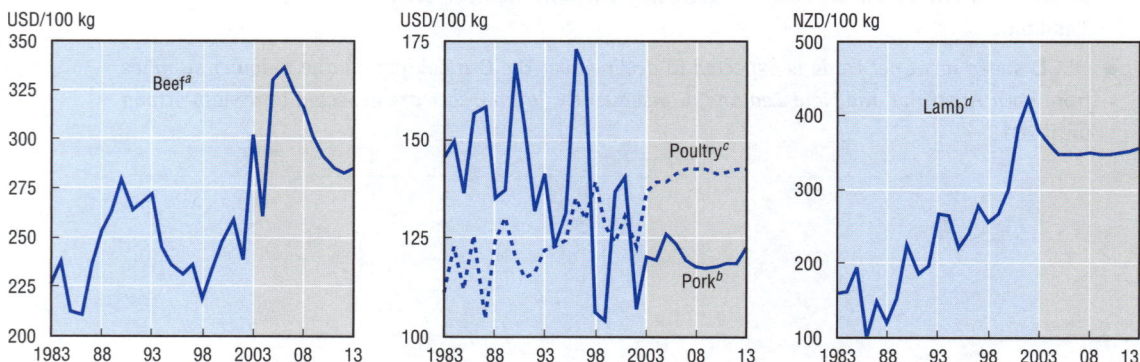

a) Choice steers, USA, dress weight Nebraska.
b) Barrows and gilts, No. 1-3 Iowa/South Minnesota, USA dress weight.
c) Wholesale weighted average broiler price, ready to cook, 12 cities, USA.
d) New Zealand lamb schedule price all grade average, dressed weight.
Source: OECD Secretariat.

partners. These bans led to major disruptions within the Canadian beef sector. Similarly the United States, which is currently the world's largest beef producer, with over 11 million tonnes cw produced annually and the second largest exporter, discovered a cow suffering from BSE in December 2003 and this resulted in the imposition of trade embargoes. The recent outbreak of avian virus in Asia, Canada and the United States appeared too late to be assessed in this year's *Outlook*, but it can be expected to have a similar impact on poultry exports as BSE on beef trade.

Following the discovery of BSE in Canada and the United States, the majority of beef importers immediately announced that they would halt all imports of North American beef. The closure of markets to US beef is of particular importance as the United States has, in recent years, been the world's largest beef trader, with exports representing around 10% of beef production. Almost two-thirds of these exports are shipped to markets outside the North American Free Trade Agreement (NAFTA) countries, mostly to Japan and Korea. The loss of access to these export markets will boost US domestic supply of beef, which, in turn, will push down US domestic prices by an estimated 14% in 2004, when compared to the previous year. However, in the short run, the loss of export markets could cause prices to decline even more. On the other hand, the Pacific market price will increase by 33% in USD terms as total export availabilities are reduced. This adjustment is reflected in a rise in market prices of those countries outside of North America which are supplying beef to Asian countries.

The resulting change in supply is difficult to assess...

The exact magnitude of the changes in supplies to Pacific beef markets is difficult to determine because trade flows are likely to shift somewhat in response to loss of export markets by Canada and the United States. However, the potential for a shift in trade flows is limited by the facts that grass fed beef from Oceania and grain fed beef from North America are imperfect substitutes. In early 2004, it is assumed that Canada, Mexico and the United States would resume their beef trade with each other, although meat would only come from animals under thirty-months of age. The time-frame for the resumption of Canadian and US beef exports to other markets now closed to their produce remains an

Figure 5.2. **Real world prices for meat to decline**

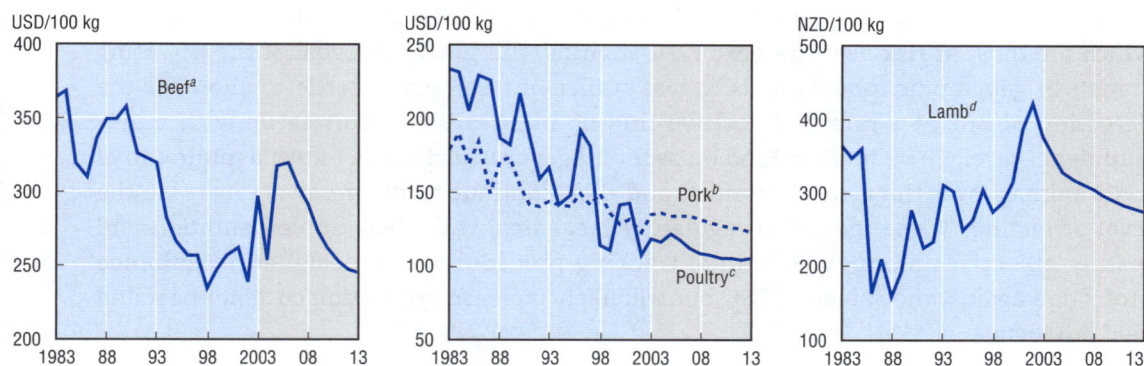

a) Choice steers, USA, dress weight Nebraska.
b) Barrows and gilts, No. 1-3 Iowa/South Minnesota, USA dress weight.
c) Wholesale weighted average broiler price, ready to cook, 12 cities, USA.
d) New Zealand lamb schedule price all grade average, dressed weight.
Source: OECD Secretariat.

open question. For the purpose of the 2004 *Agricultural Outlook*, it is assumed that the Asian market will remain closed until January 2005. To the degree that beef from North America and Oceania are substitutes, producers in Oceania will likely benefit from the lack of supply of North American beef to the Asian market, and are expected to see a temporary increase in exports estimated at 13% for Australia and 4% for New Zealand in 2004, when compared to 2003. Finally, it is also assumed that some small quantities of beef will be supplied to the Pacific market by non-traditional exporters.

As Australian and New Zealand beef producers ship more products to Asian destinations to fill the gaps left by Canadian and US exports, domestic consumers in Oceania will face higher prices. This price increase is expected to result in a fall in domestic consumption of 12%, or more then 3.3 kg, and 9%, or 1.7 kg per person, in Australia and New Zealand, respectively, in the near term. However, domestic beef consumption will likely recover thereafter, as prices are expected to fall sharply once Canada and the United States regain access to major exports markets. Per capita beef consumption is expected to reach over 25 kg in Australia and more than 19 kg in New Zealand by 2013.

Despite the occurrence of BSE in Europe, Japan and North America, beef consumption in the NMEs included in *Aglink* is set to increase over the *Outlook* period. The biggest relative increase in consumption are projected in Chile with a rise of 49%, followed by Brazil with 41%, China with 31% and even Russia with about 18%, despite the imposition of quantity limits on imports.

It is currently very difficult to assess the importance of the impact of the BSE crisis on world beef prices. The OECD's *Aglink* model estimates that the loss of export market access is expected to cause a drop in US domestic cattle prices by about 1.5%, for every 1% increase in domestic supply. This decline in price, albeit on an annual basis, is not as steep as that which took place in Canada, where the Alberta slaughter steer prices declined by about two-thirds from mid-May to mid-summer 2003. This was primarily due to Canada's high dependence on beef exports to foreign markets, which represent around 60% of Canadian production. Export markets for by-products are also very important for the beef industry, and a decline has occurred in by-product values in response to the bans on Canadian and US exports. This decline in by-product values has reduced processing margins and has negatively affected prices paid for slaughter animals.

At the time of writing this *Outlook*, the US government has decided to introduce an enhanced BSE testing programme which should test between 201 000 and 268 000 cattle which are most at risk over the next 12-18 months. Ultimately, the goal of the US, using statistical geographic modelling, is to test sufficient numbers of cattle to allow for the detection of BSE at a rate of 1 positive finding in 10 million adult cattle with a 99% confidence level. The Canadian Food Inspection Agency will also aim to test a minimum of 8 000 animals over the next 12 months, and then continue to progressively increase the level of testing. These increased regulatory measures, with their implementation and testing objectives along with animal tracking procedures, are designed to enhance protection against the spread of BSE, but will likely increase production costs for beef and beef products.

The medium term impact of BSE on North American consumption is uncertain

Changes in domestic demand, following the discovery of BSE in the US herd, may be the biggest uncertainty for the beef market *Outlook*. The European experience suggests that, although demand can be expected to decline in the short-run in response to the BSE crisis, consumption patterns will likely recover over time when consumers are reassured of government and industry efforts to guarantee food safety. However, when comparing the crises in the EU and North America, it has to be borne in mind that they are of very different orders of magnitude. Nevertheless, should any additional BSE cases be detected, or the related human form appears, this would keep the issue before consumers over an extended period of time and could lead to an on-going decline in beef demand. Thus, whether the BSE case in the United States turns out to be an isolated incident, or the first of several, will be key in determining the impact the disease will have on US consumer demand for beef. The Canadian experience of summer 2003 may be indicative for longer term developments in North America. Although there was only a single case of BSE detected in Canada, beef consumption was not affected to the same extent as observed when similar crises occurred in Europe or in Japan. In fact, overall beef consumption in Canada increased by 2.3% in 2003, when compared to the 1998-2002 average. A preliminary analysis made by Agriculture and Agri-Food Canada (AAFC), using the OECD *Aglink* model, confirmed that there was no downward shift in demand for beef in 2003 as a result of BSE. Therefore, it can be assumed that Canadian consumers were reassured by the response to the crisis and the measures taken by both government and the private sector to maintain consumer confidence in Canadian beef.

EU beef markets face brighter prospects

In contrast to North America, the European beef sector faces brighter market prospects. Given limited access to Pacific beef markets, the EU should remain relatively unaffected by the latest BSE related market disruptions. European beef production is expected to continue its long term decline, partly due to the massive culling of older livestock due to BSE measures. This effect is likely to decline in future years as the Over Thirty Months Scheme (OTMS) in the United Kingdom is assumed to finish by the end of 2004. Future development in the EU dairy quota regime will also impact on beef production. Intervention stocks in the European Union have been depleted and consumption has largely recovered from the 2001 BSE crisis. This has led to a change in the net trade position of the EU. In 2003, a year before enlargement of the Union, the EU-15 became a net beef importer again for the first time since the late 1970s. In fact, the EU-15 has witnessed significant beef imports at full duty in the previous years. Whether this is a consequence of the strong euro, the sudden recovery in internal beef demand or some other factor is uncertain. The evolution of the net trade position of the EU, which may have important implications for EU and world beef markets, is discussed in a special section at the end of this chapter.

EU enlargement should benefit the meat sector

When enlargement is taken into consideration, the *Outlook* for the EU meat market may change as several of the new member states, in particular Poland and Hungary, are important meat producers and exporters in their own right. Farmers in these new member states will have immediate and full access to some market mechanisms of the CAP, such as export refunds and intervention measures, while the Single Farm Payment (SFP) will be phased-in over a ten-year period. Details of the implementation of the reformed CAP in EU accession countries were still being discussed at the time this *Outlook* was prepared. Nevertheless, as enlargement takes place, it is expected that the impact on markets for red and white meat will differ. The level of support to beef producers in the new member states will increase throughout the accession period to catch up on the level in the existing member states. The European Union has estimated that the accession should lead to some increase in the beef surplus of the EU-25, thus allowing the enlarged Union to be a net exporter until 2007.

However, pork production is much more significant for the acceding countries. Pork remains by far the largest sector within the meat markets of the European Union. Encouraged by declining EU feed costs, resulting in part from concurrent policy reforms, greater competitiveness should help the EU to remain a major player in world pig meat export markets, the only world meat market where the EU's share of meat trade remains substantial. However, support to pig meat producers in acceding countries is likely to decline from its pre-accession level with the introduction of the CAP. Also, many of the new member states have structural difficulties to overcome, for example, in improving marketing and processing standards, before they can fully compete within and outside the European Union. In addition to this, it is expected that consumption of pork in the accession countries will increase by 16% by 2013, compared to the 2003 level, due to an increase in disposable incomes of the population. These developments will lower the growth in the EU-25 pig meat surplus over time.

Weak market prospect for pig meat...

World pig meat prices, however, remain relatively depressed over the *Outlook* period. This is partly caused by the fact that demand for pig meat shows a decline in many OECD member countries when compared with that for poultry, which becomes the most consumed meat within OECD in 2013 (see Figure 5.3). However, the shortage of beef created by reduced export supplies, and possibly even for poultry, could lead to temporary short term fluctuations in pig meat prices due to substitution effects. Some important policy measures that can affect trade flows include the quota that Russia imposed for pig meat, which will reduce its imports by about 9% by 2013, when compared to the average for 1998-2002, and the triggering of safeguard measures for pork by Japan until March 2004. As a result, pork imports into Japan are now facing a standard import price (gate price plus

Figure 5.3. **Poultry's share in total OECD meat consumption to increase**

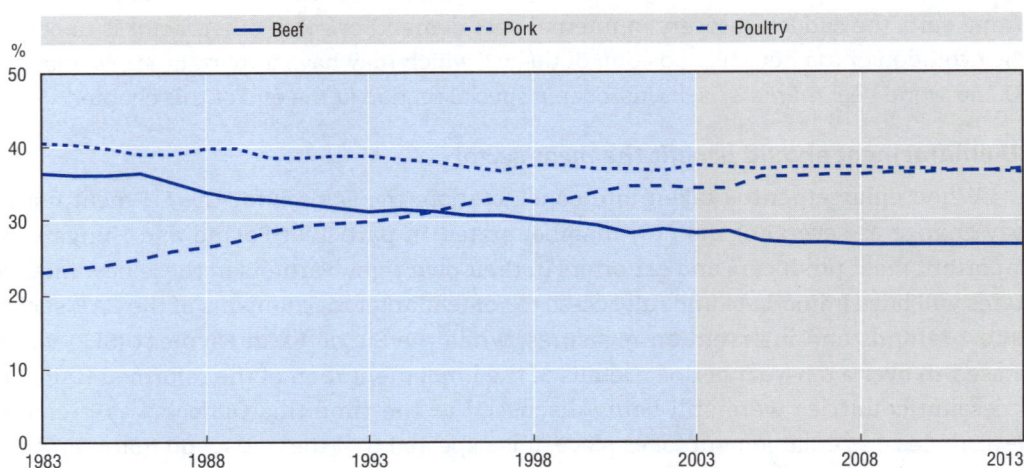

Source: OECD Secretariat.

tariff) of JPY 681.08/kg instead of JPY 546.53/kg. Finally, driven by continued structural change and productivity gains, the growing importance of Canadian and US pig meat exports in global markets continues over the *Outlook* period. When compared to the 1998-2002 average, pig meat exports (excluding live animals) by the US and Canada are expected to increase by 36% and 92%, respectively, by 2013, although in the US case the share of production that is exported remains modest at around 9% of annual production. Another development, not taken into account in this *Outlook* is the fact that Korea could start re-exporting pork to Japan, now that the foot and mouth disease (FMD) crisis of 2002 appears to be under control.

... and, to a lesser extent, the poultry meat sector

The poultry sector is also suffering from disease outbreaks. Lately, bans on imports of poultry from China and Thailand, but also the United States, have been imposed by some OECD countries following the outbreaks of avian flu. Poultry is rapidly becoming the predominant meat consumed in OECD member countries and, to a large extent, in NMEs. One aspect of poultry supply is its short production cycle as compared to other meats, allowing production units to adjust rapidly to market signals. This flexibility enables non-traditional exporters, such as Malaysia and the Philippines, to temporarily adjust their supply to respond to displaced demand from Japan, for example.

The EU poultry sector has been experiencing increased competition in its domestic market from low cost producers such as Brazil and Thailand. However, this situation may change due to enlargement, with an estimated increase in low cost production in the new member states, in particular Hungary, of 87% by 2013, when compared to the 1998-2002 average. This remains contingent on the ability of these producers to meet EU production and marketing standards. As a result, EU-25 net exports should grow by 21% in 2013, when compared to a low of 325 kt in 2006.

Under the NAFTA agreement, Mexico liberalised its poultry import market in 2003 by removing the TRQ and out-of-quota rates. But subsequently, a safeguard agreement with the United States was signed and has lead to a new duty free TRQ of 46.95 kt for chicken legs. The TRQ volume is due to increase by 1% annually, but will be phased out by 2008. As a result, Mexican imports of poultry are expected to increase by 69% by 2013, when compared to the 1998-2002 average.

The United States exports of poultry meat are projected to reach 3 million tonnes, ready-to-cook (rtc) weight basis, by 2013, an increase of 18% when compared to the 1998-2003 average. Despite this significant increase in exports, US poultry meat per capita consumption will reach almost 50 kg rtc per head by 2013, the highest level of all OECD countries.

Of the NMEs, which are part of the *Outlook*, China's net imports of poultry meat are growing rapidly, to reach 500 kt rtc by 2013. On the other hand, Brazil's net exports are falling over the *Outlook* period to a level of 1.5 million tonnes, rtc in 2013, or 21% down on the 2005 peak of 1.8 million tonnes rtc. This decline reflects a 47% increase in per capita consumption by 2013, compared to the 1998-2002 average, motivated by strong income growth and falling real prices.

Sheep meat prices remain strong

Sheep meat production in both Australia and New Zealand is projected to decline by 3% and 6%, respectively. As a result, exports from Oceania are also declining. With continued low supplies and developing demand from Europe, the Middle East and the United States, producers could expect sheep prices to remain high, when compared to the 1998-2002 average. Although the suspension of live trade with Saudi Arabia will reduce live sheep exports, increasing demand from other Middle Eastern countries should partly make up for the loss of this valuable market for Australia.

South American meat exports are an uncertainty

South American meat exports, in particular from Argentina and Brazil, are expected to grow steadily over the next ten years. Currency devaluation has led to lower export prices in USD terms and has helped in developing new markets. As a result, beef exports from both Argentina and Brazil are expected to more than double by 2013, compared to the 1998-2002 average (see Figure 5.4). Production of both pig and poultry meat in Argentina and Brazil is expected to increase rapidly over the next decade. The majority of this output will be consumed domestically. Nevertheless, Brazil's share in global exports for pig and poultry meat will increase over the *Outlook* period when compared to the 1998-2002 average. However, some aspect of the future growth in beef production and exports in Brazil and Argentina remain uncertain.

For example, Argentina has fallen from fifth to eighth place in ranking as a world beef exporter since 1995. Despite this decline, if we consider beef exports from a regional point of view, the market share of MERCOSUR has increased from 18.3% to 28.6% between 1995 and 2003, solely due to growth in Brazil and Uruguay. Argentina has transferred some of its best pasture to other agricultural activities as a result in the drop of the international price of beef. Similarly in Brazil, it has been estimated that in the next ten years, pasture area will shrink as more land will be switched over to grain crop production. With the rise in livestock production, Brazil must improve the animal carrying capacity of pasture, density of cattle per hectare, and, to this end, use more productive varieties of grass, improve irrigation of pastureland and use more fertilisers. Nevertheless, potential for growth in Brazil is huge considering that around 90 million hectares of land remains available (according to Brazilian official projections) for agricultural use. Taking this into account,

Figure 5.4. **Increasing beef exports for Argentina and Brazil**

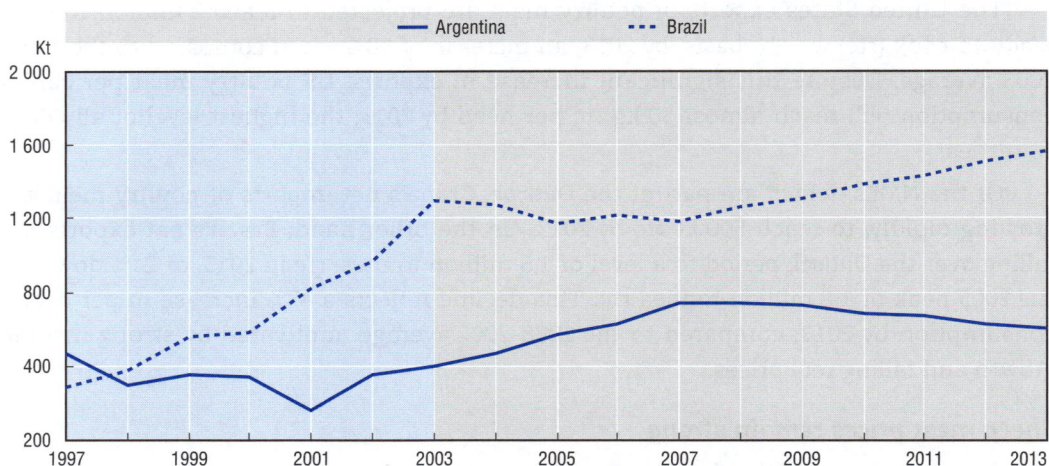

Source: OECD Secretariat.

the potential area available can be, to some degree, used as pasture which would permit Brazil to expand its beef production to a higher level than presented in this *Outlook*.

Both Argentina and Brazil can be expected to be a reliable source of quality meat in the medium term. The advantages of the regional exporting industry are based on the availability of quality raw materials, good infrastructure, a processing capacity consistent with international requirements, application of Hazard Analysis and Critical Control Point (HACCP) system, International Organization for Standardization (ISO) standards and quality protocols from production through to commercialisation. Both countries are also looking into improving traceability through cattle identification programmes. Finally, current negotiations of a regional free trade agreement with the EU and in the Doha Round of multilateral trade negotiations may also affect future market prospects for meat.

Analysis of Beef Imports by the European Union

Beef production, consumption and trade in the European Union are all greatly influenced by government programmes and policies. The purpose of this note is to trace the history of major changes in regulation of the EU beef market, with a particular focus on recent developments in EU beef imports.

The basic regulation establishing the market organisation for beef in the CAP was introduced in 1968, and, followed by a number of amendments over time, resulted in substantial changes in the European Union's beef trade position. The five main reforms which directly affected the beef sector are the Mansholt Plan of 1968, the introduction of milk quotas in 1984, the 1992 MacSharry reforms, the 1994 Uruguay Round Agreement, the 1999 Agenda 2000 CAP reforms and the subsequent Mid Term Review (MTR) or CAP reform of 2003. The MTR reforms may have further impacts on the EU beef market and trade. However, this analysis reviews the evolution in the EU's net trade position for beef over the last 30 years up to 2003, and possible impacts of the MTR are not taken into account.

The 1968 basic support system for beef provided market price support through border protection, intervention buying and export refunds. Later revisions of this system also made direct payments available through headage premiums, for male bovine animals and suckler cows, and a slaughter premium. All these mechanisms have remained dominant characteristics of the EU's beef policy, although the headage payments will be partly or completely converted to a single farm payment under the 2003 CAP reform, making this system less distorting.

1968 Mansholt Plan

The beef intervention scheme of the 1968 Mansholt Plan involved a system of set prices and levies which were determined by the European Commission rather than by the world market. A target price was established and guaranteed, set at a level above the world market price. To prevent domestic prices falling below the target price, intervention agencies were required to purchase beef and remove it from the market. The system of guaranteed prices gave incentives to beef producers, but disincentives to consumers. As a result, the EU was rapidly producing more beef than the domestic market was able to absorb. The mismatch between increasing production and lower consumption created constant surpluses of beef which in turn, generated a problem of surplus disposal. In an effort to control its growing intervention stocks, the European Commission used export refunds to dispose of the resulting beef surplus onto the world market, thus distorting world prices. Subsidised EU exports were directed to markets outside the Pacific region only, as foot and mouth disease (FMD) and later also the Andriessen Agreement with Australia both restricted access to Pacific beef markets. From being the world's largest beef importer in the early 1970s, the EU had become a net exporter of beef to world markets by 1980, with net exports continuing to trend

Figure 5.5. **EU-15 beef production and consumption (actual and trends)**

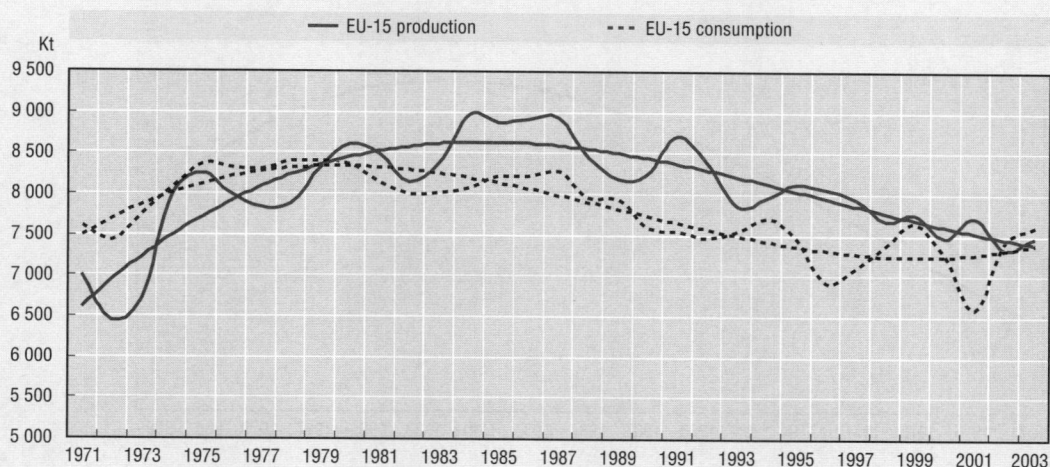

Source: OECD Secretariat.

upwards to the early 1990s. This evolution is depicted in Figure 5.5, where as a reflection of high support prices, production shows a much more strongly increasing trend in the early 1970s than does consumption.

Milk quotas introduced in 1984

In 1983, the European Commission proposed changes to the CAP, including a more restrictive price policy, followed by the introduction of milk quotas in 1984. With new limitations on the quantity of milk produced, dairy farmers gradually reduced their dairy herds and continued to improve milk yields. This resulted in a growing availability of grass land on dairy farms with low opportunity costs. Consequently, pasture based livestock activities, which could not compete with more heavily supported milk production before the introduction of milk quotas, became profitable once quotas triggered a reduction in milk cow numbers. This led to a substantial evolution of the beef cattle herd in the EU. Over the last 30 years, the total number of cows has averaged around 40% of the total herd size; but the composition of the cow herd has changed substantially. In 1981, dairy cows represented 32% of all cattle, while the beef cow herd represented only 8%. Fifteen years later, the share of dairy cows had fallen to 27%, while that for beef cows had risen to 13%. This change in the breed composition of the herd implied a growing share of animals that produce beef, consequently increasing average slaughter weights. The initial build-up of a specialised beef herd reduced the amount of beef available on the market. To overcome the shortage, a substantial increase took place in imports of live calves from third countries, which were thereafter fed domestically. This helped to increase meat supply in the EU during the second half of the 1980s.

The MacSharry Reforms and the Uruguay Round

In 1992, the MacSharry Reforms (named after the European Commissioner for Agriculture at that time) were introduced and represented a major change in the CAP. The reforms lowered the level of price support and increased the headage premia. The focus of

Figure 5.6. **EU-15 beef net trade**

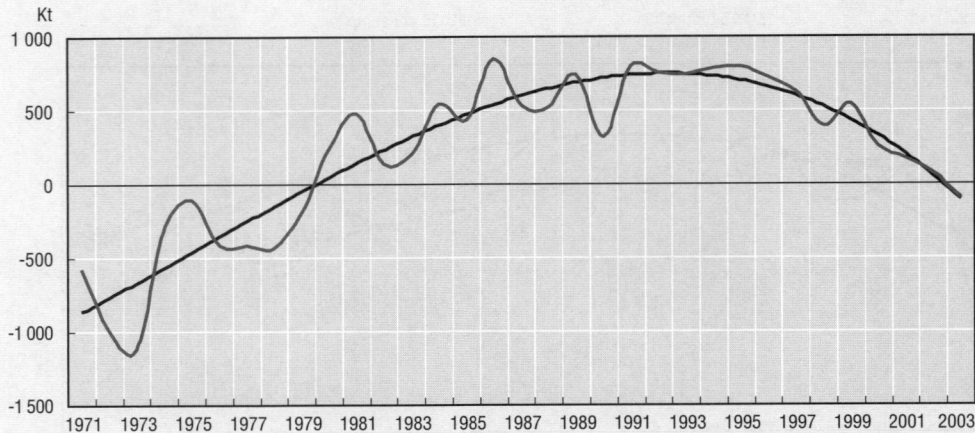

Source: OECD Secretariat.

assistance shifted from price support paid by consumers, to direct income support paid entirely by taxpayers. This reform did not substantially change incentives to beef producers, as the income support payments remained coupled to production – within certain headage limits – and support prices were not abolished. However, it did reduce domestic prices, thereby slowing the long term declining trend in beef consumption. As a result, these reforms appear to represent the turning point in the net trade position of the EU, as shown in Figure 5.6. This decline in support prices and the resulting drop in EU net exports, which started in the early 1990s, anticipated commitments undertaken under the Uruguay Round agreement of 1994, which lowered effective market support, reduced export subsidies and changed all variable levies to duties to be reduced by 36% over a six year implementation period.

Despite lower support prices, EU beef consumption continued to decline well into the first part of the 1990s, mainly due to cheaper meat alternatives like poultry and pork and changes in consumer preferences. But production declined at a faster rate, thus permitting a reduction in stocks that had accumulated in the early 1990s. The 1996 and 2001 BSE crises, as well as the 2001 outbreak of foot-and-mouth disease triggered sudden drops in per capita consumption (see Figure 5.7) during which EU beef surpluses became larger again. However, it appears that the long term declining trend in per capita beef consumption has moderated substantially. At the same time, the policy changes mentioned earlier and the destruction of all cattle over 30-months of age in the United Kingdom, is sustaining a continued decline in EU beef production, as well as in supplies available for export. The rate of decline in EU net exports accelerated even more when beef exports were banned from many markets following the BSE and FMD outbreaks and substantial quantities of beef were removed from the market to balance the drop in consumption.

The Agenda 2000

Finally, the Agenda 2000 CAP reforms of the Berlin agreement pursued the goal of lowering budget costs by reducing the intervention price for beef by 20% over a three year

Figure 5.7. **EU-15 beef consumption per capita**

Kg/person

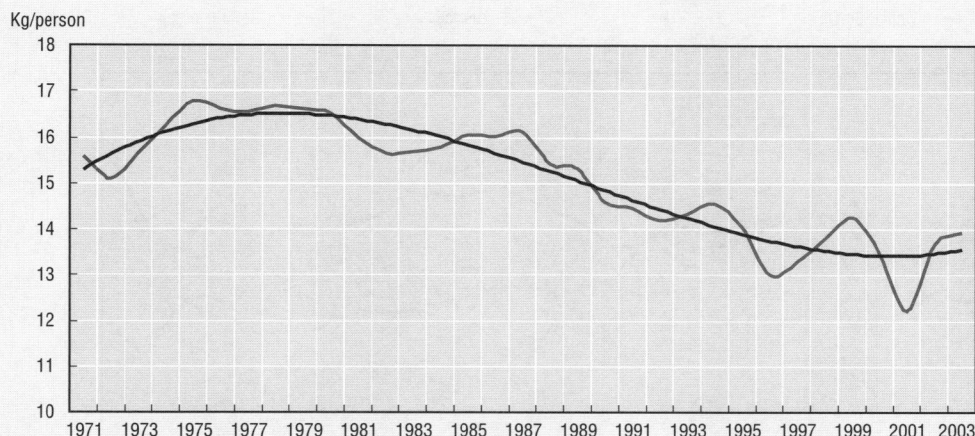

Source: OECD Secretariat.

period and by replacing public intervention by private storage aid, further reducing policy incentives for beef production.

In recent years, the additional market supply due to the depletion of stocks has not been sufficient to offset the decline in EU beef production. Coupled with increasing beef consumption since 2001, beef exports, which still depend on refunds, have continued to decline and have fallen to a level well below the WTO export subsidy limit of 821.7 kt cw. Also, the destinations of beef exports from the EU have been reduced to fewer countries, subsequent to the successive BSE outbreaks. At present, the Middle East and Russia are the only markets of importance. On the other hand, beef imports by the EU have grown in recent years, and this is becoming a significant market development. Beef imports which are limited by TRQs since 1995 as part of the URAA changes, remained stable at 400 kt cw until recently. Since 2000, in addition to some increase in in-quota imports, substantial quantities of fresh, chilled and cooked beef are entering the EU market, paying full over quota duty (see Table 5.1). Total beef imports rose from 270 kt p.w. in marketing year 2000 to 337 kt p.w. in 2002. Of this, the share of in-quota trade fell slightly from 60% to 56%. Reflecting these

Table 5.1. **EU beef imports by origin, 2000-2003**

	2000-2001			2001-2002			2002-2003		
	Inquota	Out-of quota	Cooked	Inquota	Out-of quota	Cooked	Inquota	Out-of quota	Cooked
	Tonnes	Fresh or chilled		Tonnes	Fresh or chilled		Tonnes	Fresh or chilled	
Argentina	28 812.2	3 686.9	14 775.5	28 086.8	59.23	14 698.1	52 353.2	5 342	19 321.8
Brazil	61 336.7	15 942	63 182.3	77 199	34 147.23	67 656.8	73 203.3	36 220.4	75 019.6
Uruguay	13 019.8	877.7	4 070.6	17 907.9	873	3 872.4	14 469.3	811.4	3 934
Other	60 039.7	1 129.5	3 400.8	44 665.9	1 748.94	1 388.5	50 857.1	2 150.9	2 476.6
Total	163 208.4	21 636.1	85 429.2	167 859.6	36 828.4	87 615.8	190 882.9	44 524.7	100 752

Source: EU Commission.

Figure 5.8. **Beef price comparisons**

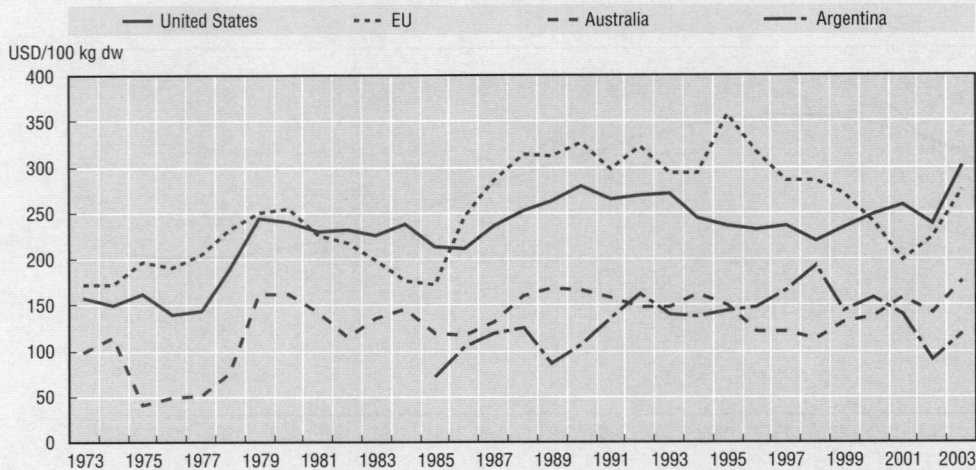

trends in trade flows, the EU became a net beef importer again in 2003, for the first time since 1979.

Against a background of growing preference in the EU for high quality South American beef, the main reasons for this increase in imports can be found in a reduction in euro denominated world market prices due to the devaluation of Peso and the Real and the appreciation of the euro. This is increasing the gap between EU and world prices such that it became profitable to import beef even at the highest over-quota tariff rate of 12.8% plus EUR 303.4 per 100kg (see Figure 5.8). Indeed, an interesting feature of recent developments in beef imports is that growth in imports at over-quota tariff levels occurred at the same time that the tariff rate quota was not completely filled. This is mostly due to problems in sanitary compliance but may also reflect quota administration rules. The decline in world prices denominated in euros helped to lift the TRQ fill rates, while at the same time raising trade outside the quota.

What will the future hold?

While factors driving the change in EU beef exports are of a long term nature, those determining the growth in imports at over-quota tariff rates are of a much more recent nature and can be explained by the recent appreciation in the euro exchange rate, in particular against the Peso and the Real, combined with a continued, if gradual, decline in EU support prices. It is difficult to say at this point in time if the EU will be an important market for South American exports over the long run. But irrespective of this, it is clear that after many years of oversupply and huge stocks induced by high support prices, recent domestic and trade policy reforms are making the EU beef market more market oriented, reducing the distortion on world markets.

The closer integration of the EU domestic market with global beef markets, also allows for adjustments to future internal EU market shocks to be shared by a larger number of participants. This can be appropriately explained by comparing the market impacts of the respective BSE outbreaks in the EU in 1996 and in Japan in 2001. In the mid-1990s, there was little integration between the EU and world beef markets. As a result, the adjustment to the

BSE outbreak was confined to the European Union's domestic beef market. Despite the fact that the introduction of sanitary measures following the first BSE crisis in the EU moderated consumer reactions during the second crisis, prices remained nevertheless depressed for a prolonged period of time. Japan's beef market, on the other hand, was by 2001 more closely linked to world markets, reflecting trade policy reforms in the early 1990s. As a result, the shock from the BSE crisis in 2001 was absorbed not only by domestic supply adjustments but also by adjustments made by beef producers and consumers in supplying countries, such as Australia, Canada and the United States.

ISBN 92-64-02008-X
OECD Agricultural Outlook: 2004-2013
© OECD 2004

Chapter 6

Dairy

Main projections – outlook in brief

- World dairy product prices firmed considerably in 2003 but are projected to increase only modestly in nominal terms over the Outlook period.

- OECD countries are expected to contribute about 22% of the projected increase in global milk production of 121 million tonnes; with most of the OECD growth occurring in Oceania.

- World production to increase for all dairy products, but in the OECD area only for WMP and cheese; OECD countries to supply more than two-thirds of the additional world cheese production.

- World trade in dairy products to remain small relative to milk production and largely regional. Global cheese exports to show the highest increase of dairy products, followed by exports of WMP.

- Consumption of dairy products in NMEs to increase by 30-40%, with growth in all dairy products. In the OECD area, consumption is expected to increase mainly for cheese and WMP.

World market trends and prospects

After a sharp recovery, world dairy prices continue a more modest long term increase

Throughout 2003 world dairy markets were steadily recovering from the 2002 price tumble which saw the lowest world dairy prices in the last decade. World prices strengthened especially during the second-half of 2003 due to limited supplies from countries such as Australia, Eastern Europe and South America. Moreover, economic recovery gathered momentum and improved import demand particularly from Asian economies. The growth in nominal terms of the US dollar denominated world dairy prices was also aided by the weakening of the US currency.

As a result of sustained import demand it is expected that world dairy prices will remain firm throughout 2004. Assuming weather conditions return to normal, fast supply response to higher prices would cause world prices to weaken slightly in 2005 and to increase only modestly in nominal terms in the following years. While in 2003 world dairy prices resulted in returns that were relatively higher for exports of butter and skim milk powder (SMP) compared to cheese, over the projection period cheese prices are expected to strengthen relative to those for other dairy products such as butter and SMP. By 2013, cheese prices are expected to rebound by 11% while butter prices would increase by 9%. Milk powder prices should see smaller gains with whole milk powder (WMP) rising by 7% and skim milk powder by 3%, when compared to the 2003 levels (see Figure 6.1). Nevertheless, in real terms, prices remain relatively flat and price increases taper off during the second half of the *Outlook* period (see Figure 6.2).

Figure 6.1. **Moderate increases in nominal world dairy prices**

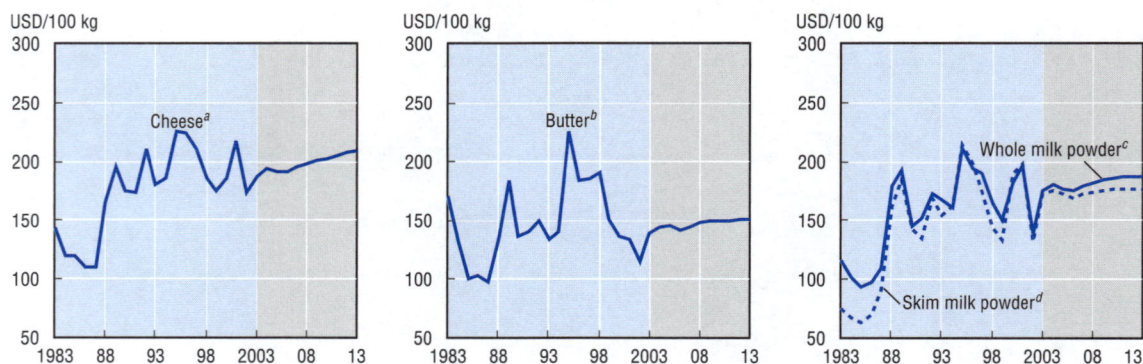

a) F.o.b. export price, cheddar cheese, 40lb blocks, Northern Europe.
b) F.o.b. export price, butter 82% butterfat, Northern Europe.
c) F.o.b. export price, WMP 26%, Northern Europe.
d) F.o.b. export price, not fat dry milk, extra grade, Northern Europe.
Source: OECD Secretariat.

Figure 6.2. **Expected softening in real dairy prices**

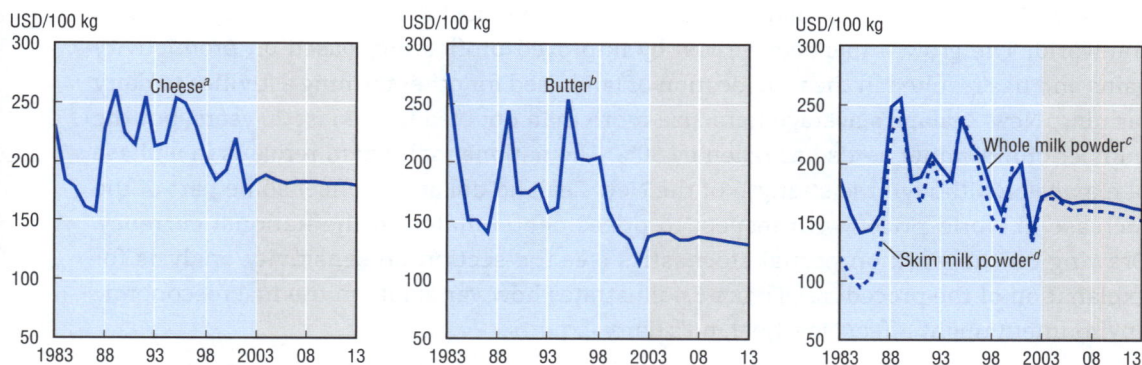

a) F.o.b. export price, cheddar cheese, 40 lb blocks, Northern Europe, deflated by US GDP 2002 = 1.
b) F.o.b. export price, butter 82% butterfat, Northern Europe, deflated by US GDP 2002 = 1.
c) F.o.b. export price, WMP 26%, Northern Europe, deflated by US GDP 2002 = 1.
d) F.o.b. export price, not fat dry milk, extra grade, Northern Europe, deflated by US GDP 2002 = 1.
Source: OECD Secretariat.

Milk production to increase mainly in the Non-member economies

Assuming that the growth in NMEs stays on trend, world milk production is expected to increase by 121 million tonnes or 20% between 2003 and 2013 with an average annual growth rate projected at 1.9%. The highest increases occur in NMEs and in that part of the OECD area not subject to production quotas. OECD member countries are contributing only about 22% or 27 million tonnes to the overall growth. The rest of the increased milk production, representing about 94 million tonnes (78% of the total), originates in the NMEs. Up until 2004 most of global milk production has been produced in the OECD area, but this is expected to change as a growing share of world milk output is produced in the NMEs (see Figure 6.3).

Figure 6.3. **NMEs to increase share of world milk production**

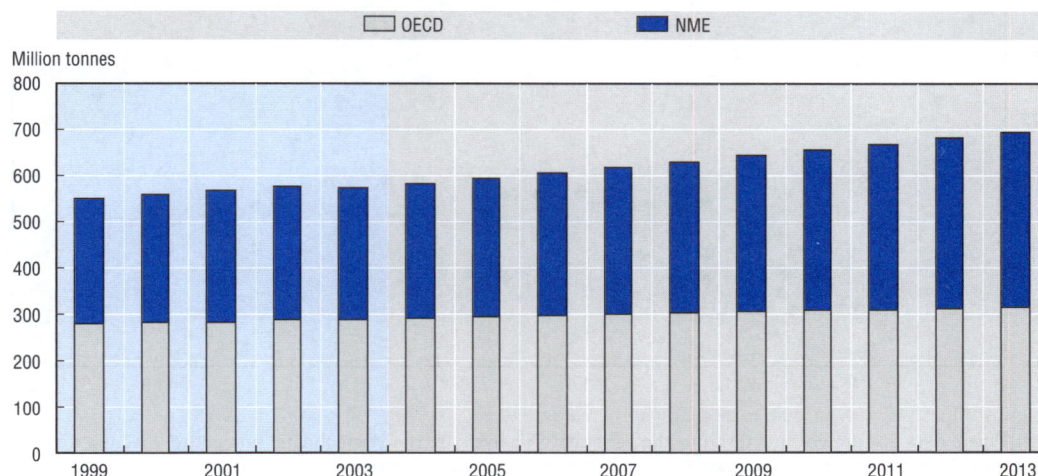

Source: OECD Secretariat.

Oceania expected to remain the most dynamic OECD milk producing region

New Zealand's dairy industry has been one of the most dynamic and fast growing in the world. The growth has been driven by improved profitability based on productivity gains and has resulted in the re-allocation of land used for other farming activities to dairy farming. New Zealand's average herd size more than doubled in the last 20 years, but the number of individual herds has fallen by 30%. The national dairy herd remains in a phase of expansion although the strength of the New Zealand dollar has offset some part of the increase in world prices with respect to prices denominated in the national currency. Drawing on the work on partial stochastics (see the section on sensitivity analysis for explanation of the procedure), Figure 6.4 illustrates how variability in the macroeconomic environment might affect New Zealand's dairy exports.

New Zealand's dairy industry is projected to continue to grow over the *Outlook* period, but at a slower pace than seen in the last decade when milk production increased by more than 80%. Between 2003 and 2013, New Zealand's milk production is projected to increase by 38% (3.3% p.a.), or an additional 4.9 millions tonnes.

Milk production in Australia fell by 8.5% in 2003 due to severe drought which forced farmers to dry off cows early and increase cow culling. This recent culling could lead to a temporary stagnation in herd growth and milk production. However, assuming weather returning to normal, herd rebuilding is expected over the *Outlook* period. By the end of the *Outlook*, Australian milk production is expected to increase by 15%, or by an annual average growth of 1.4%, relative to the reduced levels of 2003.

Growth in EU milk production remains mostly flat while US growth is trending higher

Following enlargement, the European Union has considerably strengthened its already dominant position as the world's largest dairy market and cow milk producer even though

Figure 6.4. Combinations of New Zealand exchange rates and butter exports in 2004

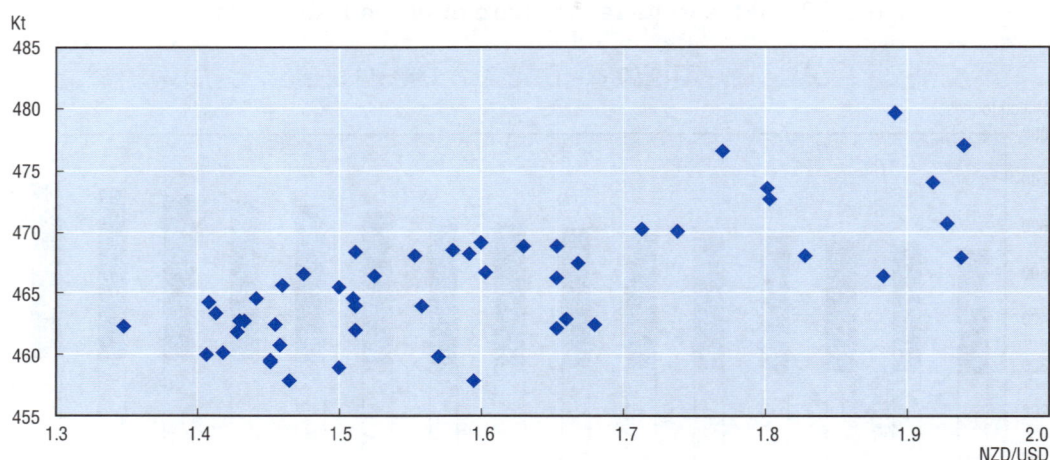

Note: These data are drawn directly from the sensitivity analysis (partially stochastic procedure) discussed earlier and only briefly summarised as follows. Fifty different combinations of macroeconomic circumstances are drawn at random and the model solved to give market outcomes associated with each macroeconomic environment. Here, the fifty values of the random New Zealand exchange rate (relative to the US dollar) are set against the corresponding value of butter exports.

Source: OECD Secretariat.

the quota allocated to the ten new EU members represents only 16% of the total EU quota. EU milk production is expected to broadly follow the evolution in milk quotas. Production would rise as the CAP quota increases are introduced, but then decline afterwards as a result of anticipated growth in the butterfat content of milk, to which the EU milk quota is tied. A reduction in subsistence production in the ten acceding countries following the expected positive developments in rural areas and improved milk quality standards, are additional elements reducing milk production. Farmers in the new member States have from the date of enlargement access to the CAP market mechanisms, such as export refunds or intervention measures. With improved product quality, technology and investments in milk production in the ten new EU members, it could be expected that their milk quotas would become increasingly binding.

In the United States, milk production remained flat in 2003 as producers adjusted to very low prices. With the expected improvement in the milk-to-feed cost price ratio and gains in yields, production growth should pick up and milk production increase by 15% over the *Outlook* period.

Recovery of milk production in Argentina – Supply surge in China

Argentina's dairy sector expanded greatly in the 1990s, reaching a record output of 10.3 million tonnes in 1999. Milk output, however, was cut by one-quarter following the economic crisis. Despite this slump, the low cost, pasture-based Argentinean dairy sector benefits from a considerable comparative cost advantage. It is therefore expected that with an improving economic situation and rising domestic demand dairy production would gradually increase beyond the 1999 record levels.

However, the highest increase in milk production over the *Outlook* period is expected in China where output is set to increase by nearly 6% annually. China's dairy sector, responding to an increase in per capita consumption of dairy products, is expected to quickly expand through the building of new structures stocked by immense imports of live breeding cows. China's government, in an attempt to match swiftly growing demand with local supplies, announced a national strategy plan called the "Advantageous Cow Milk Production Area Development Plan" for the 2003-2007 period. This is expected to help overcome the main constraints of China's dairy sector: genetics, feeding management and processing.

More milk to be channelled to the production of cheese and WMP, but away from butter and SMP...

World production is expected to increase for all dairy products with the highest growth rates expected for WMP and cheese. Increased investment in processing capacities in the NMEs is expected to enable the transformation of large milk production gains into substantial production growth of all dairy products. As a result, WMP production in the NMEs is expected to gradually over-take that in the OECD area by the end of the projection period.

Dairy production in the OECD area is projected to increase only for WMP and cheese while butter and, in particular, SMP production is likely to decline. Although the percentage growth in cheese production in the NMEs is more than twice as large as that in the OECD area (40% compared to 17% growth), OECD countries are expected to supply more than two-thirds of the additional worldwide cheese quantity. The expected strong increase in cheese production results in a gradual increase in cheese stocks throughout the *Outlook* period as growing cheese production requires a certain expansion of stocks for maturing. The production of whey, which is jointly produced with cheese, is expected to increase substantially in line with the growth in cheese production. Thus, despite the rapid increase in whey consumption for the manufacturing of foodstuffs and animal feed, the rise in whey prices will be limited due to the rapid production growth.

... with a spill over effect on of dairy product trade

World trade in dairy products will continue to represent a small share (about 5-7%) of world milk production and remains largely regional. Cheese exports, after an initial fall due to substantially lower exports from Oceania, are expected to increase more over the projection period as exports from Oceania but also Argentina quickly pick up. A sizeable increase in trade is also expected for WMP, especially in the non-OECD area where it is used for milk reconstitution, displacing SMP and condensed milk. SMP exports are anticipated to remain at relatively unchanged levels as larger exports from New Zealand and Australia are mostly offset by lower subsidised exports from the European Union and the United States. While trade in packet size butter should continue to expand (especially through imports to Russia) the overall trade in butter is expected to slightly decline. These projections consolidate an ongoing gradual trend for world trade in dairy products from supply-led trade in bulk commodities (SMP, butter) to demand-driven high value-added products, such as cheeses.

Strong growth in consumption of all dairy products in Non-member economies

The strong growth in demand for dairy products expected in the NMEs reflects not only high population and income growth, but also development of fast food, changes in lifestyles, expansion of cold storage facilities and improved product shelf life. Consumption of dairy products in the NMEs is expected to increase by 30-40% over the 2003-2013 period (see Figure 6.5). The fastest growth in per capita consumption is expected in China where milk and dairy product consumption increases by 50-60%. A large increase in consumption is also expected in countries of Latin America, partly in response to various government food assistance programmes. Despite substantial per-capita growth in the NMEs, consumption per person remains at relatively low levels in absolute terms, especially in the Asian countries.

Only modest growth in OECD consumption of cheese and WMP, some rise for butter...

In the majority of OECD countries, per capita consumption is already high and concerns with availability of food have been largely replaced by consumers' concerns about food quality and other food attributes. Cheese and WMP consumption should see an increase of 19% and 18%, respectively, over the *Outlook* period as compared to 2003 levels, while growth in SMP consumption would remain relatively flat. OECD butter consumption is expected to marginally increase by 2013, mainly due to rapid growth in Mexico and some increase in the EU. The recently stagnating EU butter consumption is expected to be stimulated by the 25% cuts in support prices under CAP reform. Given the fact that EU butter production is expected to decline over the *Outlook* period, the increase in butter consumption would be met by a decrease in net exports and a depletion of butter stocks.

... but the OECD area still dominates world cheese consumption in absolute terms...

Three quarters of world cheese consumption is attributable to the OECD area and it is expected that OECD countries will consume more than two-thirds of the additional

Figure 6.5. **Cheese consumption to show largest increase**

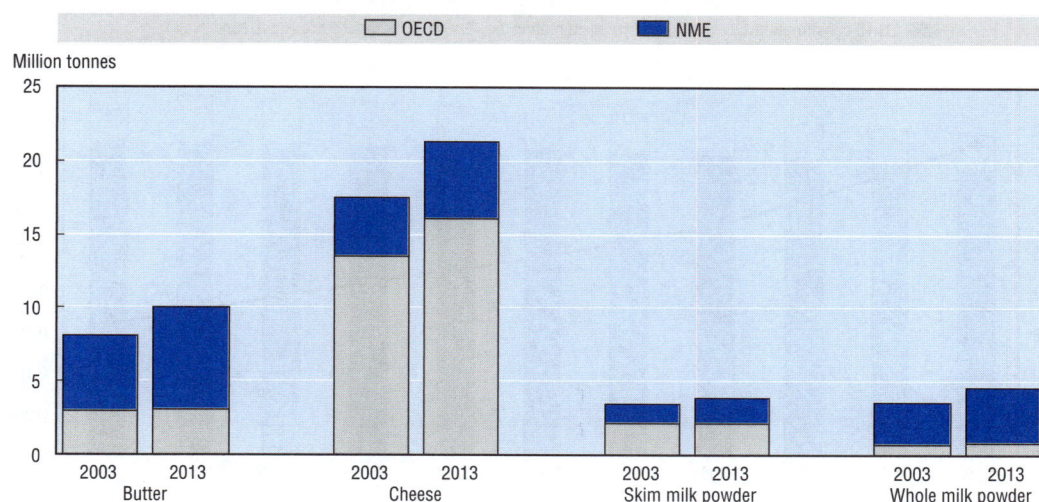

Source: OECD Secretariat.

worldwide cheese production by the end of the *Outlook* period. Cheese is an income sensitive product often seen as a substitute for meat and meat products. Many consumers are increasingly eating away from home and looking for different cheese palate experiences. As a result, cheese producers are competing for new consumers with increased variety. Producers in certain regions are also attempting to secure the approval of geographic indicators such as *appellation d'origine contrôlée* (AOC). Tension, however, may arise over brand names which have been used by cheese producers outside the specific geographic regions.

... with the European Union leading cheese per capita consumption even after enlargement

Average EU per capita consumption of dairy products is expected to decline somewhat following enlargement, as per capita consumption of dairy products in the new member states, especially SMP and cheese, is considerably lower than in the EU-15. Nevertheless, the enlarged EU would keep the world's top position with per capita cheese consumption reaching 19.2 kg per person in 2013 (This compares with 17.1 kg per person in the United States or 0.3 kg in China). The difference in per capita consumption between the EU-15 and EU-10 countries is expected to narrow rapidly over the *Outlook* period. Cheese per capita consumption in the EU-10 is expected to be boosted by more than 50% reaching 17.8 kg per person in 2013 (see Figure 6.6). The growth in cheese consumption in the EU-15 and the EU-10 will be met generally by a decline in net exports, as production is to some extent constrained by milk quotas.

Key issues and uncertainties

Market price support still remains the dominant type of support

World dairy market projections are influenced by a number of uncertainties ranging from weather conditions to global factors such as economic growth. In addition, developments in domestic dairy policies play a very important role in shaping the future of dairy markets

Figure 6.6. **European Union cheese consumption and imports to increase**

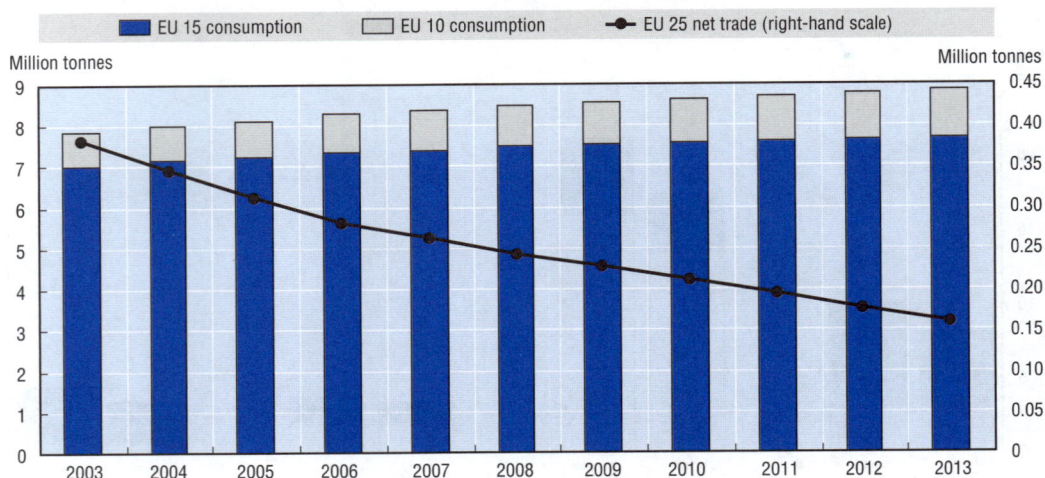

Source: OECD Secretariat.

as milk production is heavily supported in almost every OECD country and in many NMEs as well. The majority of the support is delivered by the means of market price support (MPS) which is deemed to be among the most trade distorting forms of support, requiring high border measures and often the use of export subsidies. Recently, however, several countries have signalled their willingness to reform dairy policies and reduced MPS in favour of less trade distorting measures such as direct payments.

CAP dairy reform is a positive step towards more market orientated policies

In June 2003, the EU Council agreed on a reform of the Common Agricultural Policy which endorsed the direction of the EU dairy policy decided under the Agenda 2000. The CAP reform extended the life of the milk quota system until the 2014/15 season, increased direct payments and implemented 25% and 15% cuts in butter and SMP intervention prices, respectively (for more detailed discussion see Chapter 7 entitled "Medium term market impacts of the 2003 EU Common Agricultural Policy Reform"). While this reform of the EU dairy policy is a positive step towards more market orientated and less trade distorting policies, the EU border measures, although being within the limits agreed by the WTO, remain high and continue to hinder foreign access to the largest dairy market in the world.

Milk supply management continues to be a complex and highly sensitive issue

Supply management is a powerful tool which enables government or producer groups to directly control milk production and will continue to exert a significant impact on dairy markets (see section entitled "Milk quota interaction with specific policy objectives" for more detailed analysis). Supply management systems are operated in the European Union, Canada, Switzerland,* Norway and Japan (Japan's quota is operated by cooperatives).

A non-governmental voluntary milk reduction programme was adopted in the United States. This programme, called CWT (Cooperatives Working Together), proposes to

---

* Switzerland has recently passed a new law which provides for the abolition of its milk quota from 2009 onwards.

strengthen milk producer prices and is funded by an assessment fee on member farmers' milk production. The goals of the CWT are to increase milk prices and clear inventory from the US market place. Members may participate in an export assistance programme, a herd retirement programme, and a programme to provide incentives for producers to reduce milk marketing. The accomplishment of these goals depends on the number and size of participating farms, and it remains to be seen if the program can affect the market.

Milk producers in many parts of the world seek to capture market shares and increase their efficiency by lowering costs through reaping benefits from economies of scale. These expansionary trends seem to be in contradiction with producer-led efforts in certain regions to increase producer prices through supply contraction. The positive effects of such measures, from an economic point of view, might be rather ambiguous. First, although higher prices would likely allow a certain percentage of high cost producers to survive in the short run, it will inevitably trigger further expansionary efforts of low cost producers which will only accentuate the problems of high cost producers later on. Second, of course, higher prices hurt milk and dairy product consumers. Thus, although supply management might be considered a "second best" option in a narrow sense as it alleviates surplus accumulation resulting from high market price support, it will unlikely be the long term solution in the face of increasingly competitive markets and rapid technological and structural developments throughout the world.

Concentration, consolidation and new ways of doing business

The global dairy industry has been, and is expected to be, characterised by mergers, strategic alliances, foreign direct investment and acquisitions. The main drivers behind increasing international investments are considerations of market access and input costs. Dairy companies seek out more market power, expand their brand portfolios and strengthen the security of their milk supplies. Milk producers often benefit from large dairy companies financing programs which are designed to increase milk productivity and quality. However, this high international investment and industry concentration also brings increased risk related to financial difficulties or collapse of investors. A reminder of this point was provided recently by the collapse of the seventh largest dairy company in the world Parmalat (ranking according to Rabobank, Netherlands). As Parmalat operates in more than 50 countries, the effects of the company's collapse have world wide repercussions for thousands of milk producers.

Dairy processing companies have historically had a dominant position in the dairy industry. Despite continued rapid consolidation in this sector, however, it is increasingly the retail sector that establishes the rules, shaping the structure of the processing industry and thereby influencing milk producers' production decisions. Large retailers are often setting quality standards and dictate what, when, how and how much is to be produced. Nevertheless, even these food retailers are under strict and constant pressure – the ultimate power is increasingly in the hands of consumers who are ever more demanding concerning factors such as nutrient definition, organic status, traceability of products, *etc.*

Milk Quota Interaction with Specific Policy Objectives

Introduction

It is possible to achieve the same policy objective with different tools or combinations of tools. Thus, it is important to understand the relationships and linkages between the policy tools and their respective contributions to the achievement of policy objectives. One such tool – the quota system – gives policy makers direct control over agricultural product supply. Generally, a production quota is a limit imposed on the quantity produced. It could have the character of assuring a minimum required production level where under-production might be penalised. However, typically, production quotas have a production restricting character and are combined with penalties for exceeding the quota limit. In the dairy sector, milk quotas were often introduced as a tool to control the growth of surplus production and budgetary expenditures. More than half of all OECD milk production is currently controlled by quotas.

Quota interactions with other policy objectives

A national production quota is typically set above a 100% self sufficiency level. In this case, the imposition of a quota has no direct consequence for consumers assuming that the supported domestic price is held constant. However, the level of quota has a direct influence on exports and government expenditures on subsidised exports within the objective of holding the supported price unchanged. Alternatively, the level of quota determines the cuts in domestic prices required to achieve other policy objectives such as holding exports or government expenditure on subsidised exports unchanged. The latter relationship is examined analytically and also empirically using the Aglink model.

Figure 6.7 schematically depicts the trade off between the supported domestic price and quota levels given that the policy objective is to leave the costs to taxpayers unchanged. Consider that at the initial level of support price P_S and milk quota Q^* consumers consume a quantity Q_{DS} of milk while $Q^* - Q_{DS}$ is exported with export subsidies equal to $(Q^* - Q_{DS}) \times (P_S - P_W)$. Holding the supported domestic price constant and increasing the quota to the new level Q_N^* increases taxpayers costs (export subsidies) by $(Q_N^* - Q^*) \times (P_S - P_W)$. Nevertheless, policy makers can, for a given level of quota, reduce the supported domestic price so that government expenditures on dairy exports would not be affected. Figure 6.7 illustrates that in order to keep government expenditures unchanged, the domestic price has to be reduced to a new level P_N for which the lighter shaded area equals the darker shaded area in the diagram. That is, export subsidies before and after the quota increase must be equal. Mathematically this can be expressed as follows: $(Q^* - Q_{DS}) \times (P_S - P_W) = (Q_N^* - Q_{DN}) \times (P_N - P_W)$. Note that under the new price (P_N) consumers will consume the higher quantity (Q_{DN}).[1]

In order to evaluate numerically the relationship between quota levels and other policy objectives under specific economic parameters, an empirical analysis is carried out for a

Figure 6.7. **Interaction between quota level, domestic price, exports and government expenditures**

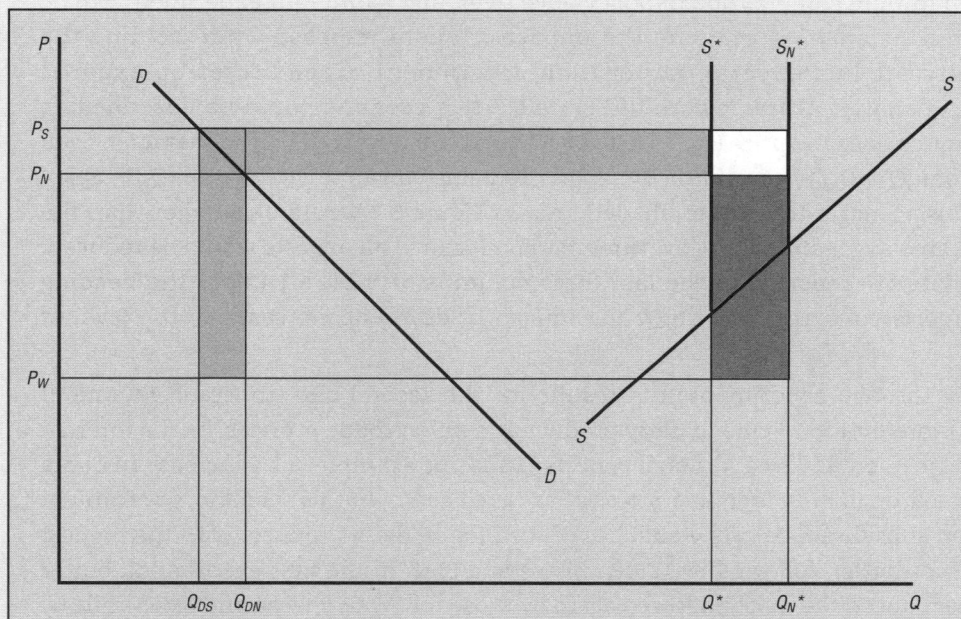

Source: OECD Secretariat.

representative OECD country, the EU, using the OECD *Aglink* model.[2] Following the analytical example of Figure 6.7 the main specific question to be addressed by the empirical experiment was "how much would domestic prices have to be lowered to accommodate a given increase in milk quota while holding government expenditures on export subsidies constant?" While Figure 6.7 depicts the analytics in terms of milk exports, in reality the milk is exported in the form of dairy products. Thus, theoretically there is a large number of permutations that would allow to adjust individual dairy product exports such that overall government expenditures on exports remain constant. For the sake of transparency, the objective of holding government expenditures on exports constant is achieved by keeping government expenditures on exports for each dairy product unchanged at the baseline level.[3]

The results for a 1%, 1.5% and 2% increase in milk production quota respectively are presented in the first three columns in Table 6.1 under the heading "Government expenditures constant". The results indicate that if the milk quota were to be increased by 1%, then the required stability in export subsidy expenditures would be achieved by a simultaneous reduction in the butter price of more than 3% and an increase in exports of butter by 5.1%. The producer price of milk in this scenario would fall by 2.4%. World prices for all dairy products would be reduced as a result of increased exports from the EU, which is a dominant player on world dairy markets. The results of the scenarios with 1.5% and 2% increase in quota are similar to those where quotas are increased by 1% only. The larger the increase in milk quota the larger is the impact on the key variables presented in Table 6.1 Butter prices would have to be reduced substantially more than SMP prices in all scenarios.

These results stem generally from the fact that in the European Union demand for fat is substantially less elastic than that for non-fat solids.

There is an infinite combination of policy objectives, and a different experiment can be constructed, for instance, to evaluate the impact of quota increase while holding the volume of dairy product exports, rather than government expenditures on exports, constant. In the analytical framework of Figure 6.7 this scenario could be described as follows: "How much does the price (P_S) have to be lowered to a new level (P_N) so that for a given increase in quota $(Q^* - Q_N^*)$ the volume of exports remains constant $(Q^* - Q_{DS}) = (Q_N^* - Q_{DN})$. Note that in this scenario the darker shaded area in Figure 6.7 would be smaller than the lighter shaded area suggesting that government expenditures on exports would be reduced. The *Aglink* results are reported in the last three columns of Table 6.1 under the heading "Volume of exports constant„ and show the impact of different increases in the level of quota.

Comparing the two experiments the results for the second one (constant volume of exports) show more profound cuts in dairy product and milk producer prices, as the internal market clearing is not aided by additional exports. Thus, for example, a 2% increase in quota would result in a drop in milk producer prices of as much as 6%. In this scenario, government expenditures on subsidised exports would be reduced for all dairy products with the highest reduction seen for butter, followed by WMP, cheese and SMP. In the first experiment, butter prices would be reduced the most, followed again by those for WMP, cheese and SMP, while by design, no change occurs in government expenditures on exports.

Table 6.1. **Impacts of quota increases on key variables**
Average changes from baseline for the EU

Variable	Product	Government expenditures constant			Volume of exports constant		
		% change	% change	% change	% change	% change	% change
Quantity	Milk	1.0	1.5	2.0	1.0	1.5	2.0
Domestic prices	Butter	−3.1	−4.6	−6.0	−4.9	−7.3	−9.7
	Cheese	−1.4	−2.0	−2.7	−1.9	−2.8	−3.7
	SMP	−0.9	−1.4	−1.9	−0.2	−0.3	−0.4
	WMP	−1.2	−1.8	−2.4	−2.1	−3.1	−4.1
	Milk	−2.4	−3.6	−4.7	−3.0	−4.4	−5.9
Subsidized exports (volume)	Butter	5.1	7.8	10.5	0	0	0
	Cheese	2.3	3.5	4.6	0	0	0
	SMP	4.5	6.8	9.2	0	0	0
	WMP	2.7	4.0	5.4	0	0	0
Government expenditures on subs. exports	Butter	0	0	0	−8.7	−12.9	−17.0
	Cheese	0	0	0	−3.4	−5.0	−6.7
	SMP	0	0	0	−0.7	−1.0	−1.4
	WMP	0	0	0	−6.6	−9.8	−12.9
World prices	Butter	−0.6	−0.9	−1.2	0.3	0.4	0.5
	Cheese	−0.3	−0.4	−0.6	−0.1	−0.1	−0.2
	SMP	−0.1	−0.2	−0.2	0.0	−0.1	−0.1
	WMP	−0.6	−0.9	−1.2	−0.2	−0.2	−0.3

Source: OECD Secretariat.

As the volume of exports is held at baseline levels, the second scenario could be expected to have a negligible impact on world dairy prices. However, the results indicate that the impact on world prices is non-trivial. This important outcome reflects that after the United Kingdom's accession to the EU, New Zealand, which is the biggest exporter of butter in the world, was granted a market access quota for butter to the EU market. Thus, as EU domestic butter prices fall substantially in the scenario, the butter incentive price in New Zealand is reduced even though New Zealand continues to fill its EU butter quota. As a result, production of other dairy products becomes relatively more profitable. Thus, channelling milk from production of butter to that of other dairy products increases New Zealand exports of these products and reduces exports of butter. As expected the world butter price increases, while those for other dairy products fall.

Conclusion

Within the context of policy objectives, the setting of the level of quota is crucial and has repercussions for markets. If the objective is to hold the volume of exports or government expenditures on exports constant for a given increase in the quota level, then this increase has to be accommodated by a change in the domestic support price. Empirical results of experiments conducted with the *Aglink* model show the amount of dairy product price cuts required for a particular increase in quota, given the policy objective. The results indicated that EU butter prices would have to be reduced substantially more than SMP prices. This is mainly due to fact that in the EU, demand for fat is substantially less elastic than that for non-fat solids. The results also show that world dairy markets would be affected even if EU exports remain unchanged. That is, changes in the domestic EU dairy market which involve a decline in support prices, impact on world markets via changes in the butter price realised by New Zealand producers from sales to the EU under the special market access agreement.

Notes

1. It should be also noted that for the simplicity the figure essentially approximates a small country case which does not have a substantial impact on world market and prices. For a large country the world price would have to be reduced in the diagram to reflect the impact of increased exports.

2. The level of the milk quota is exogenous in only one country/region in *Aglink* – the EU. The analysis is carried over using the module for fifteen member States without the new accession countries.

3. The scenario must be viewed as purely illustrative, as no account is taken of the respective WTO limits on subsidised exports and the reduction in the domestic prices will lower per unit export subsidy allowing greater export for the same amount of expenditure.

ISBN 92-64-02008-X
OECD Agricultural Outlook: 2004-2013
© OECD 2004

Chapter 7

Medium Term Market Impacts of the 2003 EU Common Agricultural Policy Reform

Introduction

The Council of Agricultural Ministers of the European Union (EU) reached agreement, in Luxembourg on 26 June 2003, on a reform of the Common Agricultural Policy (CAP), based on the Commission's proposals presented on 23 January 2003 (CEC, 2003a). It includes adjustments to the common market organisations (CMOs) for crops, beef and dairy products as well as the introduction of a Single Farm Payments largely decoupled from current production decisions. In addition, it covers Rural Development Regulation measures and a "financial discipline" mechanism to keep CAP spending in line with existing budgetary ceilings.

While a broader discussion of the CAP Reform decision is published separately, this chapter summarises the main results of the market impact analysis.[1] As such, it is restricted to the changes in the CMOs and the introduction of the Single Farm Payment (SFP) scheme. In addition, the discussion is limited to the commodities represented in the *Aglink* model: grains, oilseeds, meat (in particular beef) and dairy.

It should be noted that the impact of the CAP Reform decision strongly depends on actual implementation. EU member States have discretion to exclude some of the commodity-related payments from the SFP scheme, and/or to introduce the SFP scheme later than 2005, but no later than 2007. It is for that reason that the analysis considers two scenarios considered "Maximum Decoupling" and "Minimum Decoupling" scenarios to offer the reader a range of implications.

It should also be noted that the analysis compares impacts against projections published in the 2003 edition of the *Agricultural Outlook* report. In particular, the EU is treated as the aggregate of the 15 countries representing the EU before the 1 May 2004. Furthermore, the analysis is limited to the period 2004-2008. While policies that have been replaced by the latest reform package have never been, and will never be, implemented in the ten countries acceding to the EU in 2004, the enlargement will have an impact on market developments in all past and future member countries.

Finally, a number of limitations of this analysis should be kept in mind. As the analysis is limited to the commodities represented in the *Aglink* model, changes in other markets cannot be taken into account, nor substitutions between the commodities covered and other activities as these are not necessarily represented appropriately. This concerns, in particular, land idling. Furthermore, neither the possible regionalisation of payments nor cross-compliance are taken into account in this analysis.

Main provisions of the CAP Reform

The **intervention price** for cereals will be retained. However, the monthly increments will be reduced by 50% and intervention of rye will be abolished from 2004. The intervention price for rice will be reduced by 50% to EUR 150/tonne. The intervention price for butter – already foreseen to fall by 15% until 2007 under the earlier Berlin agreement on Agenda 2000 – is reduced by a further 10% within that period, and both the intervention price cuts for butter and skimmed milk powder (SMP) are brought forward one year compared to the earlier Berlin agreement. Intervention purchases for rice, butter and SMP are subject to ceilings.

Direct payments for cereals, oilseeds and set-aside area will be retained, but will then become part of the Single Farm Payment (SFP). Direct payments for rice will be increased from EUR 52/t to EUR 177/t in 2004 to compensate largely the intervention price reduction, based on historical yields. Of this, EUR 102/t will become part of the SFP. For beef, the maximum number of animals to receive direct payments is increased slightly for suckler cow premiums, but reduced for the special beef premiums. Headage payments will become part of the SFP, with several options for member States to exclude some of them and to keep them product specific. Compensation payments to milk producers are introduced from 2004 and will become part of the SFP once the reform is fully implemented.

Dairy quotas will be maintained until 2014, with the general quota increases decided under Agenda 2000 moved back one year to take place from 2006 onwards.

Other CMO adjustments include the reduction in the supplement for durum wheat in traditional production zones (to be included in the SFP) and a new quality premium, the phasing-out of specific durum wheat aids for other regions, a crop-specific area payment for protein crops, payments for energy crop producers outside set-aside provisions, redistribution of support in the dried fodder sector between growers and the processing industry, an increase of supplementary payments for drying aid in Nordic regions, and an annual flat rate payment for nuts.

A **Single Farm Payment** will replace most of the existing premia under different Common Market Organisations. Farmers will be allotted payment entitlements based on historical reference amounts received during the period 2000-02. The payment will be established at the farm level. The entitlement will be calculated by dividing the reference amount of the payment by the number of eligible hectares (including for forage area, which is the basis for the granting of livestock and sheep and goat premia) in the reference year.

In addition to some of the payments that will remain commodity-specific, member States may decide to maintain a proportion of direct aids to farmers in their existing form, at the national or regional level, within specified limits. This discretion concerns parts of the area payments in the arable crops sector or a share of the supplementary durum wheat premium, several options with respect to suckler cow, slaughter, and special male premiums, a share of the sheep and goats premium, and the so-called drying aid.

Dairy payments will be included in the SFP from 2006/07, once the dairy reform has been fully implemented, but member States may introduce the system earlier, from 2005, in the context of a regional implementation of the SFP. Finally, member States may put aside up to a maximum of 10% of the total SFP to encourage specific sectors (within the SFP), which are important for the environment, quality production and marketing.

In order to finance the additional Rural Development Regulation (RDR) measures, direct payments that exceed EUR 5 000 per year for any farm will be reduced by 3% in 2005, 4% in 2006 and 5% from 2007 ("**modulation**").

Medium-term implications for EU and world markets

The following results are based on the "maximum" decoupling assumption (*i.e.* the maximum amount of direct payments is included in the SFP), including the introduction of the SFP in 2005 and the exclusion only of those payments from SFP that explicitly cannot be included. A second scenario based on a "minimum" decoupling assumption (i.e. a maximum of direct payments is excluded from the SFP and remains product specific) will be discussed separately in comparison with the results given here, as well as a sensitivity analysis with respect to the degree of decoupling of the SFP. All scenarios are counterfactually compared to the projections of the 2003 Agricultural Outlook, which assumed a continuation of the Berlin agreement on Agenda 2000 provisions; as such, the baseline already includes some changes in policies, particularly in dairy markets, which were scheduled for implementation from 2005 under this agreement.

Crop market implications

The effective reduction in cereal intervention prices (due to the cut in monthly increments, as well as the abolition of rye intervention) results in corresponding declines in domestic market prices for wheat and coarse grains in the initial years of the reform implementation. Average coarse grain prices are reduced by almost 2% in 2004. With crop production adjusting and responding to changes in the payment regime, and after coarse grain stocks have declined significantly in the first implementation years, medium-term prices recover during the rest of the period. This effect is less pronounced for wheat than for coarse grains as intervention stocks of the latter are reduced significantly relative to the high levels in the Agenda 2000 run in the early years of the reform, whereas public stocks of wheat are unlikely to be significant even with unchanged Agenda 2000 intervention prices. As a result, wheat prices increase only slightly above the Agenda 2000 level, while lower intervention stocks allow coarse grain prices to increase by up to 0.8% relative to the Agenda 2000 run. Both changes become smaller towards the end of the simulation period.

Total area for cereals and oilseeds is reduced slightly as opposed to a slight increase in fallow land, pasture and fodder crops as both area payments and headage payments are converted into the SFP, with a lower incentive for production. The impact on individual crops differs substantially, with cereal production reduced more significantly in favour of oilseeds (Figure 7.1). The effective reduction of cereal intervention prices makes oilseed production more profitable relative to these grains. This effect becomes small towards the end of the simulation period when markets have adjusted to the initial change. The reduction of durum wheat specific payments is likely to have a further decreasing impact on cereal area; durum wheat is not explicitly considered in this analysis, however.

EU exports of coarse grains remain bound by WTO limits on subsidised exports, and therefore, with unsubsidised exports of malting barley assumed to be unaffected, do not change in response to policy reform. Reductions in coarse grains production allow significant reductions in coarse grain stocks (by 41% of their Agenda 2000 level by the end of the simulation period). As intervention for rye has been abolished, the composition of both exports and intervention stocks is likely to change in favour of feed barley.

Neither intervention stocks nor WTO limits on subsidised exports play a major role in the Agenda 2000 wheat projections. The combination of Agenda 2000 support policies, world market prices and exchange rates allow for significant quantities of wheat to be exported without subsidies in the Agenda 2000 run.[2] Therefore, any increase in wheat used for feed, as well as the reduction in production quantities, reflecting changed relative prices under the MTR reform, result in reduced wheat exports to third countries. Total wheat exports are found to decline by 6% on average over the period 2005-08.[3]

Oilseed production is affected directly by the shift in area payments to the SFP in this analysis, but also indirectly through substitution in land allocation due to changes in returns and demand for oilseed products, as well as to the reduction in price support for cereals. On average over the 2005-08 period, oilseed market quantities are hardly changed, whereas prices would be somewhat higher than foreseen in the Agenda 2000 projections. As larger oilseed supplies are only partly offset by higher crush quantities following the increased feed demand for oilseed meals, oilseed imports on average are slightly reduced. At the same time, imports of oilseed meals and vegetable oils are also lower than in the Agenda 2000 run.

Despite the compensation payments – 58% of which becomes part of the SFP from 2005 – rice area is found to decrease by around 9% compared to the Agenda 2000 run following the 50% reduction in the rice intervention price.[4] With some decline in rice yields, the impact on rice production is slightly larger. Consumption (up) – and hence stocks (down) – respond promptly to the price cut. The effect on domestic consumption becomes smaller at the end of the period when, as a result of the "Everything-But-Arms" Agreement (EBA), domestic consumer prices were already projected to fall under Agenda 2000 conditions. The reduction in the support price prevents intervention stocks from exploding. Rice imports to the EU are reduced vis-à-vis the Agenda 2000 results by more than 30% at the end of the projection period because EU consumers buy more domestic rice (and thereby reduce EU stock levels) as a result of lower domestic prices.[5]

Effects on world crop market prices are also shown in Figure 7.1. In general, price changes on international markets are found to be modest. At 0.6% on average, the strongest increase in world markets is found for wheat, while the increases for coarse grains and oilseeds are less pronounced at around 0.2%. Price changes are generally less pronounced in the longer run than in the early years of the implementation. Following rice

Figure 7.1. **CAP Reform impact on crop markets, 2005-2008**

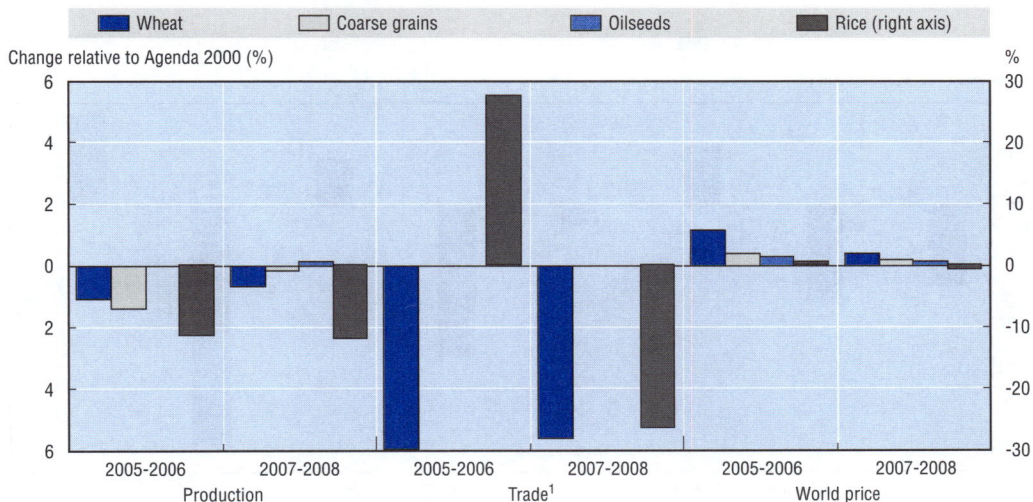

Notes: All data represent two-year averages for 2005/06-2006/07 and 2007/08-2008/09, respectively.
Rice data is represented on the right axis.
1. Trade denotes exports for wheat and coarse grains, imports for rice and oilseeds.
Source: OECD Secretariat.

imports that exceed Agenda 2000 levels in the early implementation years, but are lower in later years, world rice prices first increase *vis-à-vis* the Agenda 2000 projections, but fall below Agenda 2000 levels towards the end of the simulation period.

Even though the simulation period ends in 2008, the results suggest that decoupling the former direct payments has a smaller impact in the longer run than in the initial years of the reform with respect to the market situation which would have prevailed had the CAP Agenda 2000 reforms been maintained. Changes in feed markets due to the new regime in the livestock sector are, however, not completely taken into account within the simulation period (see next section). Lower long-term beef production on the one hand, but higher non-ruminant production on the other, are likely to impact feed use and hence export availability of wheat and coarse grains to some degree.

Meat market implications

Following the conversion of area and headage payments into the SFP in 2005, beef production receives lower incentives compared to the baseline. The impact of the conversion to the SFP is limited though by the fact that, in the baseline, increases in production were constrained by ceilings on the number of animals entitled to receive headage payments. With the SFP, although the production incentive is, on average, much lower, there are no more production limits of this nature. As a result of lower incentives to beef production, producers will gradually reduce milk and beef cow inventories. This increases the slaughter of animals and beef production in the short term, but lowers production starting in 2007 (Figure 7.2). While *Aglink* accounts for the herd dynamics, the simulation period is too short to show the longer term impact of these scenarios. This delayed, longer term impact has, however, been reported in similar analysis published by other sources. In time, the increase in domestic prices should reinforce market returns from beef production and bring an end to the decline in the beef herd.

Figure 7.2. **CAP Reform impact on meat markets, 2005-2008**

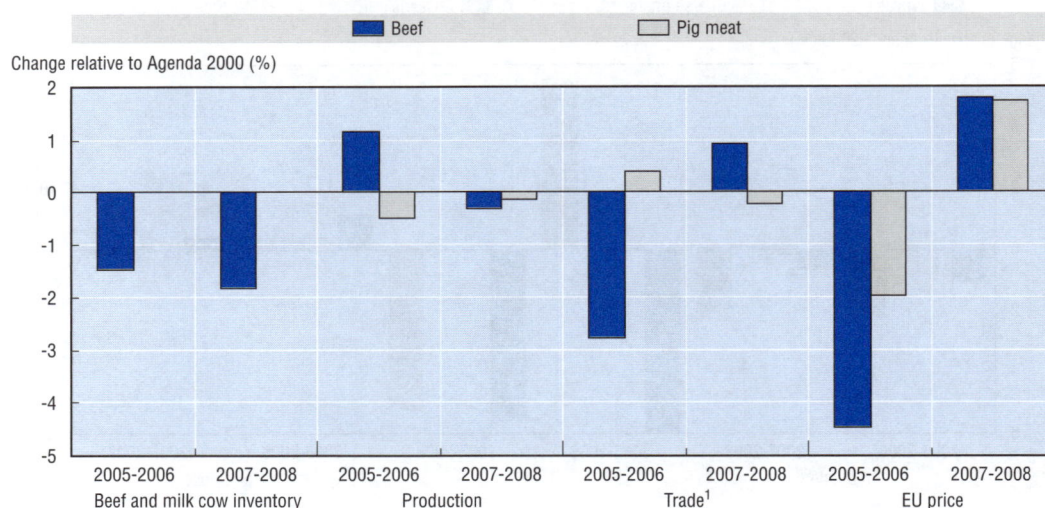

Note: All data represent two-year averages for 2005-2006 and 2007-2008, respectively.
1. Trade denotes imports for beef, exports for pig meat.
Source: OECD Secretariat.

Following the changes in beef production, EU beef prices should gain towards the end of the simulation period after being lower than with Agenda 2000 policies in the earlier years. As currently some beef imports with above quota duty are entering the EU market outside the TRQ, imports are expected to respond to price changes and would also be lower at the beginning of the reform, but higher in later years.

Due to changes in beef prices, pig meat consumption and hence prices are falling initially as well, and to recover only towards the end of the simulation period (Figure 7.2).

Dairy market implications

Compared to the Agenda 2000, which already included significant reductions in butter and SMP support prices, the CAP reform, with an additional 10% cut in the intervention price for butter, results in a decrease in the butter price by 7% by the end of the simulation period. This reduces butter production and stimulates consumption, with subsidised exports being reduced by between 16% and 19% over the simulation period. The cut in intervention prices for SMP will be implemented a year earlier relative to Agenda 2000. However, the reduction in butter production and increased demand for non fat solids in other dairy products will reduce production of SMP and drive up the SMP price by 1% to 2% over the medium term relative to the Agenda 2000 run. Cheese prices in the EU will drop only modestly (by around 2%) and production will increase by about 1% due to higher quantities of milk channelled to the relatively more profitable cheese production. WMP prices will be reduced somewhat more compared to cheese prices, with a corresponding increase in consumption and decrease in production.

While the intervention milk price equivalent is reduced by about 6% compared to the Agenda 2000 outcome, the CAP reform scenario indicates that milk producer prices would fall by less than 3% as the domestic price of butter would fall by less than the butter intervention price while the SMP price would rise slightly. All these price changes are in addition to those projected under the Agenda 2000 agreement conditions imbedded in the

Figure 7.3. **CAP Reform impact on dairy markets, average 2005-2008**

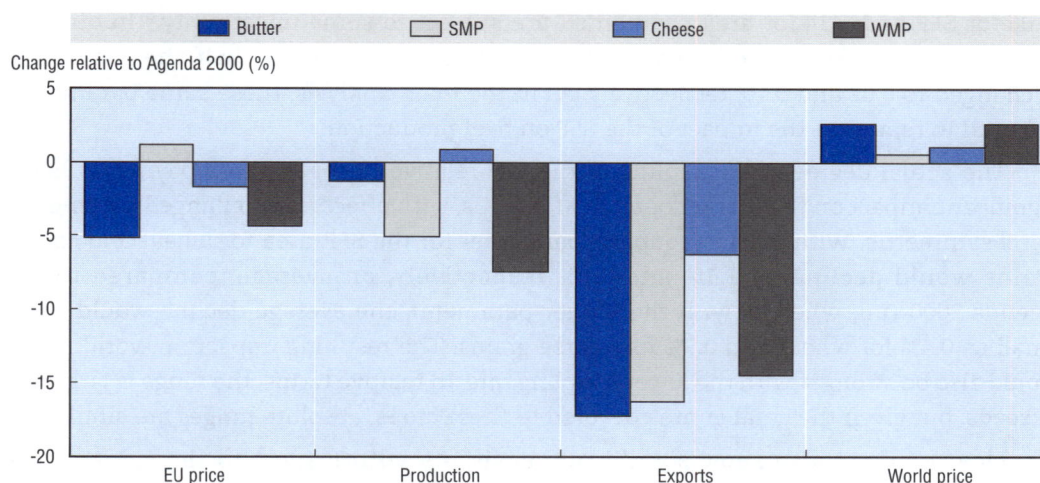

Note: All data represent four-year averages for 2005-2008.
Source: OECD Secretariat.

baseline. World dairy prices are higher than in the Agenda 2000 run on average, particularly for butter and WMP where the strongest reductions in exports are found. Prices for SMP are changed only little (Figure 7.3).

Market implications of minimum decoupling decisions

Compared to the first scenario, this second one assumes member states to opt for the maximum payments to remain linked to area use and livestock production. In particular, 25% of the area payments for crops and some 40% of beef premiums would remain commodity-specific. The new dairy payments would be included in the SFP only from 2007. Compared to the first scenario, the level of the SFP would accordingly be lower by almost 40% in 2005 and 2006, and by almost 30% in 2007 and 2008.

While the remaining area payments are incentives for maintaining crop production, any positive effect on production is offset by larger payments remaining linked in particular to beef numbers. Under the minimum decoupling scenario, total cereal and oilseed area therefore declines slightly more than with maximum decoupling in the initial years of the simulation period when the SFP is significantly reduced. From 2007, cereal and oilseed area, and hence production changes, are almost the same as in the preceding case of maximum decoupling, with slightly stronger declines for cereals, and a small decrease rather than increase for oilseeds. Consequently, the positive impact on world prices of wheat, coarse grains and oilseeds would become somewhat larger, but the differences are negligible.

In contrast, the reduction in cattle herds is estimated to be lower if a larger share of payments remains linked to production. Given the ceilings on headage payments, however, the difference is relatively small. By the end of the simulation period, beef production is therefore reduced by only 0.1% compared to 0.6% with maximum decoupling. Impacts on consumption, prices and hence other meats are less pronounced, too.

Sensitivity analysis: Implications of different assumptions on the degree of decoupling of the SFP

One of the crucial parameters in the analysis is the degree of decoupling of the SFP relative to that of area and headage payments. The parameters used in the analysis above – 0.06 for SFP and 0.14 for area payments – are subject to some uncertainty.[6] In order to increase and decrease the difference between the two parameters by half, the one for SFP is changed to 0.02 and 0.10, respectively. As in the main analysis, these same parameters are used to represent the impact of the SFP on beef production.

The actual degree of decoupling of the SFP relative to that of area payments has a significant impact on the market outcome for crops, with effects from changed parameters fairly symmetric. With a lower coupling parameter for the SFP, area for wheat and coarse grains would decline by 1.1% and 0.9% respectively, on average, compared to the Agenda 2000 run, whereas with the higher parameter, the average decline would be as small as 0.4% for wheat and 0.5% for coarse grains. The resulting impact on world prices would also be stronger with the lower coupling rate. In relative terms, the range is larger for oilseeds, but given the smaller area devoted to these crops, absolute ranges are similar.

Meat markets also show significant sensitivity with respect to the value of the coupling parameter at the end of the period. In 2008, beef production is 1% lower than in the Agenda 2000 run, if a lower coupling rate is assumed, compared to a 0.15% decrease with an assumed higher coupling rate. At the same time, beef prices are estimated to increase by close to 5% compared to the Agenda 2000 run with a lower coupling rate, and by about 1% with a higher coupling rate. The difference in impact between the lower and higher coupling rates is also significant for pig meat prices (2 percentage points in 2008).

Summary

The 2003 CAP Reform decision covers a broad range of changes on CMOs, of which only a subset were considered in this market impact analysis. Another major element of the reform considered in the analysis is the introduction of a Single Farm Payment (SFP) largely decoupled from current production and prices.

With the step towards less coupled support to farmers, grain and – in the longer run – beef production are reduced compared to Agenda 2000 conditions, with some positive, but small, implications for world prices. Oilseed markets overall are only little affected. Major changes are found for rice markets, where the more significant change in support results in substantial reductions in EU production and stocks. Imports are higher at the beginning of the implementation period, but with the increased relevance of the "Everything But Arms" Agreement and corresponding large imports as of 2007, the impact of the reform on rice trade reverses. World price effects remain, however, small. The largest effects on international markets are found for butter and WMP due to the significant reduction of EU exports after the further cut in butter intervention prices, and due to the important role the EU has in international dairy trade.

The sensitivity of the results with respect to assumptions on the actual implementation of the SFP is found to be relatively small. With the higher share of payments remaining linked to current production, the reduction of cattle herds, and hence the longer-term decline in beef production and the corresponding increase in EU meat prices would be less pronounced. In contrast, results for crop markets are found to be very similar to those with maximum decoupling decisions.

A larger sensitivity of the results was found with respect to the size of the parameters used to measure the degree of decoupling. A stronger assumption on the degree of decoupling results in more significant implications for the EU and world markets, as the above-mentioned reductions in grain and beef production and hence the increasing effects on prices are more pronounced. Stronger decoupling also results in oilseed production to be significantly reduced, while with the default decoupling factor taken from literature it was hardly changed. An assumption on a lower degree of decoupling of the SFP would yield the opposite changes in the results of the analysis: less reduction in grain and beef production and indeed some increase in oilseed area.

One should be aware of the limitations of this analysis. Given the treatment of the EU as a single block, the regionalisation option of member states cannot be taken into account, nor could a number of policy changes be considered in commodity markets that are not included in the modelling tool – Aglink. Due to the partial coverage of the agricultural sector in Aglink, substitution effects between the covered commodities and other activities, in particular land idling, may be estimated insufficiently. Furthermore, cross-compliance is not taken into account. Finally, the focus on the 15 former member states and the limited duration of the projection period to 2008 may hide some important implications of the reform.

Generally, the reform tends to result in a more extensive production particularly of livestock. The overall level of support changes only slightly, but the composition shifts significantly towards less coupled – and hence less distorting – measures. The recent CAP reform will certainly improve the performance of the EU's agricultural policies in the direction desired by Ministers as expressed in OECD reform principles (OECD, 1998). As a result of the establishment of the single farm payment based on historical entitlements, domestic market forces play a greater role in guiding the allocation of resources among a large range of commodities.

Notes

1. The discussion of both the provisions of the CAP Reform and the market impacts in this section highlights the main findings. See the main OECD (2004) study "Analysis of the 2003 CAP Reform" for details and references.

2. Note that the underlying 2003 baseline assumed a significantly weaker Euro than the one presented in this *Outlook* report. Given the sensitivity of unsubsidised grain exports with respect to the exchange rate, these results could be different with alternative assumptions.

3. It is interesting to note the initial reduction of wheat feed use in 2004 that follows the abolition of rye intervention and consequent pressures on coarse grain prices. In contrast to later years, wheat exports are therefore simulated to increase in the first year of the implementation of the reform.

4. Note that in *Aglink*, the use of land for rice production is a function of returns to rice only, without direct links to other land uses. While the reduction in land used for rice can be expected to benefit maize production to some degree, this is unlikely to have a significant impact on overall coarse grains production.

5. This result is strongly based on a peculiarity of the underlying baseline projections. With the intervention price kept at a high level, but border protection effectively eliminated due to the EBA, imported rice would increasingly serve consumer needs while domestic production would largely go to intervention stocks. Apart from some reduction in domestic supply, the cut in intervention price would therefore primarily result in lower stocks, with a consequent reduction of the large imports. With the CAP reform in place, however, rice imports in 2008 are still expected to be above current levels.

6. The decoupling parameter measures the ratio of the production impact of a Euro spent on a certain policy to that of a Euro spent on price support. In other words, the parameter 0.06 for SFP implies that a Euro spent on SFP would create a production increase equivalent to 6% of the production impact of a Euro spent on market price support. The parameters used here as well as the methodology used to derive them are from Dewbre *et al.* (2001).

ISBN 92-64-02008-X
OECD Agricultural Outlook: 2004-2013
© OECD 2004

Chapter 8

Indian Agriculture: Performance, Policy Challenges and Medium Term Market Prospects

Introduction

India, with the second largest population in the world, potentially represents one of the most important markets for bulk agricultural commodities and agricultural exports. However, in a number of cases, access to this growing market has been constrained by India's agricultural and trade policies. The overriding policy goal of self-sufficiency in food production based on high domestic production driven largely by domestic subsidies, in the form of price support and large input subsidies, and reinforced by high import tariffs have limited trade opportunities. This policy regime, however, is widely perceived as being neither sustainable nor compatible with the changed circumstances and challenges now facing the Indian agricultural economy.

The role of agriculture

Agriculture remains a very significant sector for socio-economic reasons in India, and is seen as the key driver of overall economic growth and poverty reduction. India has made considerable progress in agricultural development since it gained independence from the United Kingdom in August 1947. The food and agricultural situation in India today is markedly different from that of 50 years ago when India was plagued by famines, food shortages and was heavily reliant on external food aid to avert massive starvation. India has become largely self-sufficient in food through policies based on extensive market interventions and controls and with the adoption of improved agricultural technology. Today large-scale food imports to meet domestic shortfalls are more the exception than the rule. Despite increased food availability and even burdensome surpluses in some crops, access to food and distribution remains a problem. India still has to grapple with the paradox of persistent hunger in an apparent state of plenty, with overflowing grain bins, as millions of Indians remain chronically undernourished (Pinstrup-Anderson, 2002).

Value added in agriculture contributes about a quarter of total gross domestic product of the Indian economy today (down from 61% in 1950-51), and remains the second largest component of GDP after services. Around two-thirds of country's population continues to live in rural areas and remain dependent, directly or indirectly, on the agricultural sector for their livelihood. In line with many other developing countries, the economically active population in agriculture and closely related rural industries remains very high, employing some 57% of the country's entire workforce. The agricultural sector also contributes to about one-seventh of total Indian exports. As such, the agriculture sector of India remains an important contributor to growth in the rest of the economy and is seen by many as providing the rural poor with the most viable path to their improved well being.

Most of the agricultural production is carried out on small family owned land holdings or larger feudal type farms. The agrarian structure in India has undergone significant change with the distribution of land ownership becoming less skewed. Land fragmentation, however, has increased overtime as a result of inheritance laws with an increasing share of farm land owned by marginal and semi-medium farmers. For example, the number of operational farm holdings has increased from 50.7 million in 1961-62 to 115.58 million in 1995-96. The number of marginal holdings (*i.e.,* an operational holding of less than one hectare), have also increased from 19.8 million to an estimated 71.18 million, respectively, in the above periods. Farmers with small landholdings, averaging less than two hectares in size, are an important feature of Indian agriculture (Table 8.1). The imposition of a ceiling on land owned and the redistribution of surplus land to the landless has contributed to a changing land ownership structure in India. Ceilings on land holdings and tenancy restriction regulations have applied in some States for many years and were aimed at increasing security of tenure for land tenants.

Table 8.1. **Distribution of operational holdings – All India**

Holding size class	Number of holdings (% to total)		Area operated (%to total)		Average size of operational holding (ha)	
	1990-91	1995-96	1990-91	1995-96	1990-91	1995-96
Marginal (< 1ha)	59.40	61.60	15.00	17.30	0.39	0.40
Small (1-2 ha)	18.80	18.70	17.40	18.80	1.43	1.42
Semi-medium (2-4 ha)	13.10	12.40	23.20	23.80	2.76	2.73
Medium (4-10 ha)	7.10	6.10	27.10	25.30	5.90	5.84
Large (> 10 ha)	1.60	1.20	17.30	14.80	17.33	17.21
All holdings	100.00	100.00	100.00	100.00	1.57	1.41

Source: Agricultural Census Division, Ministry of Agriculture, New Delhi.

India faced a difficult choice after independence as the traditional markets which had dominated its external trade for decades had closed except for the limited access maintained through Commonwealth arrangements. From being a net exporter, the country had turned into a net importer of food grains to tide over the food demands of an expanding population. The food grains production in 1950-51 was 50.83 million tonnes (mt), with 17 mt of milk also produced in the agricultural economy. The net area sown was 119 million hectares (mha) with a cropping intensity of 111%. Only 18% of the net area sown was under irrigation at the time.

Agricultural output in India still depends upon the arrival of the annual monsoons, as nearly 60% of the cultivable area is rain fed. Weather conditions are an important determinant of agricultural output and consequently the growth rate of the Indian agricultural sector and the national GDP too given the importance of agriculture to the economy. The country is divided into 15 agro-climatic zones and further into 127 agro-ecological zones to facilitate planning and reflecting essentially homogeneity of outputs. India's crop year runs from July-June while the fiscal year is from April to March.

India has a rich mix of climates and soil conditions where different agricultural commodities are cultivated extensively. Crops like wheat, rice, sugarcane, pulses, oilseeds, cotton, jute etc., are grown in different areas of the country. India is the largest producer of copra, mangos, bananas, ginger, turmeric and black pepper, pulses and milk (including that from buffaloes), second largest producer of sugar, fifth largest producer of eggs and seventh largest producer of meat in the world. The main oilseeds grown in the country are groundnuts, rapeseed, mustard, sesame, sunflower, safflower, nigerseed, cottonseed and soyabeans. Production of various fruits and vegetables in recent years has been increasing and now contributes 25% towards agricultural GDP and covers about eight per cent of the cultivated areas. Being an agricultural economy, it produces more in terms of food crops, rather than commercial crops like cut flowers, on a large scale.

Over the years there has been some diversification of crops. The share of coarse grains has fallen while that of wheat and rice has increased. Similarly, the share of oilseed crops has increased. Less diversification has occurred in the livestock sector, but the phenomenon of specialisation towards milk production has been observed. The share of fruits and vegetables in the total area has also increased. There are signs that Indian agriculture is moving towards commercialisation, but cereal production still dominates the farm land of the small and marginal farmers.

As India has followed a strategy of planned development, a number of initiatives related to agriculture and rural development are formulated and implemented through different five year plans of the Central government of India (GoI). The Tenth Five Year Plan (2002-07) has set the following targets and priorities for Indian agriculture: *a*) a growth rate of 4% in the agriculture sector; *b*) the production of total coarse grains to be increased to about 43-48 mt by 2006/07, thus indicating a possibility of achieving a production level of food grains of about 245-248 mt by the end of the Tenth Plan, *i.e.*, 2006/07; *c*) a push on the commercialization of hybrid rice on a large scale; *d*) application of improved technologies in wheat to further boost food grains production; *e*) the budget allocation for the Plan period (2002-07) to the Department of Agriculture and Cooperation has been increased to INR 132 billion up from INR 91.54 billion provided in the Ninth Five Year Plan.

Table 8.2. **Production of foodgrains, oilseeds and sugar cane**

Million tonnes

Crop	2001-02	2002-03	2003-04*
Wheat	71.81	65.10	76.12
Paddy	93.08	72.66	87.94
Coarse Cereals	33.94	25.29	33.72
Pulses	13.19	11.14	14.42
Foodgrains	212.02	174.19	212.20
Summer(*kharif*) foodgrains	111.55	87.81	110.98
Winter(*rabi*) foodgrains	100.47	86.38	101.22
Summer (*kharif*) oilseeds	13.23	9.05	16.69
Winter (*rabi*) oilseeds	7.57	6.01	8.29
Total Oilseeds	20.80	15.06	24.97
Sugarcane	300.10	281.57	255.46

* Second advance estimates.

Source: Ministry of Agriculture, Government of India.

Table 8.3. **Production of meat and poultry processing In India**

'000 tonnes

Particulars	1998	1999	2000	2001	2002
Mutton and Goat meat	675	800	825	850	870
Pork	420	464	480	495	505
Chicken(Poultry meat	675	725	775	875	975
Cattle(Beef)	1 295	1 295	1 300	1 305	1 330
Buffalo(Meat)	1 210	1 250	1 270	1 300	1 365

Source: Department of Animal Husbandry and Dairying, Government of India.

The "Green Revolution"

The phrase "green revolution" is often used to describe the post 1966-67 period of rapid agricultural production growth in India. During the period of the "green revolution" production of food grains, principally rice and wheat was boosted through technology transfers and production techniques based on higher input usage. Use of high yielding cereal varieties (HYV) and improved seeds, in combination with inputs of fertilisers, pesticides and irrigation water, increased crop yields appreciably during the 1970s and 1980s. Expansion of the cropped area especially that under wheat and rice added to total food grains production. Increased food grain output was encouraged by government investments in agricultural research and extension, in irrigation and infrastructure projects and through subsidies for key inputs such as fertiliser and improved seed varieties. As a result, the output of various crops and livestock produce has increased substantially over the past several decades. The production of food grains, principally rice and wheat, increased from 89 mt in 1964-65 to 212 mt in 2001-02 and the production of oilseeds, cotton, sugarcane, fruits, vegetables and milk production also increased appreciably in the post-green revolution period. As a consequence of the increased food grain production, India was able to achieve national self-sufficiency in the 1990s which contributed to increased food security. Despite the substantial production increases that have been achieved, crop productivity and yields in India, in general, are still much lower than that of other major crop producing countries. For example, the average yield of rice in India was 3 008 kg/ha in 1999-2000, while the world average yield was 3 815 kg/ha.

Two features of Indian agriculture over the last decades have been the emphasis given to achieving food grain self-sufficiency at the expense of other crops for which India may have a comparative advantage in the global economy and a public investment emphasis on irrigated areas, and, to a lesser extent, high potential rain fed areas. This has been at the expense of underinvestment in the resource-poor rain fed areas of the country where many of India's rural poor people live (Per Pinstrup-Andersen, 2002). Due to these biases, the gains from the "green revolution" did not cover the complete agricultural economy of the country. It remained confined primarily to the irrigated areas of the north and north western regions of the country. The rain fed regions of India which constitute 60% of the arable area cover about 80% of the area under pulses and 90% of the area under oilseeds and did not benefit to the same extent from a breakthrough in technology or production as was witnessed for wheat, rice and sugarcane. Another outcome of this policy focus has been the negative effect on the environment in the green revolution areas. In one sense, the increased yields achieved in green revolution areas had a positive effect on the environment in that they preserved other more fragile dry land areas, hillsides and forests from cultivation. However, poor natural resource management and degradation through excessive use of chemicals in the green revolution areas have lead to fertiliser and pesticide contamination of waterways, leaching of soils and salination as well as lower water tables. These developments have impacted negatively on yields and crop productivity in India's intensive crop areas and thus have implications for future agricultural growth and food security.

Problems confronting the Indian agricultural sector

The Tenth Plan document produced by the Planning Commission, the GoI's planning authority very succinctly summarises the problems confronting Indian agriculture:

"… the policy of various States has been to increase production through subsidies on inputs such as power, water and fertilisers, rather than by building new capital assets in irrigation and power. These problems are particularly severe in the poorer States. Although private investment in agriculture has grown rapidly, this is hardly a substitute for lower public investment and deteriorating quality of public services in agriculture. Macroeconomic distortions are visible. For example, private investment in diesel run generating sets is increasing while power capacity is under utilized because of poor distribution and maintenance. The poor base of rural productive assets and poorer technological base because of past public/private patterns of spending has been recognised as a serious constraint in increasing production and productivity..."

Even while advocating some steps towards liberalisation in agricultural trade, the idea of being self-sufficient in food grain production continues to be paramount not only for food security but for other politico-economic reasons. But it is also accepted that over-emphasis on food self-sufficiency may result in inefficient resource allocation at the national level. The problems confronting the sector at this point in time, when markets are beginning to be opened up and there are conscious efforts to liberalise agriculture, can be stated briefly as: a) a falling growth rate of gross capital formation (GCF) in agriculture, b) a slowdown in the growth rate of total factor productivity (TFP), c) rising subsidies in the agricultural sector and d) uneconomic size of land holdings.

The gross capital formation in agriculture has been steadily declining over the last two decades. The investment in agriculture as a per cent of GDP has declined to 1.3% in 1998-99. The share of agriculture and allied sectors in total GCF declined from 14.3% in 1970-71 to 7.1% in 2000-01. Researchers have noted that the declining TFP needs to be addressed if the rate of growth in agriculture is to be sustained. The decline in growth of TFP has been viewed with concern because it is seen as sign of either a drop in the effectiveness of technology or that adoption of existing technology is no longer sufficient. In either case, this is likely to cause a slowdown in the growth of agriculture in future which India can ill afford to meet its planning targets (see Table 8.4).

Table 8.4. **Growth rate of agricultural output and Total Factor Productivity**

Period	Growth rate of agricultural output	Growth rate of total factor inputs		Contribution to agricultural growth	
		Total factor inputs	Total factor productivity	Total factor inputs	Total factor productivity
1950-51 to 1966-67	1.87	1.42	0.45	75.88	24.12
1966-67 to 1980-81	2.25	1.38	0.87	61.47	38.53
1980-81 to 1990-91	3.05	0.69	2.25	22.60	77.40
1990-91 to 1998-99	3.45	1.43	2.02	41.49	58.51

Source: Sharma, A. (2001).

Changing agricultural policy environment in the 1990s

Initially, after independence in 1947, food and agriculture policies were formulated in the context of general food shortages during the Second World War and the Bengal famines of the 1940s. As a consequence, agriculture policy in India was essentially designed to achieve self-sufficiency in production of the main food grains. Despite the transition to overall self-sufficiency in many agricultural food products, the policy reform process for agriculture has been much more tentative than for the non-agricultural sector. Government intervention in agricultural product markets in the country has, in general, been pervasive, longstanding and politically difficult to change. Policies based on controls and subsidies that were created in an era of food shortages, have continued and these have impeded the establishment of an integrated national food production and distribution system.

The central tenet of government agricultural policy over the last several decades has been essentially one of self reliance through growth in domestic production and to control trade to ensure adequate availability of essential food items to consumers at reasonable prices while protecting farmers from foreign competition. Domestic support arrangements such as support prices, statutory minimum prices, intervention purchases and use of buffer stocks as well as barriers to interstate movement of some agricultural products in some parts of the country, remain in place.

Following tentative steps taken in the 1980s, broad based economic reforms were initiated by the Central government in 1991 that included international trade liberalisation, liberalisation of direct foreign investment and financial sector deregulation and these have contributed to an improvement in macroeconomic performance in recent years. Despite concerns in some quarters that increased openness to the world economy would make India vulnerable to external shocks, the economy has managed to grow at an average rate of 5.4% per year since the second half of the 1990s, notwithstanding a number of unfavourable external factors such as the Asian economic crisis, higher energy prices and a slowing world economy. For 2004, GDP is expected to grow by about 8.1% making India one of the fastest growing and resilient economies in the world.

Minimum Support Price and Statutory Minimum Price

The most visible domestic support arrangement in the Indian agricultural economy is that of Minimum Support Prices (MSP) to producers for cereals, oilseeds and some other crops to encourage investment and production. Support prices are fixed each year, taking into account movements in input prices, domestic and world product price trends, inter crop price parity and the effect on the cost of living. The domestic price support arrangements have been an essential part of the government's agricultural marketing and price stabilisation policy since 1980. This price support policy was basically intended to protect the farmers against any sharp decline in farm prices. The government also fixes a Statutory Minimum Price (SMP) for sugarcane and cotton. This in effect is the floor price for these crops. In the case of sugarcane, the SMP is linked to the sugar recovery percentage from crushing the cane. State governments announce their own minimum price, known as the state advised prices (SAP) which are usually much higher than the SMPs, and therefore become the effective minimum prices which sugar mills have to pay to farmers for their sugarcane.

Buffer stocks

The Indian government has a policy to maintain buffer stocks to meet emergencies like drought or crop failures and to undertake open market operations (sales from stock piles) in case of a sharp rise in the domestic market price. In recent years, the actual stocks held in the central pool have been much in excess of the minimum buffer stock norms. This is the direct result of the price support policy of the government and continually raising MSPs in the context of low world prices. A food security buffer stock of 10 million tonnes (4 million tonnes of wheat and 6 million tonnes of rice) is generally considered to be adequate for most requirements.

Input subsidies

As noted earlier, production of staple cereals of rice and wheat during the period of the green revolution was boosted through technology transfers and production techniques based on improved seeds and higher input usage. This technology placed a high priority on supplying irrigation water and fertilisers to the farmers. Given the strategic importance of these inputs to the success of the government's plans to boost food grain production, the onus fell on the government to ensure that sufficient water, power and fertilisers were available, accessible and affordable to farmers. Input subsidies thus became an important instrument of domestic agricultural policy over the last three decades.

Power subsidy

The tariffs for use of electricity by the agricultural sector are generally set at rates lower than the average cost of generation and supply to farmers. The tariffs of power supply to industrial and commercial consumers are set at a higher rate to cross-subsidise the losses arising out of electricity supplied to agriculture. Some States have been supplying power free to the agriculture sector while others have been charging just one euro cent per unit of power. This means that each additional unit of subsidised power adds to the States' total burden of subsidies.

Irrigation subsidy

The subsidies given to the irrigation sector increased from USD 1.30 billion in 1993-94 to USD 3.04 billion in 2000-01 (see Figure 8.1) Furthermore, in the case agriculture, water rates are not based on the actual quantities used, which offer little incentive for efficient use. This practice may have also contributed to inappropriate cropping pattern in some regions.

Figure 8.1. **Agriculture subsidies in India during 1993-94 to 2000-01**

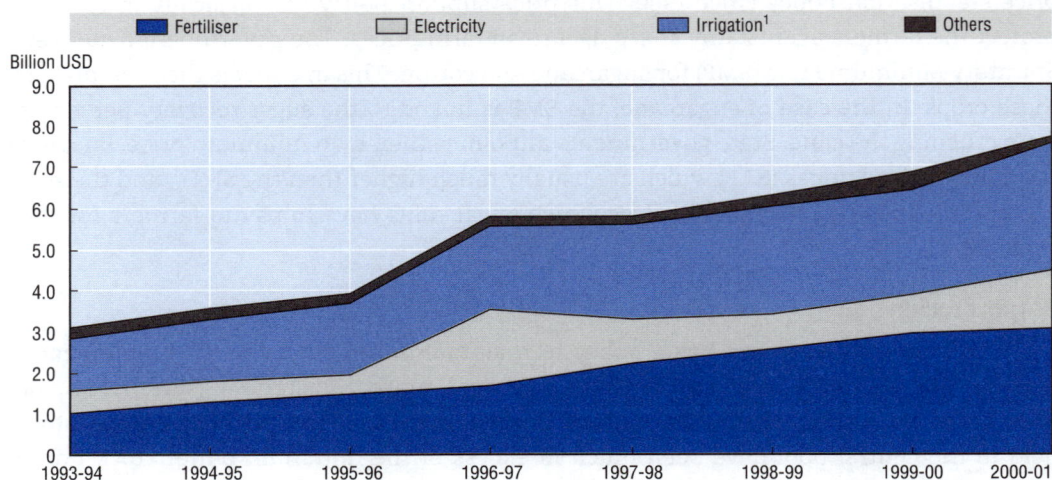

Source: Based on data central Statistical Organization, New Delhi NR45 = US$1.

Fertiliser Subsidy

A fertiliser subsidy in India is given to the fertiliser manufacturers rather than directly to farmers. Under the Retention Price Scheme (RPS), the price of fertiliser produced at each plant was fixed separately so as to give a post tax return of 12% on net worth under certain conditions. The retention price scheme has been replaced by a new pricing policy for urea units in an attempt to move towards parity with international prices. Other deregulation and market liberalisation measures are also underway and, from 1 April 2004, urea distribution and prices are to be deregulated.

In the 1990s, agricultural policy sought to further enhance production through the increased availability of input subsidies for electricity, water and fertiliser and by raising the minimum support prices, rather than building new capital assets in irrigation and other rural infrastructure. It is now widely recognized that this policy approach has outlived its utility, has not benefited the rural poor and has contributed to inefficiencies. As fertiliser subsidies are provided to fertiliser plants rather than users, these have had the effect of keeping inefficient plants in operation and at high cost to the treasury. Irrigation subsidies have promoted excessive use of agricultural water and have resulted in less funds being available for the maintenance of water facilities and led to underinvestment in new irrigation infrastructure. In addition, by setting water charges that are not based on actual water usage, this practice has led to inappropriate cropping patterns in some regions. Power subsidies for electricity use have lead to misuse of supplies designated for agricultural uses. In addition, these subsidies for fertilisers, irrigation water and power have been provided at considerable cost to the national treasury and form a large component of the Central government's budget deficit, estimated at around 5% of GDP. One consequence of the growth in subsidies and their drain on the public purse has been a squeezing out of other high priority public investments, such as for rural infrastructural developments to improve productivity.

The performance of the agricultural sector in recent years, compared to the 1980s, has been characterised by similar growth rates, an increase in non-traditional exports and importance of value-added products as well as burgeoning food grain stocks with rising MSPs. While input subsidies have helped to encourage overproduction of food grains and impose heavy storage costs on the government, other agricultural policies impede agricultural growth and diversification. For instance, restrictions on the movement and stocking of grain have kept private investment in storage and processing low. In addition, taxes on agricultural processing firms are passed back to farmers and impede expansion of higher value added exports. Such policies may have made sense in the past to assure cheap food availability but are less relevant in a situation where grain supplies are larger than can be used or stored.

A National Agriculture Policy (NAP) was adopted by the government in July 2000 with a perspective covering the next two decades which aims to attain a) a growth rate of at least 4% per annum in the agricultural sector; b) growth in demand driven by domestic markets and to maximize benefits from agricultural exports; c) growth that is technologically, environmentally and economically sustainable; d) growth with equity, i.e., growth which is spread across regions and farmers. Besides other things, the NAP aims to develop quality consciousness among farmers and agro-food processors. Better grading and standardisation of products is to be promoted for export enhancement.The Central government has indicated that agricultural sector reform should follow a holistic approach involving: *a)* a better targeting of food subsidies; and, *b)* a review of pricing and procurement operations under MSP to make them more efficient and cost effective.

Trade policy reforms

Restrictive trade policies have been implemented by the Indian government in the past to encourage import replacement and reinforce domestic market interventions, to promote exports of surplus commodities and for balance of payment reasons. In addition to quantitative import restrictions on agricultural products, tariffs were generally maintained at a high level to discourage imports except for items of mass consumption such as rice, wheat and maize. Trade policies cover the spectrum of import licensing, export restrictions, domestic market controls, tariffs, non-tariff barriers and quantitative restrictions and export subsidies. In many respects, the persistence of these policies reflect the fact that the agricultural sector was left out of the structural adjustment programme and the broad based macro – economic reforms introduced in 1991. The transformation has been more gradual for agriculture and it was only in 1994 that agriculture was included in the broader reforms. To a large extent, reform in agriculture has been as much a unilateral effort as it has been a response to external influences. Being a signatory to URAA in 1994, India committed itself to gradually liberalising its agriculture sector. In the initial years of the agreement, however, even this was limited. It was only in the late 1990s and early 2000s that more fundamental changes occurred. India's trade policy reform began in earnest with the abolition of quantitative restrictions in 2001. With the removal of some 1 400 quantitative restrictions, the customs tariff has become India's main instrument of trade policy. However, some commodities are still imported through government controlled corporations or state trading enterprises which continue to play a significant role particularly in the food grain market. In line with the lowering of trade barriers, most of the export restrictions have also been removed progressively.

In general, India bound its tariff at ceiling rates ranging from 40% for non-agricultural products to 100% or more for most agricultural products and 300% for edible oils in the URAA. As a result of India's commitments to WTO, the final average bound tariff is expected to be 50.6% in 2005, with an average of 115.7% in agriculture (HS 1-24) and 37.7% in non-agricultural products. India has tariff quotas on imports of milk powder, maize, crude sunflower-seed and safflower oil, refined rape, colza and mustard oil and it provides market access based on preferential tariff rates on commodities under various regional and bilateral agreements with different countries. Export quotas for sugar, powdered milk, ghee, and peacock tail feathers have been removed. Several agricultural exports face non-tariff measures in India's main markets, mainly for SPS/TBT reasons and as a reflection of tariff rate quota implementation rules. The commodities concerned include products such as spices, tea, tobacco, meat and poultry, groundnuts, floriculture, cereals and cereal products, fresh fruit, dairy products, and marine products.

Several autonomous bodies, within the Ministry of Commerce and Industry, aid in the implementation of trade policies. Many are holdovers from the past, and include commodity boards for tea, coffee, rubber, spices, tobacco, coir, coconut and silk which are responsible for production and market development of these commodities, respectively. Another autonomous body is the Export Inspection Council which is responsible for the enforcement of quality control and pre-shipment inspection of various commodities covered under the Export Quality Control and Inspection Act of 1963.

Agricultural exports and imports

Despite being a major agricultural producer, inward-looking trade policies have meant that India is a small player in world agricultural trade for many products. India's agricultural exports are less than 1% of total world exports and are mostly directed towards OECD countries. This performance also seems to reflect a lack of international competitiveness due to pervasive government intervention at both central and state levels in agricultural markets. The commodities where India is competitive are meat, eggs, rice, grapes, cotton, coffee, oilseed cakes and castor oil. Trade liberalisation and related domestic policy reform in agriculture will likely increase domestic prices in the short run, and in particular for exportable commodities, such as sugar, wheat and rice. Higher domestic prices in comparison with international prices will make Indian exports less competitive in global markets.

India's major partners in external trade are the OECD countries, followed by the Asian and African countries (see Tables 8.7 and 8.8). The share of exports to the United States has increased in recent years making it India's largest export destination. Since the year 1999-00, the top ten countries which imported Indian agricultural commodities are Belgium, Canada, France, Germany, Italy, Japan, Korea, the Netherlands, Spain, Switzerland, the United Kingdom and the United States. The main exports by India are still largely determined by the supply side and the availability of surpluses above domestic requirements. This is

Table 8.5. **India's export of rice**

Quantity '000 mt value USD million

Description	1998-99 Quantity	1998-99 Value	1999-00 Quantity	1999-00 Value	2000-01 Quantity	2000-01 Value	2001-02 Quantity	2001-02 Value
Rice in husk (Paddy)	1.88	0.30	2.51	0.38	5.39	1.51	42.33	7.37
Husked rice (Brown)	0.00	0.23	0.70	0.10	0.26	0.03	0.25	0.03
Rice parboiled	2 009.71	444.67	737.25	172.48	381.99	96.01	723.73	133.41
Other rice	2 280.60	519.52	465.95	117.90	293.40	74.79	751.23	151.32
Broken rice	73.65	14.14	51.33	8.16	1.72	0.44	23.95	3.73
Total non-Basmati rice	4 365.84	978.63	1 257.74	299.02	682.76	173.00	1 541.49	295.86
Basmati rice	597.76	417.09	638.38	395.63	851.72	481.32	667.07	409.50
Grand total	4 963.60	1 395.72	1 896.12	694.65	1 534.48	654.10	2 208.56	705.37

Source: Ministry of Agriculture, Government of India.

Table 8.6. **Projected demand and supply of food grains in India**

Million tonnes

SN Supply scenario	Total supply (net of seed and waste)	Demand (food + feed) Per capita income grows by: 3.70%	Demand (food + feed) Per capita income grows by: 6.00%
Total demand		**296.2**	**374.7**
1. 1962/65-1993 Trend extrapolated	**321.1**	24.9	−53.6
2. Reasonable increase in fertlizer and irrigation use	**232.2**	−64.0	−142.5
3. (2) + genetic and efficiency improvements	**259.9**	−36.3	−114.8
4. (3) + additional land degradation	**242.1**	−54.1	−132.6

Source: Bhalia, G.S. *et al.* (1999).

one of the major limitations, and provides a critical context for the Central government's current projections which call for growth in agricultural exports of 9% per annum.

India may not have consistent surpluses of average quality food grains in future years that can be relied on for exports because of increasing domestic consumption due to population growth. Other commodities like basmati rice and many fruits and vegetables, as well as certain processed foods are expected to form part of a continuing export basket. India's exports of basmati rice have been steady at around 0.7 million tonnes per annum, with most being shipped in bulk form (Table 8.5). The major export destinations for this type of rice are the countries of the Middle East such as Kuwait, Saudi Arabia and UAE. However, OECD member countries like France, Germany, Italy, the United Kingdom and the United States together form the largest destination. Exports of non-basmati rice have witnessed wide annual fluctuations. The importers of basmati and parboiled rice are mostly African and Asian countries which are traditionally very price sensitive.

The contribution of agricultural exports to foreign exchange earnings is critical for India. Every one per cent rise in agricultural exports could mean INR 85 billion worth of additional revenue for the agricultural sector. Following a government directive, the Food Corporation of India, has discontinued subsidised allocations for export since August 2003. The majority of agricultural exports are net foreign exchange earners for India as there is little or no import content. This contrasts with the case for many manufactured and industrial products, where import content in terms of raw materials and machinery use is very high. Under the auspices of the WTO, many new opportunities for trade have opened up for the developing countries and India has ambitions to achieve a substantial growth of exports in the coming years. Nevertheless, some estimates indicate that in the absence of significant technological changes in the agricultural sector there would be a demand-supply gap of 115-142 million tonnes of food grains if India's per capita income continues to grow at 5.5% per annum and even if there is a slowdown in the economy (per capita income grows at 2% per annum) there would be a demand-supply gap of 25 million tonnes (Table 8.6).

Future policy directions

In the agricultural sector, the holdover of policies formulated in a period of scarcity such as controls and subsidies, have impeded the creation of an integrated national food system and remain a major barrier to India exploiting its comparative advantage in agriculture in the global economy. As discussed above, one of the main structural problems faced by Indian agriculture is the fragmentation of farmland into a large number of small and marginal farm holdings. Technology may not be able to change static agriculture, where producers continue to grow staple crops using traditional methods without responding much to market demand for newer crops and varieties. This contrasts with a more dynamic agriculture in which farmers are more responsive to changing market realities and have access to all the necessary farm inputs. One of the first things to be addressed in facilitating such a transition is to find ways for removing or at least increasing the ceiling limit on land holdings and to change tenancy laws. Provisions made for contract farming in recent years is one such step in this direction. If India is to exploit the potential available in terms of rich and diverse resources, it has to build a modern outward-looking and market orientated agricultural marketing system. To realise the potential gains from the external trade it would be essential to carry out reforms in the domestic sector. However, without speedy domestic market reforms, the opportunity to become more fully integrated in the world trading system and to capture markets may well be missed, and with reduced trade barriers could be transformed into a threat to India's future growth.

Conclusions

Agriculture in India has significant potential and will remain a crucial sector in the next and coming decades. Promoting more rapid agricultural growth at a rate of 4% growth per year, as set out in the government's latest Five-Year Plan, will be important for achieving strong economic performance for the country as a whole and improving the well being of millions of poor farm households in rural areas. The current agricultural environment while helping in keeping the production of many crops viable in the near term, is also eroding the foundation of agricultural production (land and water resources) and thus threatening the longer term viability and growth prospects of the sector. India has started to reform its agricultural sector in recent years to become more outward-looking. A continuation of this process could include the adoption of a more market orientated agricultural policy with a focus on encouraging investment from both public and private sources and less so on poorly targeted subsidies and restrictive trade barriers. In that way, world market signals could exert a stronger influence on the orientation of agricultural production and market outcomes and help promote India's comparative advantage in

Table 8.7. **Destination of Indian exports, 1995-96 to 2001-02**

Group/Country	Exports (USD million)						
	95-96	96-97	97-98	98-99	99-00	00-0	01-02
I. OECD Countries	17 705	18 601	19 485	19 264	21 107	23 474	21 622.1
A. EU, of which:	8 708.3	8 655.3	9 144.6	8 946.6	9 382.4	10 411	9 845.9
1. Belgium	1 120.4	1 092.7	1 215.5	1 287.9	1 367.7	1 470.6	1 390.6
2. France	747	716.2	759.6	829.7	897.3	1 020	945
3. Germany	1 977.4	1 893.1	1 923.7	1 851.9	1 738.4	1 907.6	1 788.4
4. Italy	1 014	933.7	1 115.2	1 055	1 119.8	1 308.8	1 206.5
5. Netherlands	769	852.4	803.8	763.5	885.8	880.1	863.9
6. United Kingdom	2 010.8	2 046.9	2 140.8	1 855.4	2 034.8	2 298.7	2 160.9
B. North America, of which:	5 825.8	6 908.4	7 236.1	7 672.6	8 973.8	9 961.6	9 098.2
1. Canada	305.4	353	433.2	473	578.3	656.5	584.8
2. United States	5 520.4	6 555.4	6 802.9	7 199.6	8 395.5	9 305.1	8 513.3
C. Asia and Oceania, of which:	2 651.9	2 456.9	2 408.7	2 096.2	2 153	2 263.6	1 990.7
1. Australia	375.7	385.4	438.3	387.4	403.3	405.9	418
2. Japan	2 215.6	2 005.9	1 898.5	1 652	1 685.4	1 794.5	1 510.4
D. Other OECD countries, of which:	519.1	580.7	695.5	548.6	597.4	837.6	687.3
1. Switzerland	281.6	299.9	367.5	319.1	353.7	437.7	409.1
II. O P E C	3 080.8	3 232.5	3 535.1	3 560	3 902.4	4 864.4	5 224.5
III. Eastern Europe, of which:	1 340	1 098.5	1 283.3	1 052.9	1 292.9	1 317.8	1 254.8
IV. Developing countries, of which:	9 196.6	10 033	10 304	9 212	10 453	12 998	13 535.5
A. Asia	7 307.8	8 133.9	7 972.4	6 844.5	8 205.5	10 038	10 332.7
a) S A A R C	1 720.6	1 701.6	1 610.9	1 679.2	1 394.6	1 928.5	2 026
b) Other Asian developing countries	5 587.2	6 432.3	6 361.5	5 165.3	6 810.9	8 109.4	8 306.6
B. Africa	1 512.4	1 420	1 634.5	1 757.2	1 550.3	1 951.5	2 260.9
V. Others	18.8	22.5	45.9	40.4	37.1	76.7	89.9
Total trade	31 795	33 470	35 006	33 219	36 822	44 560	43 826.7

Note: Exports of petroleum products are taken in account in total exports, but are not included in country details.
Source: Handbook of Statistics on Indian Economy, Reserve Bank of India, 2003.

Table 8.8. **Origin of Indian imports, 1995-96 to 2001-02**

Group/Country	Imports (USD million)						
	95-96	96-97	97-98	98-99	99-00	00-01	01-02
I. OECD Countries	17 705	18 601	19 485	19 264	21 107	23 474	21 622.1
A. EU, *of which:*	8 708.3	8 655.3	9 144.6	8 946.6	9 382.4	10 411	9 845.9
1. Belgium	1 120.4	1 092.7	1 215.5	1 287.9	1 367.7	1 470.6	1 390.6
2. France	747	716.2	759.6	829.7	897.3	1 020	945
3. Germany	1 977.4	1 893.1	1 923.7	1 851.9	1 738.4	1 907.6	1 788.4
4. Italy	1 014	933.7	1 115.2	1 055	1 119.8	1 308.8	1 206.5
5. Netherlands	769	852.4	803.8	763.5	885.8	880.1	863.9
6. United Kingdom	2 010.8	2 046.9	2 140.8	1 855.4	2 034.8	2 298.7	2 160.9
B. North America, *of which:*	5 825.8	6 908.4	7 236.1	7 672.6	8 973.8	9 961.6	9 098.2
1. Canada	305.4	353	433.2	473	578.3	656.5	584.8
2. United States	5 520.4	6 555.4	6 802.9	7 199.6	8 395.5	9 305.1	8 513.3
C. Asia and Oceania, *of which:*	2 651.9	2 456.9	2 408.7	2 096.2	2 153	2 263.6	1 990.7
1. Australia	375.7	385.4	438.3	387.4	403.3	405.9	418
2. Japan	2 215.6	2 005.9	1 898.5	1 652	1 685.4	1 794.5	1 510.4
D. Other OECD countries, *of which:*	519.1	580.7	695.5	548.6	597.4	837.6	687.3
1. Switzerland	281.6	299.9	367.5	319.1	353.7	437.7	409.1
II. O P E C	3 080.8	3 232.5	3 535.1	3 560	3 902.4	4 864.4	5 224.5
III. Eastern Europe, *of which:*	1 340	1 098.5	1 283.3	1 052.9	1 292.9	1 317.8	1 254.8
IV. Developing countries, *of which:*	9 196.6	10 033	10 304	9 212	10 453	12 998	13 535.5
A. Asia	7 307.8	8 133.9	7 972.4	6 844.5	8 205.5	10 038	10 332.7
a) S A A R C	1 720.6	1 701.6	1 610.9	1 679.2	1 394.6	1 928.5	2 026
b) Other Asian developing countries	5 587.2	6 432.3	6 361.5	5 165.3	6 810.9	8 109.4	8 306.6
B. Africa	1 512.4	1 420	1 634.5	1 757.2	1 550.3	1 951.5	2 260.9
V. Others	18.8	22.5	45.9	40.4	37.1	76.7	89.9
Total trade	31 795	33 470	35 006	33 219	36 822	44 560	43 826.7

Source: Handbook of Statistics on Indian Economy, Reserve Bank of India, 2003.

competing in the global economy. In addition, reducing the emphasis put on inefficient input subsidies will free scarce public sector resources for investment that could contribute to more permanent productivity gains and output growth and help raise rural incomes. Emerging structural problems such as limits on farm size, infrastructure inadequacies, and constrained access to better inputs also remain a drag on development. More investment in the less favoured areas would also contribute to the productivity and competitiveness of the entire sector. There may also be a need for well targeted safety nets to ensure that the rural poor are not left worse off in the transition from controls and subsidies to a more open and market orientated agricultural policy.

Bibliography

Babu, S.C. (2004), More Supportive Policies, The Hindu Survey of Indian Agriculture 2004, Chennai.

Bhalla, G.S. *et al.* (1999),Prospects for India's Cereal supply and demand to 2020, International Food Policy Research Institute, Washington.

Bhalla, G.S. *et al.* (1999), Prospects for India's Cereal Supply and Demand to 2020, International Food Policy Research Institute, Washington, Discussion Paper 29.

Bhatia, M.S. (1994), Prices, Marketing, Agro-Processing and International Trade, *Indian Journal of Agricultural Economics*, Volume 49, No. 3, July – December.

Bathla, S. (2003), Agriculture Market Intervention Policies: Trends and Implications in a New Regime, Institute of Economic Growth, New Delhi, Processed.

Dandekar, V.M. (1994),The Indian Economy 1947-92 , Volume I: Agriculture, Sage Publications, New Delhi.

Debroy, B. (2000), India's Agricultural Exports in the Post Uruguay Round Setup – Implications, Prospects and Policies, NCAER, New Delhi.

Dohlman, Erik *et al.* (2003), India's Edible Oil Sector: Imports Fill Rising Demand, USAD, No. OCS-0903-01.

Government of India, (2000), National Agricultural Policy, New Delhi.

Gulati, Ashok and Tim Kelly (1999), Trade Liberalization and Indian Agriculture Cropping Pattern Changes and Efficiency Gains in Semi-Arid Tropics, Oxford University Press, New Delhi.

Lagh, Y.K. (1999), Agriculture Trade and Sustainable Development, *Indian Journal of Agricultural Economics*, Volume 54, No. 1, January-March.

Kondaiah, N. (1997), Export of Meat and Meat Products: Prospects and Constraints in Promotion of Agricultural Exports, Indian Institute of Foreign Trade, New Delhi.

Kumar, P. (1998), Food demand and supply projections for India, IARI, New Delhi. Agricultural Economics Policy Paper 98-01.

Pachauri, R.K. and Pooja Mehrotra (2002), Vision 2020: Sustainability of India's Material Resources, Planning Commission, New Delhi.

Parikh, K.S. (2002), Hungry Under Food Mountain, *Ind Jour. Agril.Mktng.*, 16(1).

Pinstrup-Anderson (2002) Reshaping Indian Food and Agricultural Policy to Meet the Challenges and Opportunities of Globalisation, Exim Bank Commencement Day Annual Lecture, Mumbai India, April 22.

Planning Commission (2002), Tenth Plan Document, Planning Commission, New Delhi, Vol. II.

Planning Commission (2002) Report of the Steering Group on Agriculture and Allied Sectors for the Tenth Five Year Plan 2002-2007, Planning Commission, New Delhi.

Radhakrishna, R. and K.V. Reddy (2002), Food Security and Nutrition: Vision 2020, Planning Commission, New Delhi.

Reddy, V.R. and K.B. Narayanan (1992), Trade Experience of Indian Agriculture : Behaviour of Net Export Supply Functions for Dominant Commodities, *Indian Journal of Agricultural Economics*, Volume 47, No. 1, January – March.

Satyasai , K.J.S. and K.U. Viswanathan (1996), Diversification of Indian Agriculture and Food Security, *Indian Journal of Agricultural Economics*, Volume 51, No. 4, Oct.-Dec.

Sharma , A. (2001), The Agricultural Sector in *Economic and Policy Reforms in India*, NCAER, New Delhi.

Subramanian, S. (1993), Agricultural Trade Liberalization and India, OECD, Paris.

The Hindu Business Line (2004), Rewoking Grain Exports, Editorial, Vol. 11, No. 60.

Trade Policy Review Body (2002), Trade Policy Review – India, No. WT/TPR/S/100, World Trade Organization, Geneva.

Virmani, A. and P.V. Rajeev (2002), Excess Food Stocks, PDS and Procurement Policy, Working Paper No. 5/2002-PC, Planning Commission, New Delhi.

ISBN 92-64-02008-X
OECD Agricultural Outlook: 2004-2013
© OECD 2004

Methodology

The projections presented and analysed in this document are the result of a process that brings together information from member countries and a number of other sources. Consistency in this process is ensured by the use of the OECD's *Aglink* model. A large amount of expert judgement, however, is applied at various stages of the *Outlook* process. The *OECD Agricultural Outlook* presents a single assessment, judged by the Secretariat to be plausible given the underlying assumptions, the procedure of information exchange outlined below and the information to which it had access as of 7 April 2004.

The starting point of the *Outlook* process is the reply by member countries (and some Non-member economies) to an annual questionnaire circulated by the Secretariat at mid-year. Through these questionnaires, the Secretariat obtains information from member countries on future commodity market developments and on the evolution of agricultural policies in OECD countries. This information is supplemented by that obtained from external sources, such as the FAO, the World Bank or the IMF, to establish a view of the main forces determining market developments in the Non-member economies. This part of the process is aimed at creating a first insight into possible market developments and at establishing the key assumptions which condition the *Outlook*. The main economic and policy assumptions are indicated in the chapter on Economic and Policy Assumptions, and in specific commodity tables of the present report. The assumed *Outlook* period developments in main macroeconomic variables are based on December 2003 medium term projections of the OECD's Economics Department for member countries, and September 2003 *Global Economic Prospects* projections of the World Bank. While sometimes different from the macroeconomic assumptions provided through the questionnaire replies, it has been judged preferable to use just two consistent sources for these variables.

As a next step, the OECD's *Aglink* model is used to facilitate a consistent integration of this information and to derive an initial set of global market projections (baseline). *Aglink* is a dynamic economic and policy specific model of major temperate-zone agricultural commodity markets. It currently consists of complete agricultural sector modules for major OECD agricultural producing and trading members, namely Australia, Canada, the European Union, Japan, Korea, Mexico, New Zealand, and the United States, in addition to certain Non-member economies, namely Argentina, Russia, China and Brazil, as well as a beef sector module for other MERCOSUR countries. A revised standalone sugar model has also been developed to produce a set of long term baseline projections for world and OECD sugar markets, covering raw and white (or refined) sugar. The modules are all developed by the Secretariat in conjunction with experts in member countries and Non-member economies and, in some cases, with assistance from other national administrations. The initial baseline results are compared with those obtained from the questionnaire replies and any emerging issues are sometimes discussed in bilateral exchanges with country experts. On the basis of these discussions and of updated information, a second baseline is produced.

In addition to quantities produced, consumed and traded, the baseline also includes projections for nominal prices for the commodities concerned. Unless otherwise stated, prices referred to in the text are also in nominal terms.

The information generated is used to prepare reports presenting *Outlook* assessments for cereals, oilseeds, meats, dairy products and sugar. These reports are discussed at the annual meetings of the *Working Group on Meat and Dairy Products* and the *Working Group on Cereals, Animal Feeds and Sugar* of the OECD *Committee for Agriculture*. The *Outlook* discussions in the *Working Groups* focus on key issues emerging from the replies to the questionnaires and any adjustments which have to be made to member country projections in order to derive a coherent global baseline. Subsequent to the meetings of the commodity *Working Groups* and final data revisions, a revised baseline is produced and its sensitivity to major uncertainties evaluated. The revised projections form the basis of a draft of the present OECD *Agricultural Outlook* publication, which is normally discussed by the *Working Party on Agricultural Policies and Markets of the Committee for Agriculture*, prior to publication.

The above procedure implies that the baseline projections presented in this report are heavily conditioned by those developed by member countries and participating Non-member economies. It also reconciles inconsistencies between individual country projections through the use of a formal modelling framework and highlights the sensitivity of the outcomes to key assumptions. The review process ensures that the judgement of country experts is applied to the projections and related analyses. However, the final responsibility for the projections and their interpretation rests with the OECD Secretariat.

ISBN 92-64-02008-X
OECD Agricultural Outlook: 2004-2013
© OECD 2004

ANNEX I

Statistical Tables

Annex Table 1. **ECONOMIC ASSUMPTIONS**

Calendar year[a]		Average 1998-02	2003 est.	2004	2005	2006	2007	2008	2009	2010	2011	2012	2013
REAL GDP[b]													
Australia	%	3.8	2.4	3.7	4.0	3.8	3.6	3.4	3.2	3.2	3.2	3.2	3.2
Canada	%	4.0	1.8	2.8	3.2	3.2	3.3	3.1	3.1	3.1	3.1	3.1	3.1
EU-15	%	2.4	0.5	1.8	2.5	2.5	2.4	2.2	2.1	1.9	2.1	2.1	2.1
Japan	%	0.5	2.7	1.8	1.8	1.5	1.5	1.5	1.5	1.5	1.5	1.5	1.5
Korea	%	4.6	2.7	4.7	5.5	5.7	5.3	5.3	5.4	5.4	5.3	5.3	5.3
Mexico	%	3.2	1.5	3.6	4.2	4.5	4.4	4.3	4.3	4.1	4.2	4.2	4.2
New Zealand	%	2.9	2.7	3.1	2.9	3.1	3.2	3.2	3.1	3.1	3.1	3.1	3.1
United States	%	3.0	2.9	4.2	3.8	3.2	3.0	2.9	2.9	3.0	2.9	2.9	2.9
OECD[c, e]	%	2.5	1.9	2.9	3.1	2.9	2.7	2.6	2.6	2.6	2.6	2.6	2.6
Argentina	%	−3.1	4.0	4.0	3.8	3.6	3.3	3.4	3.3	3.3	5.7	5.7	5.7
Brazil	%	1.7	1.8	3.6	3.9	4.8	4.5	4.4	3.8	3.6	4.1	4.1	4.1
China	%	7.6	7.2	7.5	7.3	8.6	8.3	8.2	7.2	7.0	8.1	8.1	8.1
Russia	%	3.8	5.2	4.5	3.2	2.7	2.5	2.6	2.3	1.7	1.1	1.1	1.1
Rest of world[d]	%	2.8	2.9	4.0	4.1	3.9	4.0	4.1	3.9	3.7	3.7	3.6	3.4
CPI[b]													
Australia	%	2.8	2.8	2.0	2.3	2.5	2.4	2.4	2.5	2.4	2.4	2.4	2.4
Canada	%	4.0	2.8	1.4	2.1	2.0	2.0	2.0	2.0	2.0	2.0	2.0	2.0
EU-15	%	1.8	2.0	1.5	1.4	1.5	1.6	1.6	1.6	1.6	1.6	1.6	1.6
Japan	%	−0.4	−0.2	−0.2	−0.2	−0.1	0.0	0.0	0.0	0.0	0.0	0.0	0.0
Korea	%	3.5	3.5	2.7	3.0	3.0	3.0	3.0	3.0	3.0	3.0	3.0	3.0
Mexico	%	10.6	4.6	3.4	3.1	3.2	3.2	3.2	3.2	3.2	3.2	3.2	3.2
New Zealand	%	1.6	0.6	2.0	2.3	2.3	2.3	2.3	2.3	2.3	2.3	2.3	2.3
United States	%	2.3	2.3	1.7	1.8	1.8	1.8	1.8	1.8	1.8	1.8	1.8	1.8
OECD[c, e]	%	3.0	2.3	1.7	1.6	1.6	1.6	1.6	1.6	1.6	1.6	1.6	1.6
Argentina	%	5.4	16.2	14.4	10.4	7.9	6.3	5.2	4.2	3.9	3.9	3.9	3.9
Brazil	%	5.3	13.4	7.4	6.3	5.4	5.5	4.9	4.7	4.6	4.8	4.8	4.8
China	%	0.2	1.0	1.7	1.7	1.9	2.4	2.5	2.9	3.1	2.8	2.8	2.8
Russia	%	3.8	13.2	12.1	5.5	5.8	3.8	3.1	3.5	2.3	3.0	3.0	3.0
POPULATION													
Australia	million	19.2	19.8	20.0	20.2	20.4	20.5	20.7	20.8	21.0	21.1	21.2	21.3
Canada	million	30.8	31.7	31.9	32.1	32.3	32.5	32.7	32.8	33.0	33.1	33.2	33.3
EU-25	million	452.3	455.1	455.8	456.6	457.3	457.9	458.5	459.1	459.6	460.0	460.4	460.7
Japan	million	127.0	127.5	127.5	127.5	127.5	127.4	127.3	127.0	126.8	126.4	126.1	125.7
Korea	million	47.0	48.0	48.3	48.6	48.9	49.2	49.4	49.7	49.9	50.1	50.3	50.4
Mexico	million	98.2	102.6	104.1	105.6	106.8	108.1	109.4	110.7	112.1	113.4	114.7	116.1
United States	million	282.4	291.6	293.7	295.7	297.7	299.7	301.7	303.7	305.8	307.8	309.8	311.8
OECD[c]	million	1 139.7	1 162.6	1 168.6	1 174.4	1 179.9	1 185.2	1 190.4	1 195.5	1 200.5	1 205.2	1 209.8	1 214.4
Argentina	million	37.0	38.4	38.8	39.2	39.5	39.9	40.2	40.6	40.9	41.3	41.7	42.0
Brazil	million	170.2	176.4	178.4	180.4	182.5	184.5	186.5	188.5	190.5	192.4	194.4	196.3
China	million	1 259.5	1 288.2	1 297.0	1 305.6	1 313.8	1 322.0	1 330.4	1 338.7	1 347.1	1 355.4	1 363.8	1 372.1
Russia	million	146.1	144.8	144.3	143.8	143.2	142.6	142.1	141.5	140.8	140.2	139.6	138.9
Rest of world[d]	million	3 166.1	3 345.7	3 402.7	3 459.3	3 515.1	3 571.0	3 626.7	3 682.5	3 738.3	3 794.1	3 849.8	3 905.6

For notes, see end of the table.

Annex Table 1. **ECONOMIC ASSUMPTIONS** *(cont.)*

Calendar year[a]		Average 1998-02	2003 est.	2004	2005	2006	2007	2008	2009	2010	2011	2012	2013
EXCHANGE RATE													
Australia	*AUD/USD*	1.73	1.54	1.41	1.41	1.41	1.41	1.41	1.41	1.41	1.41	1.41	1.41
Canada	*CAD/USD*	1.51	1.40	1.33	1.33	1.33	1.33	1.33	1.33	1.33	1.33	1.33	1.33
EU-15	*EUR/USD*	1.02	0.89	0.87	0.87	0.87	0.87	0.87	0.87	0.87	0.87	0.87	0.87
Japan	*JPY/USD*	119.7	118.0	116.4	116.4	114.5	111.4	108.5	105.6	102.8	100.1	97.5	94.9
Korea	*'000 KRW/USD*	1.25	1.19	1.19	1.19	1.19	1.19	1.19	1.19	1.19	1.19	1.19	1.19
Mexico	*MXN/USD*	9.43	10.75	10.99	10.99	10.99	10.99	10.99	10.99	10.99	10.99	10.99	10.99
New Zealand	*NZD/USD*	2.10	1.73	1.63	1.63	1.63	1.63	1.63	1.63	1.63	1.63	1.63	1.63
Argentina	*ARS/USD*	1.41	3.08	3.33	3.55	3.68	3.76	3.82	3.85	3.85	3.95	4.06	4.17
Brazil	*BRL/USD*	2.02	3.09	2.99	3.13	3.27	3.38	3.49	3.59	3.72	3.85	3.98	4.11
Russia	*RUR/USD*	24.6	33.0	35.7	36.3	37.5	38.5	39.4	40.9	41.7	42.3	43.0	43.6
China	*CNY/USD*	8.28	8.18	8.03	7.95	8.06	8.09	8.06	8.27	8.41	8.56	8.72	8.89

a) For OECD member countries, historical data for real GDP, population and exchange rate were obtained from the *OECD Economic Outlook No. 74*, December 2003, and for CPI from the *OECD Main Economic Indicators*, December 2003. For Non-member economies, historical macroeconomic data were obtained from the World Bank, September 2003. Assumptions for the projection period draw on the recent medium term macroeconomic projections of the OECD Economics Department, projections of the World Bank, and responses to a questionnaire sent to member country agricultural experts. Data for the EU-15 are euro area aggregates.

b) Annual per cent change. For Non-member economies the price index used is the Private Consumption Expenditure deflator.

c) Excludes Iceland.

d) Excludes OIS, Argentina, China, Brazil and Russia. Source: World Bank, September 2003.

e) Annual weighted average real GDP and CPI growth rates in OECD countries are based on weights using 1995 GDP and purchasing power parities (PPPs).

est.: estimate.

Source: OECD Secretariat.

Annex Table 2. **WORLD PRICES**[a]

		Average 98/99-02/03	03/04 est.	04/05	05/06	06/07	07/08	08/09	09/10	10/11	11/12	12/13	13/14
WHEAT													
Price[b]	USD/t	126.4	152.7	148.1	154.0	156.2	156.8	156.4	154.7	154.8	153.6	153.5	152.9
COARSE GRAINS													
Price[c]	USD/t	93.5	104.9	106.2	111.5	112.6	113.6	114.1	114.5	114.5	113.7	113.8	113.7
RICE													
Price[d]	USD/t	218.0	203.8	231.5	252.0	263.7	274.9	286.3	291.0	298.8	302.9	310.2	316.3
OILSEEDS													
Price[e]	USD/t	218.0	316.8	252.5	245.6	257.6	256.1	254.9	252.7	249.9	253.3	253.3	254.1
OILSEED MEALS													
Price[f]	USD/t	165.6	237.3	192.2	180.2	183.0	181.3	180.7	181.0	179.1	179.5	179.4	179.5
VEGETABLE OILS													
Price[g]	USD/t	429.3	582.6	570.2	568.1	582.3	593.4	599.0	593.1	589.0	595.0	599.4	602.8
SUGAR													
Price, raw sugar[h]	USD/t	176.7	149.9	160.3	200.0	164.6	165.3	150.0	160.0	185.1	155.8	168.4	176.4
Price, refined sugar[i]	USD/t	224.9	196.2	190.0	229.3	193.8	194.3	178.8	188.3	213.4	184.1	196.2	203.6
BEEF AND VEAL													
Price, EU[j]	EUR/100 kg dw	245.4	245.4	242.6	240.7	245.3	249.2	248.9	248.4	247.5	247.9	247.8	247.9
Price, USA[k]	USD/100 kg dw	239.3	302.0	260.7	329.5	336.1	323.9	315.6	301.0	291.3	284.8	282.6	284.7
Price, Argentina[l]	ARS/100 kg dw	181.7	359.7	379.3	411.0	408.3	409.4	407.0	394.8	378.3	381.4	385.1	391.1
PIG MEAT													
Price, EU[m]	EUR/100 kg dw	127.2	124.9	122.8	129.5	133.3	135.3	133.4	132.8	130.3	133.6	134.4	134.9
Price, USA[n]	USD/100 kg dw	118.9	120.4	119.5	126.1	123.6	119.4	117.6	117.4	117.6	118.5	118.6	122.7
Price, Brazil[o]	BRL/100 kg dw	122.8	156.6	160.6	172.9	174.1	186.6	199.0	208.9	213.6	226.1	234.5	251.9
POULTRY MEAT													
Price, EU[p]	EUR/100 kg rtc	99.1	103.6	102.9	98.5	97.9	97.8	97.5	97.4	97.3	97.0	96.8	96.7
Price, USA[q]	USD/100 kg rtc	128.8	136.5	139.0	139.3	141.2	142.5	142.4	142.4	141.3	141.7	142.5	142.5
SHEEP MEAT													
Price, New Zealand[r]	NZD/100 kg dw	325.2	379.4	362.9	347.4	345.9	346.8	348.5	346.4	347.2	349.0	352.2	355.5
BUTTER													
Price[s]	USD/100 kg	145.2	139.2	143.9	146.0	142.1	143.9	148.6	149.2	149.6	150.1	151.1	151.6
CHEESE													
Price[t]	USD/100 kg	187.6	187.7	193.9	191.1	191.4	195.3	198.6	200.8	202.5	205.0	207.2	209.0

For notes, see end of the table.

Annex Table 2. **WORLD PRICES**[a] *(cont.)*

		Average 98/99-02/03	03/04 est.	04/05	05/06	06/07	07/08	08/09	09/10	10/11	11/12	12/13	13/14
SKIM MILK POWDER													
Price[u]	*USD/100 kg*	159.4	173.3	176.0	173.4	168.8	172.4	174.1	175.9	176.6	177.5	177.2	177.2
WHOLE MILK POWDER													
Price[v]	*USD/100 kg*	166.9	175.2	181.5	177.1	176.1	179.7	181.6	184.7	186.0	187.2	187.1	187.4
WHEY POWDER													
Wholesale price, USA[w]	*USD/100 kg*	47.0	45.8	47.1	45.9	45.8	46.0	47.1	47.7	47.7	47.5	47.5	47.6
CASEIN													
Price[x]	*USD/100 kg*	434.4	359.6	396.3	411.6	425.9	440.1	436.5	437.1	435.1	434.1	435.3	433.0

a) This table is a compilation of price information presented in the detailed commodity tables further in this annex. Prices for crops are on marketing year basis and those for meat and dairy products on calendar year basis (e.g. 00/01 is calendar year 2000).
b) No. 2 hard red winter wheat, ordinary protein, USA f.o.b. Gulf Ports (June/May).
c) No. 2 yellow corn, US f.o.b. Gulf Ports (September/August).
d) Milled, 100%, grade b, Nominal Price Quote, NPQ, f.o.b. Bangkok (August/July).
e) Weighted average oilseed price, European port.
f) Weighted average meal price, European port.
g) Weighted average price of oilseed oils and palm oil, European port.
h) Raw sugar world price, New York No. 11, f.o.b. stowed Caribbean port (including Brazil), bulk spot price.
i) Refined sugar price, London No. 5, f.o.b. Europe, spot.
j) Producer price.
k) Choice steers, 1100-1300 lb lw, Nebraska – lw to dw conversion factor 0.63.
l) Buenos Aires wholesale price linier, young bulls.
m) Pig producer price.
n) Barrows and gilts, No. 1-3, 230-250 lb lw, Iowa/South Minnesota – lw to dw conversion factor 0.74.
o) Producer price.
p) Weighted average farmgate live fowls, top quality, (lw to rtc conversion of 0.75), EU-15 starting in 1995.
q) Wholesale weighted average broiler price 12 cities.
r) Lamb schedule price, all grade average.
s) F.o.b. export price, butter, 82% butterfat, Northern Europe.
t) F.o.b. export price, cheddar cheese, 40 lb blocks, Northern Europe.
u) F.o.b. export price, nonfat dry milk, extra grade, Northern Europe.
v) F.o.b. export price, WMP 26% butterfat, Northern Europe.
w) Edible dry whey, Wisconsin, plant.
x) World price, New Zealand.
est.: estimate.

Source: OECD Secretariat

Annex Table 3. **MAIN POLICY ASSUMPTIONS FOR CEREAL MARKETS**

Crop year[a]		Average 98/99-02/03	03/04 est.	04/05	05/06	06/07	07/08	08/09	09/10	10/11	11/12	12/13	13/14
ARGENTINA													
Crops export tax	%	4	20	20	20	20	20	20	20	20	20	20	20
Rice export tax	%	2	10	10	10	10	10	10	10	10	10	10	10
CANADA													
Tariff-quotas[b]													
wheat	kt	339	350	350	350	350	350	350	350	350	350	350	350
in-quota tariff	%	1.3	1.1	1.1	1.1	1.1	1.1	1.1	1.1	1.1	1.1	1.1	1.1
out-of-quota tariff	%	65	62	62	62	62	62	62	62	62	62	62	62
barley	kt	380	399	399	399	399	399	399	399	399	399	399	399
in-quota tariff	%	0.8	0.7	0.7	0.7	0.7	0.7	0.7	0.7	0.7	0.7	0.7	0.7
out-of-quota tariff	%	59	58	58	58	58	58	58	58	58	58	58	58
EUROPEAN UNION[c, d]													
Cereal support price[e]	EUR/t	110	101	101	101	101	101	101	101	101	101	101	101
Cereal compensation[f, g]	EUR/ha	280	290	290	0	0	0	0	0	0	0	0	0
Rice support price[h]	EUR/t	302	298	150	150	150	150	150	150	150	150	150	150
Compulsory set-aside rate	%	9	10	5	10	10	10	10	10	10	10	10	10
Set-aside payment[g]	EUR/ha	297	290	290	0	0	0	0	0	0	0	0	0
Direct payment for rice	EUR/ha	329	329	1 120	475	475	475	475	475	475	475	475	475
Wheat tariff-quota[b]													
EU-15	kt	350	3 332	3 332	3 332	3 332	3 332	3 332	3 332	3 332	3 332	3 332	3 332
EU-10	kt	..	448	448	448	448	448	448	448	448	448	448	448
Coarse grain tariff-quota[b]													
EU-15	kt	2 622	3 122	3 122	3 122	3 122	3 122	3 122	3 122	3 122	3 122	3 122	3 122
EU-10	kt	..	347	347	347	347	347	347	347	347	347	347	347
Subsidised export limits[b, i]													
wheat	mt	16.3	15.6	15.6	15.6	15.6	15.6	15.6	15.6	15.6	15.6	15.6	15.6
EU-15	mt	15.1	14.4	14.4	14.4	14.4	14.4	14.4	14.4	14.4	14.4	14.4	14.4
EU-10	mt	1.2	1.1	1.1	1.1	1.1	1.1	1.1	1.1	1.1	1.1	1.1	1.1
coarse grains[j]	mt	..	10.8	10.8	10.8	10.8	10.8	10.8	10.8	10.8	10.8	10.8	10.8
EU-15	mt	10.9	10.4	10.4	10.4	10.4	10.4	10.4	10.4	10.4	10.4	10.4	10.4
EU-10	mt	0.3	0.4	0.4	0.4	0.4	0.4	0.4	0.4	0.4	0.4	0.4	0.4
JAPAN													
Rice land diversion program	'000ha	982	1 010	1 010	1 010	1 010	1 010	1 010	1 010	1 010	1 010	1 010	1 010
Wheat support price[k]	'000 JPY/t	147	145	145	145	145	145	145	145	145	145	145	145
Barley support price[l]	'000 JPY/t	127	125	125	125	125	125	125	125	125	125	125	125
Wheat tariff-quota	kt	5 719	5 740	5 740	5 740	5 740	5 740	5 740	5 740	5 740	5 740	5 740	5 740
in-quota tariff	%	9.5	9.5	9.5	9.5	9.5	9.5	9.5	9.5	9.5	9.5	9.5	9.5
out-of-quota tariff	%	496	488	488	488	488	488	488	488	488	488	488	488
Barley tariff-quota	kt	1 364	1 369	1 369	1 369	1 369	1 369	1 369	1 369	1 369	1 369	1 369	1 369
in-quota tariff	%	0	0	0	0	0	0	0	0	0	0	0	0
out-of-quota tariff	%	348	352	352	352	352	352	352	352	352	352	352	352
Rice tariff-quota[m]	kt	659	682	682	682	682	682	682	682	682	682	682	682
in-quota tariff	%	5	5	5	5	5	5	5	5	5	5	5	5
out-of-quota tariff	%	1 484	1 689	1 689	1 689	1 689	1 689	1 689	1 689	1 689	1 689	1 689	1 689

For notes, see end of the table.

Annex Table 3. **MAIN POLICY ASSUMPTIONS FOR CEREAL MARKETS** (cont.)

Crop year[a]		Average 98/99-02/03	03/04 est.	04/05	05/06	06/07	07/08	08/09	09/10	10/11	11/12	12/13	13/14
KOREA													
Wheat tariff	%	8.9	6.3	5.4	5.4	5.4	5.4	5.4	5.4	5.4	5.4	5.4	5.4
Maize tariff-quota	kt	6 102	6 102	6 102	6 102	6 102	6 102	6 102	6 102	6 102	6 102	6 102	6 102
in-quota tariff	%	1.9	1.7	1.7	1.7	1.7	1.7	1.7	1.7	1.7	1.7	1.7	1.7
out-of-quota tariff	%	422	408	404	404	404	404	404	404	404	404	404	404
Barley tariff-quota	kt	49	53	54	54	54	54	54	54	54	54	54	54
in-quota tariff	%	23	23	23	23	23	23	23	23	23	23	23	23
out-of-quota tariff	%	376	363	359	359	359	359	359	359	359	359	359	359
Rice quota[m]	kt	134	188	205	205	205	205	205	205	205	205	205	205
in-quota tariff	%	5	5	5	5	5	5	5	5	5	5	5	5
MERCOSUR													
Wheat tariff	%	12	12	12	10	10	10	10	10	10	10	10	10
Coarse grain tariff	%	8	8	8	8	8	8	8	8	8	8	8	8
Rice tariff	%	13	12	12	10	10	10	10	10	10	10	10	10
MEXICO													
Cereal income payment[n]	MXN/ha	763	913	944	973	1 004	1 036	1 069	1 103	1 139	1 175	1 213	1 252
Wheat NAFTA tariff	%	4.5	0.0	0.0	0.0	0.0	0.0	0.0	0.0	0.0	0.0	0.0	0.0
Fidelist social program	MXN mn	1 403	1 330	1 827	2 320	2 635	2 776	2 926	3 070	3 216	3 362	3 507	3 653
Tortilla consumption subsidy	MXN mn	751	0	0	0	0	0	0	0	0	0	0	0
Maize tariff-quota	kt	2 501	2 501	2 501	2 501	2 501	2 501	2 501	2 501	2 501	2 501	2 501	2 501
in-quota tariff	%	50	50	50	50	50	50	50	50	50	50	50	50
out-of-quota tariff	%	202	196	194	194	194	194	194	194	194	194	194	194
Barley tariff-quota	kt	5	5	5	5	5	5	5	5	5	5	5	5
in-quota tariff	%	50	50	50	50	50	50	50	50	50	50	50	50
out-of-quota tariff	%	120	116	115	115	115	115	115	115	115	115	115	115
UNITED STATES													
Wheat loan rate	USD/t	96.4	102.9	101.0	101.0	101.0	101.0	101.0	101.0	101.0	101.0	101.0	101.0
Maize loan rate	USD/t	75.1	77.9	76.8	76.8	76.8	76.8	76.8	76.8	76.8	76.8	76.8	76.8
Prod. flex. contract payment													
wheat	USD/t	20.7	16.9	16.9	16.9	16.9	16.9	16.9	16.9	16.9	16.9	16.9	16.9
maize	USD/t	12.6	10.3	10.3	10.3	10.3	10.3	10.3	10.3	10.3	10.3	10.3	10.3
CRP areas[o]	mha	..	6.9	6.9	7.1	7.1	7.4	7.4	7.4	7.4	7.4	7.4	7.4
wheat	mha	3.3	3.6	3.6	3.6	3.6	3.6	3.6	3.6	3.6	3.6	3.6	3.6
coarse grains	mha	..	3.3	3.3	3.5	3.5	3.8	3.8	3.8	3.8	3.8	3.8	3.8
Subsidised export limits[b]													
wheat	mt	15.2	14.5	14.5	14.5	14.5	14.5	14.5	15.5	16.5	17.5	18.5	19.5
coarse grains	mt	1.6	1.6	1.6	1.6	1.6	1.6	1.6	1.6	1.6	1.6	1.6	1.6
Wheat EEP payment[p]	USD/t	0.0	0.0	0.0	0.0	0.0	0.0	0.0	0.0	0.0	0.0	0.0	0.0

For notes, see end of the table.

Annex Table 3. **MAIN POLICY ASSUMPTIONS FOR CEREAL MARKETS** (cont.)

Crop year[a]		Average 98/99-02/03	03/04 est.	04/05	05/06	06/07	07/08	08/09	09/10	10/11	11/12	12/13	13/14
CHINA													
Wheat support price	CNY/t	666	704	731	763	797	837	879	920	961	1 004	1 048	1 094
Coarse grains support price	CNY/t	572	614	641	672	709	748	790	831	872	915	960	1 008
Rice support price	CNY/t	1 692	1 093	1 147	1 209	1 282	1 359	1 417	1 426	1 430	1 434	1 438	1 442
Wheat tariff-quota	kt	3 270	9 198	9 636	9 636	9 636	9 636	9 636	9 636	9 636	9 636	9 636	9 636
in-quota tariff	%	..	2.3	2.3	2.3	2.3	2.3	2.3	2.3	2.3	2.3	2.3	2.3
out-of-quota tariff	%	70.2	67.3	65.0	65.0	65.0	65.0	65.0	65.0	65.0	65.0	65.0	65.0
Coarse grains tariff	%	12	2	2	2	2	2	2	2	2	2	2	2
Maize tariff-quota	kt	2 205	6 525	7 200	7 200	7 200	7 200	7 200	7 200	7 200	7 200	7 200	7 200
in-quota tariff	%	..	3.7	3.7	3.7	3.7	3.7	3.7	3.7	3.7	3.7	3.7	3.7
out-of-quota tariff	%	53.4	45.4	41.7	41.7	41.7	41.7	41.7	41.7	41.7	41.7	41.7	41.7
Rice tariff-quota	%	1 463	4 655	5 320	5 320	5 320	5 320	5 320	5 320	5 320	5 320	5 320	5 320
in-quota tariff	%	..	2.3	2.3	2.3	2.3	2.3	2.3	2.3	2.3	2.3	2.3	2.3
out-of-quota tariff	%	61.8	54.2	51.7	51.7	51.7	51.7	51.7	51.7	51.7	51.7	51.7	51.7

a) Beginning crop marketing year – see Glossary of Terms for definitions.
b) Year beginning 1 July.
c) Prices and payments in market euro – see Glossary of Terms.
d) EU farmers also benefit from the Single Farm Payment (SFP) Scheme, which provides flat-rate payments independent from current production decisions and market developments. The total amount spent under the SFP scheme, before modulation, is assumed to increase from 26.9 billion euro in 2005 to 28.4 billion euro in 2008 for the total of the 15 former member States. The final number is equivalent to 233 euro per hectare of eligible farm land on average. For the accession countries, payments are phased in with the assumption of maximum top-ups from national budgets. Due to modulation, between 2.7% and 4.6% of the total SFP will go to rural development spending rather than directly to the farmers.
e) Common intervention price for soft wheat, barley, maize and sorghum.
f) Compensatory area payments.
g) Actual payments made per hectare based on program yields.
h) Subject to a purchase limit of 75 000 tonnes per year.
i) The export volume for coarse grain excludes 0.4 mt of exported potato starch.
j) The original limit on subsidised exports is 10.8 mt; the figure given here is used to take into account subsidised exports for potato starch.
k) Government purchase price, domestic wheat.
l) Government purchase price, barley, 2nd grade, 1st class.
m) Husked rice basis.
n) Applies to producers of wheat, maize and sorghum.
o) Includes wheat, barley, maize, oats and sorghum.
p) Average per tonne of total exports.
Note: The source for tariffs and Tariff Rate Quotas is AMAD (Agricultural market access database). The tariff and TRQ data are based on Most Favoured Nation rates scheduled with the WTO and exclude those under preferential or regional agreements, which may be substantially different. Tariffs are simple averages of several product lines. Specific rates are converted to ad valorem rates using world prices in the Outlook. Import quotas are based on global commitments scheduled in the WTO rather than those allocated to preferential partners under regional or other agreements. For Mexico, the NAFTA in-quota tariff on maize and barley is zero, while the tariff-rate quota becomes unlimited in 2003 for barley and 2008 for maize.
est.: estimate.

Annex Table 4. **WORLD CEREAL PROJECTIONS**

Crop year[a]		Average 98/99-02/03	03/04 est.	04/05	05/06	06/07	07/08	08/09	09/10	10/11	11/12	12/13	13/14
WHEAT													
OECD[b]													
Production	mt	243.6	240.8	253.5	250.4	256.0	259.9	263.8	267.0	269.4	272.8	275.9	279.3
Consumption	mt	190.2	189.7	193.1	193.3	194.3	195.1	196.8	198.8	200.3	202.1	203.8	205.6
feed use	mt	68.3	67.7	69.6	68.8	68.8	68.7	69.3	70.2	70.6	71.2	71.8	72.5
Closing stocks	mt	61.0	54.6	58.4	59.0	60.8	61.9	63.9	66.1	68.1	70.3	72.0	73.9
Non-OECD													
Production	mt	333.3	310.5	329.1	338.4	343.8	348.5	352.8	358.1	363.2	368.0	372.8	378.3
Consumption	mt	391.3	391.3	397.4	400.8	405.1	411.7	417.1	423.3	429.9	436.1	442.5	449.1
feed use	mt	35.1	33.3	35.9	35.8	36.7	37.9	38.7	39.5	40.5	41.3	42.1	43.1
Net trade[d]	mt	−53.4	−47.9	−58.4	−58.3	−61.6	−65.4	−66.8	−67.7	−68.9	−70.4	−72.1	−73.6
Closing stocks	mt	160.3	109.5	97.8	91.8	90.5	91.0	91.6	92.4	92.7	93.2	93.8	94.8
WORLD[c]													
Production	mt	576.9	551.2	582.6	588.8	599.8	608.4	616.6	625.1	632.6	640.8	648.7	657.6
Consumption	mt	581.5	581.0	590.5	594.1	599.3	606.8	614.0	622.1	630.2	638.2	646.3	654.6
feed use	mt	103.4	100.9	105.4	104.6	105.5	106.6	108.0	109.7	111.1	112.5	113.9	115.6
Closing stocks	mt	221.3	164.1	156.2	150.8	151.3	152.9	155.5	158.5	160.8	163.4	165.8	168.7
Price[e]	USD/t	126.4	152.7	148.1	154.0	156.2	156.8	156.4	154.7	154.8	153.6	153.5	152.9
COARSE GRAINS													
OECD[b]													
Production	mt	470.5	469.6	490.8	498.4	508.6	514.9	520.1	525.0	530.6	535.0	537.8	541.1
Consumption	mt	444.5	457.5	456.8	467.6	474.8	479.4	481.8	484.3	487.9	490.4	493.9	497.3
feed use	mt	..	328.1	333.6	343.4	349.4	352.7	354.1	355.5	357.9	359.2	361.7	363.6
Closing stocks	mt	96.3	78.6	82.9	83.7	83.9	84.0	84.0	84.3	85.1	86.3	85.3	83.1
Non-OECD													
Production	mt	390.8	405.4	419.0	428.7	436.0	443.0	450.0	456.4	463.5	470.9	478.6	484.6
Consumption	mt	..	445.6	454.7	462.2	471.5	478.8	488.1	496.5	505.0	513.9	523.0	530.4
feed use	mt	..	272.6	278.6	282.8	290.4	296.3	303.6	310.3	316.7	323.3	330.2	335.4
Net trade[d]	mt	−27.1	−17.6	−29.6	−30.1	−33.6	−35.5	−38.2	−40.5	−41.8	−43.4	−44.9	−46.0
Closing stocks	mt	92.9	44.2	38.2	34.8	32.9	32.5	32.7	33.1	33.5	33.9	34.3	34.5
WORLD[c]													
Production	mt	861.2	875.0	909.7	927.1	944.6	957.9	970.1	981.4	994.1	1 005.9	1 016.4	1 025.7
Consumption	mt	..	903.1	911.5	929.7	946.3	958.2	969.9	980.8	992.9	1 004.3	1 016.9	1 027.7
feed use	mt	..	600.7	612.2	626.2	639.8	649.0	657.7	665.8	674.5	682.5	691.9	699.0
Closing stocks	mt	189.2	122.9	121.1	118.5	116.8	116.5	116.7	117.3	118.6	120.1	119.6	117.6
Price[f]	USD/t	93.5	104.9	106.2	111.5	112.6	113.6	114.1	114.5	114.5	113.7	113.8	113.7
RICE													
OECD[b]													
Production	mt	23.5	21.1	23.0	22.9	23.0	22.9	23.0	23.1	23.2	23.4	23.5	23.6
Consumption	mt	22.7	22.2	22.8	22.6	22.6	22.7	23.0	23.1	23.3	23.5	23.7	23.9
Closing stocks	mt	7.6	6.2	5.9	5.8	6.0	6.1	6.2	6.4	6.5	6.7	7.0	7.2
Non-OECD													
Production	mt	372.8	372.2	386.4	393.9	398.8	402.4	406.0	410.8	414.4	417.6	420.7	424.5
Consumption	mt	377.7	392.3	391.8	395.6	398.7	402.6	406.3	410.6	414.2	417.3	420.1	423.6
Net trade[d]	mt	−0.9	−0.3	−0.5	−0.4	−0.2	0.0	0.1	0.2	0.3	0.4	0.5	0.6
Closing stocks	mt	129.4	84.6	79.7	78.4	78.7	78.5	78.0	78.1	78.0	77.9	78.0	78.3
WORLD[c]													
Production	mt	396.3	393.2	409.4	416.8	421.7	425.3	429.0	433.9	437.6	441.0	444.2	448.1
Consumption	mt	400.4	414.5	414.6	418.2	421.3	4 25.3	429.3	433.8	437.6	440.8	443.9	447.5
Closing stocks	mt	137.1	90.8	85.6	84.2	84.7	84.6	84.3	84.4	84.5	84.7	85.0	85.5
Price[g]	USD/t	218.0	203.8	231.5	252.0	263.7	274.9	286.3	291.0	298.8	302.9	310.2	316.3

a) Beginning crop marketing year – see Glossary of Terms for definitions.
b) Excludes Iceland but includes Cyprus, Estonia, Latvia, Lithuania, Malta and Slovenia.
c) Source of historic data is USDA.
d) Non-OECD net exports (imports) equal OECD net imports (exports).
e) No. 2 hard red winter wheat, ordinary protein, USA f.o.b. Gulf Ports (June/May).
f) No. 2 yellow corn, US f.o.b. Gulf Ports (September/August).
g) Milled, 100%, grade b, Nominal Price Quote, NPQ, f.o.b. Bangkok (August/July).
est.: estimate.

Source: OECD Secretariat.

Annex Table 5. **MAIN POLICY ASSUMPTIONS FOR OILSEED MARKETS**

Crop year[a]		Average 98/99-02/03	03/04 est.	04/05	05/06	06/07	07/08	08/09	09/10	10/11	11/12	12/13	13/14
ARGENTINA													
Oilseed export tax	%	7.5	23.5	23.5	23.5	23.5	23.5	23.5	23.5	23.5	23.5	23.5	23.5
Oilseed meal export tax	%	4.0	20.0	20.0	20.0	20.0	20.0	20.0	20.0	20.0	20.0	20.0	20.0
Oilseed oil export tax	%	4.0	20.0	20.0	20.0	20.0	20.0	20.0	20.0	20.0	20.0	20.0	20.0
AUSTRALIA													
Tariffs													
soyabean oil	%	8.2	8.0	8.0	8.0	8.0	8.0	8.0	8.0	8.0	8.0	8.0	8.0
rapeseed oil	%	8.2	8.0	8.0	8.0	8.0	8.0	8.0	8.0	8.0	8.0	8.0	8.0
CANADA													
Tariffs													
rapeseed oil	%	6.8	6.4	6.4	6.4	6.4	6.4	6.4	6.4	6.4	6.4	6.4	6.4
EUROPEAN UNION[c, d]													
Oilseed compensation[e, f]	EUR/ha	280	290	290	0	0	0	0	0	0	0	0	0
Compulsory set-aside rate	%	10.0	10.0	5.0	10.0	10.0	10.0	10.0	10.0	10.0	10.0	10.0	10.0
Set-aside payment[f]	EUR/ha	296.9	290.1	290.1	0.0	0.0	0.0	0.0	0.0	0.0	0.0	0.0	0.0
Tariffs													
soyabean oil	%	6.4	6.0	6.0	6.0	6.0	6.0	6.0	6.0	6.0	6.0	6.0	6.0
rapeseed oil	%	6.4	6.0	6.0	6.0	6.0	6.0	6.0	6.0	6.0	6.0	6.0	6.0
JAPAN													
Deficiency payments													
soyabeans	bn. JPY	12.1	12.9	12.9	12.9	12.9	12.9	12.9	12.9	12.9	12.9	12.9	12.9
Tariffs													
soyabean oil	%	11.5	10.9	10.9	10.9	10.9	10.9	10.9	10.9	10.9	10.9	10.9	10.9
rapeseed oil	%	27.6	28.4	28.4	28.4	28.4	28.4	28.4	28.4	28.4	28.4	28.4	28.4
KOREA													
soyabean tariff-quota	kt	1 032	1 032	1 032	1 032	1 032	1 032	1 032	1 032	1 032	1 032	1 032	1 032
in-quota tariff	%	5	5	5	5	5	5	5	5	5	5	5	5
out-of-quota tariff	%	509	492	487	487	487	487	487	487	487	487	487	487
Soyabean (for food) mark up	'000 KRW/t	170	145	146	147	145	145	144	144	143	143	143	143
MEXICO													
Soyabeans income payment[g]	MXN/ha	763	913	944	973	1 004	1 036	1 069	1 103	1 139	1 175	1 213	1 252
Tariffs													
soyabeans	%	34.5	33.4	33.0	33.0	33.0	33.0	33.0	33.0	33.0	33.0	33.0	33.0
soyabean meal	%	29.3	25.1	23.8	23.8	23.8	23.8	23.8	23.8	23.8	23.8	23.8	23.8
soyabean oil	%	47.0	45.5	45.0	45.0	45.0	45.0	45.0	45.0	45.0	45.0	45.0	45.0
UNITED STATES													
Soyabeans loan rate	USD/t	191.4	183.7	183.7	183.7	183.7	183.7	183.7	183.7	183.7	183.7	183.7	183.7
CRP area													
soyabeans	mha	1.9	2.2	2.2	2.2	2.2	2.2	2.2	2.2	2.2	2.2	2.2	2.2
Tariffs													
rapeseed	%	3.0	3.0	3.0	3.0	3.0	3.0	3.0	3.0	3.0	3.0	3.0	3.0
soyabean meal	%	2.3	2.2	2.2	2.2	2.2	2.2	2.2	2.2	2.2	2.2	2.2	2.2
rapeseed meal	%	1.2	1.2	1.2	1.2	1.2	1.2	1.2	1.2	1.2	1.2	1.2	1.2
soyabean oil	%	13.0	12.7	12.7	12.7	12.7	12.7	12.7	12.7	12.7	12.7	12.7	12.7
rapeseed oil	%	3.2	3.2	3.2	3.2	3.2	3.2	3.2	3.2	3.2	3.2	3.2	3.2
Subsidised export limits[b]													
oilseed oils	kt	194.8	141.0	141.0	141.0	141.0	141.0	141.0	141.0	141.0	141.0	141.0	141.0

For notes, see end of the table.

OECD AGRICULTURAL OUTLOOK: 2004-2013 – ISBN 92-64-02008-X – © OECD 2004

Annex Table 5. **MAIN POLICY ASSUMPTIONS FOR OILSEED MARKETS** (cont.)

Crop year[a]		Average 98/99-02/03	03/04 est.	04/05	05/06	06/07	07/08	08/09	09/10	10/11	11/12	12/13	13/14
CHINA													
Soyabeans support price	CNY/t	751.4	0.0	0.0	0.0	0.0	0.0	0.0	0.0	0.0	0.0	0.0	0.0
Tariffs[b]													
soyabeans	%	59.4	2.4	2.4	2.4	2.4	2.4	2.4	2.4	2.4	2.4	2.4	2.4
soyabean meal	%	13.0	6.3	6.3	6.3	6.3	6.3	6.3	6.3	6.3	6.3	6.3	6.3
soyabean oil in-quota tariff	%	..	9.0	9.0	9.0	9.0	9.0	9.0	9.0	9.0	9.0	9.0	9.0
Vegetable oil tariff-quota	kt	2 150.8	6 436.6	6 944.6	7 998.1	7 998.1	7 998.1	7 998.1	7 998.1	7 998.1	7 998.1	7 998.1	7 998.1

a) Beginning crop marketing year – see Glossary of Terms for definitions.
b) Calendar year, except for China and subsidised export limit in USA, beginning 1 July.
c) Prices and payments in market euro – see Glossary of Terms.
d) EU farmers also benefit from the Single Farm Payment (SFP) Scheme, which provides flat-rate payments independent from current production decisions and market developments. The total amount spent under the SFP scheme, before modulation, is assumed to increase from 26.9 billion euro in 2005 to 28.4 billion euro in 2008 for the total of the 15 former member States. The final number is equivalent to 233 euro per hectare of eligible farm land on average. For the accession countries, payments are phased in with the assumption of maximum top-ups from national budgets. Due to modulation, between 2.7% and 4.6% of the total SFP will go to rural development spending rather than directly to the farmers.
e) Compensatory area payments, before penalties.
f) Payments made per hectare based on region.
g) Weighted average of autumn/winter and spring/summer.
Note: The source for tariffs and Tariff Rate Quotas is AMAD (Agricultural market access database). The tariff and TRQ data are based on Most Favoured Nation rates scheduled with the WTO and exclude those under preferential or regional agreements, which may be substantially different. Tariffs are simple averages of several product lines. Specific rates are converted to ad valorem rates using world prices in the Outlook. Import quotas are based on global commitments scheduled in the WTO rather than those allocated to preferential partners under regional or other agreements. For Mexico, the NAFTA tariffs on soyabeans, oilmeals and soyabean oil are zero after 2003.
est.: estimate.
Source: OECD Secretariat.

Annex Table 6. WORLD OILSEED PROJECTIONS

Marketing year[a]		Average 98/99-02/03	03/04 est.	04/05	05/06	06/07	07/08	08/09	09/10	10/11	11/12	12/13	13/14
OILSEEDS													
OECD[b]													
Production	mt	105.2	92.7	111.8	111.8	110.7	112.4	113.0	113.5	114.6	115.7	117.4	118.3
Consumption	mt	106.3	99.0	105.9	109.1	111.0	112.9	115.0	116.8	118.4	119.6	120.8	122.0
crush	mt	95.4	90.4	95.9	98.9	100.8	102.7	104.6	106.3	107.8	108.8	110.0	111.1
Closing stocks	mt	13.5	9.2	12.5	13.2	12.7	13.1	13.3	13.5	13.7	13.5	13.6	13.6
Non-OECD													
Production	mt	123.4	151.4	157.5	159.4	163.3	168.9	174.5	180.4	185.5	189.4	194.4	199.8
Consumption	mt	122.1	147.7	157.0	161.0	163.0	167.4	172.1	176.5	181.1	185.5	190.6	195.8
crush	mt	105.0	129.0	136.8	140.1	141.9	145.8	150.0	154.0	158.1	162.1	166.9	171.7
Net trade[d]	mt	0.9	5.2	−2.6	−2.0	−0.3	1.0	2.2	3.6	4.1	3.7	3.5	3.7
Closing stocks	mt	7.6	7.0	10.1	10.4	11.0	11.5	11.8	12.1	12.4	12.7	13.0	13.4
WORLD[c]													
Production	mt	228.6	244.2	269.2	271.2	274.0	281.3	287.5	293.9	300.0	305.1	311.7	318.1
Consumption	mt	228.4	246.7	262.9	270.1	274.0	280.3	287.0	293.3	299.5	305.0	311.3	317.8
crush	mt	200.4	219.4	232.7	239.0	242.7	248.4	254.6	260.3	265.8	271.0	276.9	282.8
Closing stocks	mt	21.1	16.2	22.5	23.6	23.6	24.6	25.1	25.6	26.1	26.2	26.6	27.0
Price[e]	USD/t	218.0	316.8	252.5	245.6	257.6	256.1	254.9	252.7	249.9	253.3	253.3	254.1
OILSEED MEALS													
OECD[b]													
Production	mt	70.6	66.9	71.4	73.9	75.4	76.7	78.3	79.5	80.7	81.5	82.3	83.2
Consumption	mt	89.2	92.9	96.8	99.1	99.8	101.1	102.5	103.8	105.1	106.2	107.5	108.6
Closing stocks	mt	2.7	2.6	3.0	2.9	2.9	2.9	3.0	3.0	3.0	3.0	3.0	3.0
Non-OECD													
Production	mt	75.8	93.5	99.0	101.3	102.6	105.4	108.5	111.4	114.4	117.4	120.8	124.4
Consumption	mt	57.2	68.2	72.8	76.0	78.2	81.0	84.2	87.0	89.9	92.5	95.7	98.9
Net trade[d]	mt	18.6	26.0	25.8	25.1	24.4	24.4	24.3	24.3	24.5	24.8	25.1	25.4
Closing stocks	mt	3.1	2.7	3.2	3.5	3.4	3.5	3.6	3.6	3.7	3.7	3.8	3.8
WORLD[c]													
Production	mt	146.5	160.4	170.4	175.2	178.0	182.1	186.8	190.9	195.1	198.8	203.2	207.5
Consumption	mt	146.4	161.1	169.6	175.0	178.0	182.0	186.7	190.9	195.0	198.8	203.1	207.5
Closing stocks	mt	5.7	5.4	6.2	6.4	6.4	6.5	6.5	6.6	6.7	6.7	6.8	6.8
Price[f]	USD/t	165.6	237.3	192.2	180.2	183.0	181.3	180.7	181.0	179.1	179.5	179.4	179.5
VEGETABLE OILS													
OECD[b]													
Production	mt	22.7	21.3	22.6	23.2	23.5	23.9	24.3	24.7	25.0	25.3	25.6	25.8
Consumption	mt	24.9	26.4	26.8	27.3	27.6	28.0	28.4	28.9	29.4	29.8	30.3	30.8
Closing stocks	mt	2.2	1.6	1.8	1.9	1.9	2.0	2.0	2.0	2.0	2.1	2.1	2.1
Non-OECD													
Production	mt	44.7	51.4	53.9	55.3	56.5	58.5	60.7	62.9	65.1	67.2	69.5	71.9
Consumption	mt	42.3	46.6	48.9	50.9	52.3	54.2	56.4	58.5	60.6	62.7	64.7	66.8
Net trade[d]	mt	2.3	4.9	4.4	4.3	4.2	4.1	4.1	4.2	4.4	4.5	4.7	4.9
Closing stocks	mt	5.5	5.1	5.7	5.8	5.8	5.9	6.1	6.2	6.4	6.4	6.5	6.7
WORLD[c]													
Production	mt	67.4	72.7	76.5	78.5	80.0	82.4	85.0	87.6	90.1	92.5	95.1	97.7
of which palm oil	mt	20.1	22.5	23.1	23.7	24.5	25.5	26.8	28.1	29.4	30.6	31.9	33.1
Consumption	mt	67.2	73.0	75.7	78.2	80.0	82.2	84.9	87.4	90.0	92.4	95.0	97.6
Closing stocks	mt	7.7	6.7	7.5	7.7	7.8	7.9	8.1	8.3	8.4	8.5	8.6	8.7
Oil price[g]	USD/t	429.3	582.6	570.2	568.1	582.3	593.4	599.0	593.1	589.0	595.0	599.4	602.8

a) Beginning crop marketing year – see Glossary of Terms for definitions.
b) Excludes Iceland but includes Cyprus, Estonia, Latvia, Lithuania, Malta and Slovenia.
c) Source of historic data is USDA.
d) Non-OECD net exports (imports) equal OECD net imports (exports).
e) Weighted average oilseed price, European port.
f) Weighted average meal price, European port.
g) Weighted average price of oilseed oils and palm oil, European port.
est.: estimate.
Source: OECD Secretariat.

OECD AGRICULTURAL OUTLOOK: 2004-2013 – ISBN 92-64-02008-X – © OECD 2004

Annex Table 7. **MAIN POLICY ASSUMPTIONS FOR MEAT MARKETS**

		Average 1998-02	2003 est.	2004	2005	2006	2007	2008	2009	2010	2011	2012	2013
ARGENTINA													
Beef export tax	%	1	5	5	5	5	5	5	5	5	5	5	5
CANADA													
Beef tariff-quota	kt pw	76	76	76	76	76	76	76	76	76	76	76	76
in-quota tariff	%	0	0	0	0	0	0	0	0	0	0	0	0
out-of-quota tariff	%	28	27	27	27	27	27	27	27	27	27	27	27
Poultry meat tariff-quota	kt pw	45	45	45	45	45	45	45	45	45	45	45	45
in-quota tariff	%	3	2	2	2	2	2	2	2	2	2	2	2
out-of-quota tariff	%	201	197	197	197	197	197	197	197	197	197	197	197
EUROPEAN UNION[a, b]													
Beef basic price[c, d, e]	EUR/kg dw	3.09	2.22	2.22	2.22	2.22	2.22	2.22	2.22	2.22	2.22	2.22	2.22
Beef buy-in price[c, f]	EUR/kg dw	..	1.56	1.56	1.56	1.56	1.56	1.56	1.56	1.56	1.56	1.56	1.56
Pig meat basic price[d]	EUR/kg dw	1.51	1.51	1.51	1.51	1.51	1.51	1.51	1.51	1.51	1.51	1.51	1.51
Sheep meat basic price	EUR/kg dw	5.04	5.04	5.04	5.04	5.04	5.04	5.04	5.04	5.04	5.04	5.04	5.04
Sheep basic rate[g]	EUR/head	..	21.00	21.0	21.0	21.0	21.0	21.0	21.0	21.0	21.0	21.0	21.0
Male bovine premium[h]	EUR/head	183	229	229	0	0	0	0	0	0	0	0	0
Adult bovine slaughter premium[i]	EUR/head	45	102	102	0	0	0	0	0	0	0	0	0
Calf slaughter premium	EUR/head	20	50	50	0	0	0	0	0	0	0	0	0
Suckler cow premium	EUR/head	167	200	200	0	0	0	0	0	0	0	0	0
Beef tariff-quota													
EU-15	kt pw	164	164	164	164	164	164	164	164	164	164	164	164
EU-10	kt pw	..	52	52	52	52	52	52	52	52	52	52	52
Pig meat tariff-quota													
EU-15	kt pw	56	67	67	67	67	67	67	67	67	67	67	67
EU-10	kt pw	..	101	101	101	101	101	101	101	101	101	101	101
Poultry meat tariff-quota													
EU-15	kt pw	28	30	30	30	30	30	30	30	30	30	30	30
EU-10	kt pw	..	66	66	66	66	66	66	66	66	66	66	66
Sheep meat tariff-quota													
EU-15	kt cwe	285	285	285	285	285	285	285	285	285	285	285	285
EU-10	kt cwe	..	1	1	1	1	1	1	1	1	1	1	1
Subsidised export limits[d]													
beef[j]													
EU-15	kt cwe	860	822	822	822	822	822	822	822	822	822	822	822
EU-10	kt pw	..	106	106	106	106	106	106	106	106	106	106	106
pig meat[j]													
EU-15	kt cwe	456	444	444	444	444	444	444	444	444	444	444	444
EU-10	kt pw	..	142	142	142	142	142	142	142	142	142	142	142
poultry meat													
EU-15	kt cwe	304	286	286	286	286	286	286	286	286	286	286	286
EU-10	kt pw	..	158	158	158	158	158	158	158	158	158	158	158
JAPAN[k]													
Beef stabilisation prices													
upper price	JPY/kg dw	1 024	1 010	1 010	1 010	1 010	1 010	1 010	1 010	1 010	1 010	1 010	1 010
lower price	JPY/kg dw	789	780	780	780	780	780	780	780	780	780	780	780
Beef tariff	%	..	45	41	39	39	39	39	39	39	39	39	39
Pig meat stabilisation prices													
upper price	JPY/kg dw	489	480	480	480	480	480	480	480	480	480	480	480
lower price	JPY/kg dw	369	365	365	365	365	365	365	365	365	365	365	365
Pig meat import system[l]													
tariff	%	4	4	4	4	4	4	4	4	4	4	4	4
standard import price	JPY/kg dw	416	410	410	410	410	410	410	410	410	410	410	410
Poultry meat tariff	%	8	7	7	7	7	7	7	7	7	7	7	7

For notes, see end of the table.

Annex Table 7. **MAIN POLICY ASSUMPTIONS FOR MEAT MARKETS** (cont.)

		Average 1998-02	2003 est.	2004	2005	2006	2007	2008	2009	2010	2011	2012	2013
KOREA													
Beef tariff	%	42	40	40	40	40	40	40	40	40	40	40	40
Beef mark-up	%	6	0	0	0	0	0	0	0	0	0	0	0
Pig meat tariff	%	25	23	22	22	22	22	22	22	22	22	22	22
Poultry meat tariff	%	22	21	21	21	21	21	21	21	21	21	21	21
MEXICO													
Pig meat tariff	%	47	46	45	45	45	45	45	45	45	45	45	45
Pig meat NAFTA tariff	%	6	0	0	0	0	0	0	0	0	0	0	0
Poultry meat tariff-quota	kt pw	41	41	41	41	41	41	41	41	41	41	41	41
in-quota tariff	%	50	50	50	50	50	50	50	50	50	50	50	50
out-of-quota tariff	%	238	230	228	228	228	228	228	228	228	228	228	228
RUSSIA													
Beef tariff-quota	kt pw	..	315	420	420	420	420	420	420	420	420	420	420
in-quota tariff	%	15	15	15	15	15	15	15	15	15	15	15	15
out-of-quota tariff	%	..	60	60	60	60	60	60	60	60	60	60	60
Pigmeat tariff-quota	kt pw	..	335	450	450	450	450	450	450	450	450	450	450
in-quota tariff	%	15	15	15	15	15	15	15	15	15	15	15	15
out-of-quota tariff	%	..	80	80	80	80	80	80	80	80	80	80	80
Poultry meat tariff-quota	kt pw	..	774	1 050	1 050	1 050	1 050	1 050	1 050	1 050	1 050	1 050	1 050
in-quota tariff	%	28	25	25	25	25	25	25	25	25	25	25	25
UNITED STATES													
Beef tariff-quota	kt pw	673	697	697	697	697	697	697	697	697	697	697	697
in-quota tariff	%	5	5	5	5	5	5	5	5	5	5	5	5
out-of-quota tariff	%	27	26	26	26	26	26	26	26	26	26	26	26
CHINA													
Beef tariff	%	40	20	16	16	16	16	16	16	16	16	16	16
Pig meat tariff	%	19	17	16	16	16	16	16	16	16	16	16	16
Sheep meat tariff	%	22	16	15	15	15	15	15	15	15	15	15	15
Poultry meat tariff	%	20	19	19	19	19	19	19	19	19	19	19	19

a) Prices and payments in market euro's – see Glossary of Terms.
b) EU farmers also benefit from the Single Farm Payment (SFP) Scheme, which provides flat-rate payments independent from current production decisions and market developments. The total amount spent under the SFP scheme, before modulation, is assumed to increase from 26.9 billion euro in 2005 to 28.4 billion euro in 2008 for the total of the 15 former member States. The final number is equivalent to 233 euro per hectare of eligible farm land on average. For the accession countries, payments are phased in with the assumption of maximum top-ups from national budgets. Due to modulation, between 2.7% and 4.6% of the total SFP will go to rural development spending rather than directly to the farmers.
c) Price for R3 grade male cattle.
d) Year beginning 1 July, except for E10 which is calendar year. Poland has a commitment on export subsidies on unspecified meat.
e) Ending 1 July 2002, replaced by basic price for storage.
f) Starting 1 July 2002.
g) A supplementary payment of 7 euro per head is provided for Less Favoured Areas.
h) Weighted average of all bull and steers payments.
i) Includes national envelopes for beef.
j) Includes live trade.
k) Year beginning 1 April.
l) Pig carcass imports. Emergency import procedures triggered from November 1995 to March 1996, from July 1996 to June 1997, from August 2001 to March 2002, from August 2002 to March 2003 and from August 2003 to March 2004.
Note: The source for tariffs and Tariff Rate Quotas (excluding Russia) is AMAD (Agricultural market access database). The tariff and TRQ data are based on Most Favoured Nation rates scheduled with the WTO and exclude those under preferential or regional agreements, which may be substantially different. Tariffs are simple averages of several product lines. Specific rates are converted to ad valorem rates using world prices in the Outlook. Import quotas are based on global commitments scheduled in the WTO rather than those allocated to preferential partners under regional or other agreements. For Mexico, the NAFTA in-quota tariff on poultry meat is zero and the tariff-rate quota is unlimited from 2003.
est.: estimate.
Source: OECD Secretariat.

Annex Table 8. **OECD MEAT PROJECTIONS**[a]

Calendar year		Average 1998-02	2003 est.	2004	2005	2006	2007	2008	2009	2010	2011	2012	2013
BEEF AND VEAL[b]													
Production	kt cwe	27 092	26 565	26 894	26 149	26 172	26 608	26 789	27 108	27 461	27 786	28 081	28 333
Net trade	kt cwe	560	−69	−59	−78	−77	53	138	252	344	389	443	498
Consumption	kt cwe	26 479	26 800	27 006	26 326	26 311	26 551	26 645	26 851	27 112	27 396	27 641	27 842
Ending stocks	kt cwe	968	1 058	999	893	826	824	823	821	820	818	817	815
Per capita consumption	kg rwt	16.3	16.1	16.2	15.7	15.6	15.7	15.7	15.7	15.8	15.9	16.0	16.0
Price, Australia[c]	AUD/100 kg dw	238	272	330	272	277	267	260	248	241	235	233	235
Price, EU[d]	EUR/100 kg dw	245	245	243	241	245	249	249	248	247	248	248	248
Price, USA[e]	USD/100 kg dw	239	302	261	329	336	324	316	301	291	285	283	285
Price, Argentina[f]	ARS/100 kg dw	182	360	379	411	408	409	407	395	378	381	385	391
PIG MEAT[g]													
Production	kt cwe	..	37 104	37 293	37 769	38 207	38 588	39 029	39 318	39 482	39 719	39 978	40 073
Net trade	kt cwe	..	1 059	1 182	1 245	1 225	1 245	1 239	1 189	1 181	1 220	1 201	1 187
Consumption	kt cwe	34 081	35 829	35 868	36 327	36 770	37 144	37 585	37 928	38 091	38 292	38 569	38 705
Ending stocks	kt cwe	620	702	738	732	737	728	722	710	703	695	690	660
Per capita consumption	kg rwt	23.3	24.0	23.9	24.1	24.3	24.4	24.6	24.7	24.7	24.8	24.9	24.9
Price, EU[h]	EUR/100 kg dw	127	125	123	129	133	135	133	133	130	134	134	135
Price, USA[i]	USD/100 kg dw	119	120	120	126	124	119	118	117	118	118	119	123
POULTRY MEAT													
Production	kt rtc	32 780	34 289	34 838	36 691	37 197	37 621	38 167	38 726	39 219	39 727	40 141	40 763
Net trade	kt rtc	2 273	1 804	1 880	1 853	1 833	1 885	1 921	1 920	1 950	1 984	2 002	2 028
Consumption	kt rtc	30 577	32 608	32 923	34 839	35 364	35 735	36 245	36 806	37 268	37 742	38 137	38 734
Stock changes	kt rtc	−71	−123	34	0	0	1	1	1	1	1	1	1
Per capita consumption	kg rwt	23.6	24.7	24.8	26.1	26.4	26.5	26.8	27.1	27.3	27.6	27.7	28.1
Price, EU[j]	EUR/100 kg rtc	99	104	103	98	98	98	97	97	97	97	97	97
Price, USA[k]	USD/100 kg rtc	129	137	139	139	141	143	142	142	141	142	143	143
SHEEP MEAT													
Production	kt cwe	..	2 751	2 665	2 672	2 662	2 673	2 675	2 665	2 654	2 637	2 620	2 602
Net trade	kt cwe	..	415	313	327	307	315	313	305	299	285	270	250
Consumption	kt cwe	..	2 324	2 349	2 341	2 352	2 354	2 358	2 356	2 353	2 350	2 349	2 352
Stock changes	kt cwe	..	12	4	4	3	4	5	6	7	8	9	10
Per capita consumption	kg rwt	..	1.8	1.8	1.8	1.8	1.7	1.7	1.7	1.7	1.7	1.7	1.7
Price, Australia[l]	AUD/100 kg dw	232	388	250	252	265	273	282	290	299	308	318	327
Price, Australia[m]	AUD/100 kg dw	106	193	125	120	123	127	131	135	139	143	147	152
Price, New Zealand[n]	NZD/100 kg dw	325	379	363	347	346	347	348	346	347	349	352	355
TOTAL MEAT													
Per capita consumption	kg rwt	..	66.6	66.7	67.7	68.0	68.4	68.8	69.3	69.6	70.0	70.3	70.7

a) Excludes Iceland but includes Cyprus, Estonia, Latvia, Lithuania, Malta and Slovenia. Carcass weight to retail weight conversion factors of 0.7 for beef and veal, 0.78 for pig meat and 0.88 for sheep meat. Rtc to retail weight conversion factor 0.88 for poultry meat.
b) Do not balance due to statistical differences in New Zealand.
c) Weighted average price of cows 201-260 kg, steers 301-400 kg, yearling < 200 kg dw.
d) Producer price.
e) Choice steers, 1100-1300 lb lw, Nebraska – lw to dw conversion factor 0.63.
f) Buenos Aires wholesale price linier, young bulls.
g) Do not balance due to consumption in Canada which excludes non-food parts.
h) Pig producer price.
i) Barrows and gilts, No. 1-3, 230-250 lb lw, Iowa/South Minnesota – lw to dw conversion factor 0.74.
j) Weighted average farmgate live fowls, top quality, (lw to rtc conversion of 0.75), EU-15 starting in 1995.
k) Wholesale weighted average broiler price 12 cities.
l) Saleyard price, lamb, 16-20 kg dw.
m) Saleyard price, wethers, < 22 kg dw.
n) Lamb schedule price, all grade average.
est.: estimate.

Source: OECD Secretariat.

Annex Table 9. **MAIN POLICY ASSUMPTIONS FOR DAIRY MARKETS**

		Average 1998-02	2003 est.	2004	2005	2006	2007	2008	2009	2010	2011	2012	2013
ARGENTINA													
Dairy export tax	%	1	5	5	5	5	5	5	5	5	5	5	5
AUSTRALIA[a]													
Domestic support payment[b]	AUDc/kg	1.0	0.0	0.0	0.0	0.0	0.0	0.0	0.0	0.0	0.0	0.0	0.0
CANADA													
Milk target price[b]	CADc/litre	57	61	64	65	66	66	67	68	69	70	70	71
Butter support price	CAD/t	5 604	6 096	6 252	6 346	6 440	6 536	6 634	6 732	6 833	6 934	7 038	7 142
SMP support price	CAD/t	4 672	5 153	5 430	5 464	5 571	5 607	5 676	5 725	5 793	5 847	5 896	5 962
Dairy subsidy	CADc/hltr	1.77	0.00	0.00	0.00	0.00	0.00	0.00	0.00	0.00	0.00	0.00	0.00
Cheese tariff-quota	kt pw	20	20	20	20	20	20	20	20	20	20	20	20
in-quota tariff	%	1	1	1	1	1	1	1	1	1	1	1	1
out-of-quota tariff	%	250	246	246	246	246	246	246	246	246	246	246	246
Subsidised export limits[c]													
cheese	kt pw	9	9	9	9	9	9	9	9	9	9	9	9
SMP	kt pw	46	45	45	45	45	45	45	45	45	45	45	45
EUROPEAN UNION[d, e, f]													
Milk quota[g]	mt pw	..	139	139	139	140	140	141	141	141	141	141	141
Butter intervention price	EUR/t	3 282	3 282	3 167	2 938	2 708	2 528	2 462	2 462	2 462	2 464	2 464	2 464
SMP intervention price	EUR/t	2 055	2 055	2 004	1 901	1 798	1 747	1 747	1 747	1 747	1 747	1 747	1 747
Butter tariff-quotas													
EU-15	kt pw	86	87	87	87	87	87	87	87	87	87	87	87
EU-10	kt pw	..	3	3	3	3	3	3	3	3	3	3	3
Cheese tariff-quota[h]													
EU-15	kt pw	88	102	102	102	102	102	102	102	102	102	102	102
EU-10	kt pw	..	1	1	1	1	1	1	1	1	1	1	1
SMP tariff-quota													
EU-15	kt pw	63	68	68	68	68	68	68	68	68	68	68	68
EU-10	kt pw	..	3	3	3	3	3	3	3	3	3	3	3
Subsidised export limits[a]													
butter													
EU-25	kt pw	410	399	399	399	399	399	399	399	399	399	399	399
cheese													
EU-15	kt pw	334	321	321	321	321	321	321	321	321	321	321	321
EU-10	kt pw	3	2	2	2	2	2	2	2	2	2	2	2
SMP													
EU-15	kt pw	280	273	273	273	273	273	273	273	273	273	273	273
EU-10	kt pw	97	95	95	95	95	95	95	95	95	95	95	95
other milk products													
EU-15	kt pw	1 129	1 098	1 098	1 098	1 098	1 098	1 098	1 098	1 098	1 098	1 098	1 098
EU-10	kt pw	144	140	140	140	140	140	140	140	140	140	140	140

For notes, see end of the table.

Annex Table 9. **MAIN POLICY ASSUMPTIONS FOR DAIRY MARKETS** (cont.)

		Average 1998-02	2003 est.	2004	2005	2006	2007	2008	2009	2010	2011	2012	2013
JAPAN[d]													
Direct payments[i]	JPY/kg	..	11	11	11	11	11	11	11	11	11	11	11
Deficiency/direct payment ceiling[j]	kt pw	2 334	2 100	2 100	2 100	2 100	2 100	2 100	2 100	2 100	2 100	2 100	2 100
Cheese tariff[k]	%	33	30	30	30	30	30	30	30	30	30	30	30
Tariff-quotas													
Butter	kt pw	2	2	2	2	2	2	2	2	2	2	2	2
in-quota tariff	%	35	35	35	35	35	35	35	35	35	35	35	35
out-of-quota tariff	%	664	733	733	733	733	733	733	733	733	733	733	733
SMP	kt pw	116	116	116	116	116	116	116	116	116	116	116	116
in-quota tariff	%	16	16	16	16	16	16	16	16	16	16	16	16
out-of-quota tariff	%	231	210	210	210	210	210	210	210	210	210	210	210
WMP	kt pw	0	0	0	0	0	0	0	0	0	0	0	0
in-quota tariff	%	24	24	24	24	24	24	24	24	24	24	24	24
out-of-quota tariff	%	325	316	316	316	316	316	316	316	316	316	316	316
KOREA													
Tariff-quotas													
Butter	kt pw	0.3	0.4	0.4	0.4	0.4	0.4	0.4	0.4	0.4	0.4	0.4	0.4
in-quota tariff	%	40	40	40	40	40	40	40	40	40	40	40	40
out-of-quota tariff	%	89	89	89	89	89	89	89	89	89	89	89	89
SMP	kt pw	0.9	1.0	1.0	1.0	1.0	1.0	1.0	1.0	1.0	1.0	1.0	1.0
in-quota tariff	%	20	20	20	20	20	20	20	20	20	20	20	20
out-of-quota tariff	%	176	176	176	176	176	176	176	176	176	176	176	176
WMP	kt pw	0.5	0.5	0.6	0.6	0.6	0.6	0.6	0.6	0.6	0.6	0.6	0.6
in-quota tariff	%	40	40	40	40	40	40	40	40	40	40	40	40
out-of-quota tariff	%	176	176	176	176	176	176	176	176	176	176	176	176
MEXICO													
Butter tariff	%	6	0	0	0	0	0	0	0	0	0	0	0
Tariff-quotas													
cheese	kt pw	9	9	9	9	9	9	9	9	9	9	9	9
in-quota tariff	%	50	50	50	50	50	50	50	50	50	50	50	50
out-of-quota tariff	%	131	126	125	125	125	125	125	125	125	125	125	125
SMP	kt pw	90	90	90	90	90	90	90	90	90	90	90	90
in-quota tariff	%	0	0	0	0	0	0	0	0	0	0	0	0
out-of-quota tariff	%	131	126	125	125	125	125	125	125	125	125	125	125
Liconsa social program	MXN mn	3 300	3 395	3 380	3 364	3 349	3 334	3 319	3 304	3 289	3 274	3 259	3 244
RUSSIA													
Butter tariff	%	20	20	20	20	20	20	20	20	20	20	20	20
Cheese tariff	%	15	15	15	15	15	15	15	15	15	15	15	15

For notes, see end of the table.

Annex Table 9. **MAIN POLICY ASSUMPTIONS FOR DAIRY MARKETS** (cont.)

		Average 1998-02	2003 est.	2004	2005	2006	2007	2008	2009	2010	2011	2012	2013
UNITED STATES[l]													
Milk support price[b]	USDc/litre	23	22	22	22	22	22	22	22	22	22	22	22
Target price[m]	USDc/litre	..	38.5	38.5	38.5	0.0	0.0	0.0	0.0	0.0	0.0	0.0	0.0
Butter support price	USD/t	1 596	2 315	2 315	2 315	2 315	2 315	2 315	2 315	2 315	2 315	2 315	2 315
SMP support price	USD/t	2 155	1 764	1 764	1 764	1 764	1 764	1 764	1 764	1 764	1 764	1 764	1 764
Butter tariff-quota	kt pw	12	13	13	13	13	13	13	13	13	13	13	13
in-quota tariff	%	10	10	10	10	10	10	10	10	10	10	10	10
out-of-quota tariff	%	105	112	112	112	112	112	112	112	112	112	112	112
Cheese tariff-quota	kt pw	134	135	135	135	135	135	135	135	135	135	135	135
in-quota tariff	%	12	12	12	12	12	12	12	12	12	12	12	12
out-of-quota tariff	%	90	87	87	87	87	87	87	87	87	87	87	87
Subsidised export limits[a]													
butter	kt pw	24	21	21	21	21	21	21	21	21	21	21	21
SMP	kt pw	73	68	68	68	68	68	68	68	68	68	68	68

a) Year ending 30 June.
b) For manufacturing milk.
c) The effective volume of cheese and SMP subsidized export will be lower reflecting the binding nature of subsidized export limits in value terms.
d) Year beginning 1 April.
e) Prices and payments in market euro's – see Glossary of Terms.
f) EU farmers also benefit from the Single Farm Payment (SFP) Scheme, which provides flat-rate payments independent from current production decisions and market developments. The total amount spent under the SFP scheme, before modulation, is assumed to increase from 26.9 billion euro in 2005 to 28.4 billion euro in 2008 for the total of the 15 former member States. The final number is equivalent to 233 euro per hectare of eligible farm land on average. For the accession countries, payments are phased in with the assumption of maximum top-ups from national budgets. Due to modulation, between 2.7% and 4.6% of the total SFP will go to rural development spending rather than directly to the farmers.
g) Total quota, EU-25 starting in 1997.
h) Calendar year minimum access for Australia, New Zealand and Canada before 1995.
i) In addition to direct payments, a compensation payment is paid – equal to 80% difference between the market price and the base price (the average price of the past three years).
j) Manufacturing milk eligible for deficiency/direct payments.
k) Excludes processed cheese.
l) Year beginning 1 January.
m) The counter-cyclical payment is determined as a 45% difference between the target price and the Boston class I price.
Note: The source for tariffs and Tariff Rate Quotas (except Russia) is AMAD (Agricultural market access database). The tariff and TRQ data are based on Most Favoured Nation rates scheduled with the WTO and exclude those under preferential or regional agreements, which may be substantially different. Tariffs are simple averages of several product lines. Specific rates are converted to ad valorem rates using world prices in the Outlook. Import quotas are based on global commitments scheduled in the WTO rather than those allocated to preferential partners under regional or other agreements.
est.: estimate.
Source: OECD Secretariat.

Annex Table 10. **WORLD DAIRY PROJECTIONS (BUTTER AND CHEESE)**

Calendar year[a]		Average 1998-02	2003 est.	2004	2005	2006	2007	2008	2009	2010	2011	2012	2013
BUTTER													
OECD[b, f]													
Production	kt pw	..	3 708	3 701	3 611	3 602	3 616	3 640	3 653	3 675	3 683	3 699	3 702
Imports	kt pw	..	207	232	234	266	268	272	276	280	283	287	290
Exports	kt pw	..	822	900	860	845	846	861	864	836	832	834	832
Consumption	kt pw	..	3 009	3 050	3 085	3 113	3 113	3 120	3 119	3 121	3 125	3 128	3 124
Closing stocks	kt pw	423	569	538	427	337	263	193	138	135	144	168	202
Non-OECD													
Production	kt pw	..	4 422	4 607	4 795	4 985	5 143	5 338	5 530	5 733	5 946	6 169	6 409
Consumption	kt pw	..	5 038	5 274	5 420	5 562	5 720	5 926	6 118	6 289	6 494	6 716	6 951
Net trade[d]	kt pw	..	−615	−668	−626	−579	−577	−589	−588	−556	−549	−547	−542
Closing stocks	kt pw	55	45	47	49	50	50	50	50	50	50	50	50
WORLD[f]													
Production[c]	kt pw	7 423	8 130	8 308	8 406	8 587	8 759	8 977	9 183	9 407	9 629	9 868	10 111
Consumption	kt pw	..	8 047	8 324	8 505	8 675	8 832	9 046	9 237	9 410	9 619	9 844	10 075
Closing stocks	kt pw	478	614	585	475	387	313	243	188	185	194	218	252
Price[e]	USD/100 kg	145	139	144	146	142	144	149	149	150	150	151	152
CHEESE													
OECD[b]													
Production	kt pw	..	13 925	14 217	14 469	14 743	14 961	15 228	15 438	15 662	15 887	16 106	16 302
Imports	kt pw	..	751	800	853	934	967	1 007	1 036	1 063	1 091	1 119	1 147
Exports	kt pw	..	1 167	1 115	1 168	1 233	1 270	1 294	1 322	1 345	1 368	1 394	1 413
Consumption	kt pw	..	13 518	13 869	14 083	14 391	14 602	14 908	15 131	15 368	15 609	15 842	16 060
Closing stocks	kt pw	659	814	846	914	965	1 018	1 049	1 068	1 078	1 077	1 065	1 040
Non-OECD													
Production	kt pw	..	3 598	3 727	3 841	3 989	4 111	4 245	4 387	4 538	4 698	4 863	5 036
Consumption	kt pw	..	4 008	4 042	4 156	4 287	4 414	4 533	4 672	4 819	4 975	5 137	5 302
Net trade[d]	kt pw	..	−416	−314	−315	−299	−303	−288	−286	−282	−277	−275	−266
Closing stocks	kt pw	72	75	76	76	76	77	77	77	78	78	78	79
WORLD													
Production[c]	kt pw	16 352	17 523	17 945	18 310	18 731	19 072	19 473	19 825	20 200	20 585	20 969	21 338
Consumption	kt pw	..	17 527	17 910	18 239	18 678	19 017	19 441	19 803	20 187	20 584	20 979	21 362
Closing stocks	kt pw	732	889	922	990	1 041	1 095	1 126	1 145	1 156	1 155	1 143	1 118
Price[g]	USD/100 kg	188	188	194	191	191	195	199	201	203	205	207	209

a) Year ending 30 June for Australia and 31 May for New Zealand in OECD aggregate.
b) Excludes Iceland but includes Cyprus, Estonia, Latvia, Lithuania, Malta and Slovenia.
c) Source of data is FAO.
d) Non-OECD net exports (imports) equals OECD net imports (exports).
e) F.o.b. export price, butter, 82% butterfat, northern Europe.
f) Do not balance due to statistical differences in New Zealand.
g) F.o.b. export price, cheddar cheese, 40 lb blocks, Northern Europe.
est.: estimate.

Source: OECD Secretariat.

Annex Table 11. **WORLD DAIRY PROJECTIONS (POWDERS AND CASEIN)**

Calendar year[a]		Average 1998-02	2003 est.	2004	2005	2006	2007	2008	2009	2010	2011	2012	2013
SKIM MILK POWDER													
OECD[b, f]													
Production	kt pw	..	3 005	2 895	2 855	2 831	2 848	2 825	2 820	2 818	2 818	2 811	2 803
Imports	kt pw	..	257	288	291	295	300	306	312	319	328	337	348
Exports	kt pw	..	1 013	1 015	1 024	1 004	1 001	1 006	1 023	1 036	1 025	1 020	1 015
Consumption	kt pw	..	2 115	2 181	2 192	2 206	2 190	2 157	2 136	2 103	2 111	2 114	2 122
Closing stocks	kt pw	603	1 017	991	918	830	783	747	716	710	716	726	736
Non-OECD													
Production	kt pw	..	634	676	727	761	807	846	884	931	980	1 032	1 098
Consumption	kt pw	..	1 401	1 403	1 460	1 471	1 508	1 546	1 595	1 648	1 678	1 714	1 765
Net trade[d]	kt pw	..	−756	−727	−733	−709	−701	−701	−711	−717	−698	−683	−667
Closing stocks	kt pw	86	60	60	60	60	60	60	60	60	60	60	60
WORLD[f]													
Production[c]	kt pw	3 402	3 639	3 571	3 582	3 592	3 655	3 671	3 704	3 749	3 798	3 842	3 902
Consumption	kt pw	..	3 516	3 584	3 652	3 676	3 698	3 703	3 731	3 752	3 789	3 828	3 887
Closing stocks	kt pw	689	1 077	1 050	978	890	843	807	776	770	776	786	796
Price[e]	USD/100 kg	159	173	176	173	169	172	174	176	177	177	177	177
WHOLE MILK POWDER													
OECD[b]													
Production	kt pw	1 789	1 901	1 914	1 998	2 033	2 074	2 115	2 149	2 181	2 213	2 249	2 291
Imports	kt pw	82	79	79	76	74	72	70	68	66	64	62	60
Exports	kt pw	1 169	1 305	1 319	1 399	1 415	1 433	1 460	1 474	1 487	1 500	1 517	1 540
Consumption	kt pw	702	686	663	674	691	711	724	742	759	776	793	810
Non-OECD													
Production	kt pw	1 521	1 657	1 740	1 799	1 854	1 908	1 971	2 037	2 108	2 178	2 249	2 326
Consumption	kt pw	..	2 883	2 981	3 123	3 195	3 270	3 361	3 443	3 528	3 614	3 703	3 806
Net trade[d]	kt pw	−1 086	−1 225	−1 241	−1 323	−1 341	−1 362	−1 390	−1 406	−1 421	−1 436	−1 455	−1 480
WORLD													
Production[c]	kt pw	3 310	3 558	3 654	3 798	3 887	3 982	4 087	4 186	4 288	4 391	4 497	4 617
Consumption	kt pw	..	3 569	3 644	3 797	3 886	3 981	4 085	4 185	4 287	4 390	4 496	4 616
Price[g]	USD/100 kg	167	175	181	177	176	180	182	185	186	187	187	187
WHEY POWDER													
Non-OECD													
Net trade	kt pw	−268	−356	−306	−287	−281	−275	−271	−265	−258	−251	−243	−234
Wholesale price, USA[h]	USD/100 kg	47	46	47	46	46	46	47	48	48	48	48	48
CASEIN													
Price[i]	USD/100 kg	434	360	396	412	426	440	436	437	435	434	435	433

a) Year ending 30 June for Australia and 31 May for New Zealand in OECD aggregate.
b) Excludes Iceland but includes Cyprus, Estonia, Latvia, Lithuania, Malta and Slovenia.
c) Source of data is FAO.
d) Non-OECD net exports (imports) equal OECD net imports (exports).
e) F.o.b. export price, nonfat dry milk, extra grade, Northern Europe.
f) Do not balance due to stastitical differences in New Zealand.
g) F.o.b. export price, WMP 26% butterfat, Northern Europe.
h) Edible dry whey, Wisconsin, plant.
i) World price, New Zealand.
est.: estimate.

Source: OECD Secretariat.

Annex Table 12. **OECD TRADE PROJECTIONS**[a]

		Average 1998-02	2003 est.	2004	2005	2006	2007	2008	2009	2010	2011	2012	2013
EXPORTS													
Wheat	kt	78 058	77 012	83 853	84 125	87 478	91 386	92 811	93 909	95 148	96 823	98 706	100 299
Coarse grains	kt	80 847	70 373	80 676	80 219	84 067	86 614	89 724	93 145	94 329	95 368	96 376	97 265
Rice	kt	4 240	3 838	4 341	4 466	4 478	4 375	4 415	4 469	4 463	4 524	4 489	4 487
Sugar	kt	11 653	11 521	11 797	11 765	11 975	11 980	11 982	12 087	12 228	12 185	12 426	12 591
Beef[b]	kt	4 995	4 330	4 046	4 993	5 082	5 120	5 276	5 474	5 754	5 924	6 058	6 106
Pig meat[b]	kt	3 074	3 441	3 582	3 770	3 871	3 962	4 027	4 099	4 184	4 249	4 292	4 306
Poultry meat	kt	3 759	3 598	3 799	3 863	3 943	4 053	4 164	4 266	4 367	4 468	4 551	4 645
Sheep meat[b, d]	kt	..	415	313	327	307	315	313	305	299	285	270	250
Butter	kt	..	822	900	860	845	846	861	864	836	832	834	832
Cheese	kt	..	1 167	1 115	1 168	1 233	1 270	1 294	1 322	1 345	1 368	1 394	1 413
Skim milk powder	kt	..	1 013	1 015	1 024	1 004	1 001	1 006	1 023	1 036	1 025	1 020	1 015
Whole milk powder	kt	1 169	1 305	1 319	1 399	1 415	1 433	1 460	1 474	1 487	1 500	1 517	1 540
Whey powder[c]	kt	268	332	280	266	254	248	242	236	230	223	214	205
IMPORTS													
Wheat	kt	24 615	29 066	25 484	25 858	25 833	25 943	26 037	26 188	26 288	26 404	26 572	26 725
Coarse grains	kt	53 786	52 784	51 035	50 145	50 444	51 162	51 501	52 655	52 481	51 977	51 521	51 241
Rice	kt	3 337	3 520	3 813	4 044	4 258	4 372	4 534	4 625	4 752	4 879	5 001	5 133
Sugar	kt	8 447	8 951	8 368	8 619	8 704	8 764	9 203	9 269	9 710	9 905	10 172	10 457
Oilseeds[d]	kt	..	5 166	−2 644	−1 955	−254	955	2 167	3 551	4 060	3 688	3 504	3 705
Oilseed meals[d]	kt	..	26 012	25 777	25 146	24 424	24 404	24 258	24 350	24 457	24 779	25 121	25 431
Vegetable oils[d]	kt	..	4 882	4 387	4 266	4 151	4 148	4 111	4 236	4 358	4 497	4 699	4 932
Beef[b]	kt	4 088	4 086	3 858	4 809	4 891	4 789	4 849	4 929	5 112	5 228	5 300	5 287
Pig meat[b]	kt	2 064	2 400	2 414	2 545	2 668	2 741	2 814	2 937	3 032	3 062	3 126	3 155
Poultry meat	kt	1 486	1 794	1 918	2 011	2 110	2 168	2 244	2 347	2 417	2 484	2 548	2 617
Butter	kt	..	207	232	234	266	268	272	276	280	283	287	290
Cheese	kt	..	751	800	853	934	967	1 007	1 036	1 063	1 091	1 119	1 147
Skim milk powder	kt	..	257	288	291	295	300	306	312	319	328	337	348
Whole milk powder	kt	82	79	79	76	74	72	70	68	66	64	62	60

a) Excludes Iceland but includes Cyprus, Estonia, Latvia, Lithuania, Malta and Slovenia. For meats, year are calendar year; for grains, meals and oils products, year are crop or marketing year; for dairy products, year are calendar year but year ends 30 June for Australia and 31 May for New Zealand in the OECD aggregate.
b) Includes trade of live animals.
c) Net exports.
d) Net imports.
est.: estimate.
Source: : OECD Secretariat

Annex Table 13. **WHEAT PROJECTIONS**

Crop year[a]		Average 98/99-02/03	03/04 est.	04/05	05/06	06/07	07/08	08/09	09/10	10/11	11/12	12/13	13/14
AUSTRALIA													
Production	mt	20.5	23.9	23.8	23.7	23.9	24.1	24.3	24.5	24.7	24.9	25.1	25.3
Consumption	mt	4.8	5.0	5.5	5.7	5.8	5.8	5.8	5.8	5.9	5.9	6.0	6.0
Exports	mt	15.1	15.2	16.3	17.5	18.1	18.3	18.5	18.7	18.8	19.0	19.1	19.3
Price[b]	AUD/t	229	228	227	235	239	240	239	237	238	236	236	236
CANADA													
Production	mt	23.7	23.6	23.4	22.9	22.9	22.6	22.6	22.8	23.0	23.6	23.8	24.2
Consumption	mt	7.6	5.7	6.3	6.5	6.7	6.8	6.7	6.9	6.9	6.9	6.9	6.9
Exports	mt	16.3	16.4	16.7	15.9	16.1	16.1	15.9	15.8	16.1	16.6	16.9	17.3
Closing stocks	mt	7.4	7.1	7.5	8.1	8.2	8.0	8.0	8.2	8.2	8.3	8.3	8.4
Price[c]	CAD/t	206	180	175	182	182	183	183	181	181	179	179	178
EU-25													
Production	mt	120.1	109.1	125.0	123.0	126.9	129.8	132.5	134.3	135.8	137.3	139.3	141.2
EU-15	mt	99.5	92.4	103.1	101.3	104.6	107.2	109.7	111.7	113.0	114.4	116.2	117.9
EU-10	mt	..	16.6	21.9	21.8	22.2	22.6	22.8	22.6	22.8	22.9	23.1	23.3
Consumption	mt	109.2	111.9	114.1	113.7	114.2	114.7	115.9	117.0	117.8	119.0	120.1	121.4
EU-15	mt	90.2	93.3	94.4	94.1	94.4	94.7	95.6	96.5	97.2	98.2	99.2	100.2
EU-10	mt	..	18.7	19.7	19.7	19.8	20.0	20.2	20.4	20.6	20.8	21.0	21.1
Exports[d]	mt	17.1	9.2	17.5	16.3	18.9	21.0	21.6	22.3	22.5	23.2	24.0	24.8
Closing stocks	mt	22.9	21.6	21.5	21.1	21.5	22.2	23.9	25.6	27.7	29.4	31.2	32.8
Price[e]	EUR/t	120	113	104	108	109	109	109	108	108	107	107	106
JAPAN													
Production	mt	0.7	0.7	0.8	0.8	0.8	0.8	0.8	0.8	0.8	0.8	0.8	0.8
Consumption	mt	6.2	6.1	6.1	6.1	6.1	6.1	6.1	6.1	6.1	6.1	6.1	6.1
Imports	mt	5.5	5.4	5.4	5.3	5.3	5.3	5.3	5.3	5.3	5.3	5.3	5.3
Closing stocks	mt	1.9	1.7	1.7	1.7	1.7	1.7	1.7	1.7	1.7	1.7	1.7	1.7
Price[f]	'000 JPY/t	22	25	25	25	25	25	24	23	23	22	22	21
KOREA													
Consumption	mt	3.7	3.9	4.2	4.3	4.3	4.4	4.5	4.6	4.6	4.7	4.8	4.9
Imports	mt	3.7	3.9	4.3	4.2	4.4	4.4	4.5	4.6	4.6	4.7	4.8	4.9
Price[g]	'000 KRW/t	163	194	187	195	198	199	198	196	196	195	194	194
MEXICO													
Production	mt	3.1	2.8	2.8	2.8	2.8	2.9	2.9	3.0	3.1	3.2	3.2	3.2
Consumption	mt	5.6	5.4	5.3	5.5	5.4	5.4	5.5	5.6	5.7	5.8	6.0	6.1
Imports	mt	2.8	3.0	2.9	3.2	3.0	3.0	3.0	3.0	3.1	3.1	3.2	3.3
Price[h]	MXN/t	1 325	1 287	1 377	1 396	1 521	1 593	1 632	1 659	1 679	1 711	1 729	1 758
NEW ZEALAND													
Production	mt	0.3	0.3	0.3	0.4	0.4	0.4	0.4	0.4	0.4	0.4	0.4	0.4
Consumption	mt	0.5	0.6	0.6	0.6	0.6	0.6	0.6	0.6	0.6	0.7	0.7	0.7
Imports	mt	0.2	0.3	0.3	0.3	0.3	0.3	0.3	0.3	0.3	0.3	0.3	0.3
Price[i]	NZD/t	274	287	270	281	285	287	286	283	283	281	281	280
UNITED STATES													
Production	mt	58.0	63.6	60.3	59.5	60.8	61.4	62.0	62.6	62.7	63.5	63.8	64.3
Consumption	mt	34.5	32.7	32.4	32.1	32.0	32.1	32.2	32.5	32.6	32.7	32.8	33.0
Imports	mt	2.6	2.0	2.0	2.2	2.2	2.3	2.3	2.3	2.3	2.3	2.3	2.3
Exports	mt	27.3	31.5	28.5	29.6	29.5	31.0	31.7	32.0	32.5	32.8	33.3	33.5
Closing stocks	mt	22.0	14.9	16.3	16.3	17.7	18.3	18.7	19.1	19.0	19.3	19.3	19.5
Price[j]	USD/t	104	127	123	128	129	130	130	128	128	127	127	127

For notes, see end of the table.

OECD AGRICULTURAL OUTLOOK: 2004-2013 – ISBN 92-64-02008-X – © OECD 2004

Annex Table 13. **WHEAT PROJECTIONS** (cont.)

Crop year[a]		98/99-02/03	03/04 est.	04/05	05/06	06/07	07/08	08/09	09/10	10/11	11/12	12/13	13/14
OTHER OECD[k]													
Production	mt	17.1	16.7	17.1	17.4	17.7	18.0	18.3	18.6	18.9	19.2	19.5	19.8
Consumption	mt	17.9	18.3	18.5	18.8	19.0	19.3	19.6	19.7	20.0	20.2	20.4	20.7
Net trade	mt	−0.6	−1.5	0.3	0.4	0.5	0.5	0.6	0.7	0.7	0.8	0.9	0.9
ARGENTINA													
Production	mt	14.3	14.6	15.1	15.9	16.9	17.1	17.4	17.6	17.9	18.2	18.3	18.4
Consumption	mt	5.5	5.4	5.3	5.3	5.4	5.5	5.6	5.8	5.8	5.9	6.1	6.2
Exports	mt	8.8	9.4	9.7	10.6	11.5	11.6	11.8	11.8	12.0	12.2	12.2	12.2
Price[l]	ARS/t	174	450	457	513	524	533	527	516	531	537	536	551
BRAZIL													
Production	mt	2.5	4.1	4.1	4.3	4.3	4.4	4.5	4.5	4.7	4.8	4.9	5.1
Consumption	mt	9.5	10.4	10.7	10.9	11.1	11.4	11.7	11.9	12.1	12.3	12.6	12.8
Net trade	mt	−6.9	−6.4	−6.6	−6.6	−6.9	−7.0	−7.2	−7.3	−7.4	−7.5	−7.6	−7.7
Price	BRL/t	237	419	390	406	434	454	468	479	494	508	523	539
CHINA													
Production	mt	101.4	86.7	88.3	88.6	88.9	88.4	87.8	88.4	88.5	88.3	88.3	89.0
Consumption	mt	108.0	102.9	99.7	98.8	97.9	97.9	97.2	97.1	97.5	97.4	97.3	97.0
Imports	mt	0.6	0.0	1.6	4.7	7.1	9.6	9.6	9.6	9.6	9.6	9.6	9.1
Closing stocks	mt	106.2	63.7	52.8	46.2	43.3	42.4	41.8	41.7	41.6	41.3	41.2	41.5
Price[m]	CNY/t	705	780	743	765	786	799	829	839	833	837	839	848
RUSSIA													
Production	mt	37.5	35.0	40.0	41.6	41.6	42.6	43.4	43.9	44.6	45.1	45.6	45.9
Consumption	mt	37.8	36.9	39.6	40.1	40.7	41.4	42.1	42.4	43.0	43.4	43.9	44.5
Net trade	mt	0.1	2.2	0.2	1.5	0.9	1.1	1.3	1.5	1.6	1.7	1.6	1.3
Price	RUR/t	1 444	1 541	1 597	1 629	1 719	1 774	1 811	1 864	1 886	1 906	1 925	1 947
OIS													
Production	mt	36.5	29.5	36.8	37.4	37.6	38.3	39.0	39.7	40.4	41.1	41.8	42.6
Consumption	mt	30.4	32.0	32.2	32.4	32.6	33.2	33.8	34.4	35.0	35.6	36.2	36.8
Net trade	mt	4.9	2.7	4.7	4.8	4.8	4.9	5.1	5.2	5.3	5.4	5.5	5.6
Closing stocks	mt	7.9	9.7	9.5	9.7	9.9	10.0	10.2	10.4	10.5	10.7	10.9	11.0
REST OF WORLD													
Production	mt	141.2	140.7	144.8	150.6	154.5	157.8	160.6	163.9	167.1	170.5	173.8	177.3
Consumption	mt	200.0	203.6	209.8	213.2	217.3	222.3	226.9	231.7	236.5	241.5	246.5	251.7
Net trade	mt	−60.8	−57.0	−64.2	−63.1	−64.2	−65.7	−67.3	−68.4	−69.8	−71.6	−73.2	−75.0
Closing stocks	mt	41.7	33.7	32.9	33.3	34.7	35.9	36.9	37.5	37.8	38.4	38.9	39.4

a) Beginning crop marketing year – see the Glossary of Terms for definitions.
b) AWB net pool return, ASW 10.
c) CWB final producer price, No. 1 CWRS, in store Thunder Bay or Vancouver. From 1995 in store St Lawrence or Vancouver.
d) Excludes intra-EU-25 trade.
e) Weighted average producer price, common and durum wheat, year ended 31 December.
f) Average import price c.i.f., all wheat, year ended 31 December.
g) Import price.
h) Average producer price.
i) Indicative wheat price.
j) Average price received by farmers.
k) Includes Norway, Switzerland and Turkey. Excludes Iceland.
l) Export price f.o.b., Argentinean ports.
m) Free market price.
est.: estimate.

Source: OECD Secretariat.

Annex Table 14. **COARSE GRAINS PROJECTIONS**

Crop year[a]		Average 98/99-02/03	03/04 est.	04/05	05/06	06/07	07/08	08/09	09/10	10/11	11/12	12/13	13/14
AUSTRALIA													
Production	mt	10.2	12.6	11.4	11.2	11.3	11.4	11.4	11.5	11.6	11.6	11.7	11.7
Consumption	mt	5.8	6.6	6.6	6.5	6.4	6.4	6.5	6.6	6.8	6.9	7.0	7.1
Exports	mt	4.7	5.8	4.7	4.6	4.6	4.7	4.7	4.7	4.7	4.6	4.6	4.6
Price[b]	AUD/t	191	227	248	254	254	255	255	255	255	253	253	252
CANADA													
Production	mt	24.0	26.1	27.3	26.3	27.6	27.8	28.4	28.8	29.4	30.1	30.6	31.0
Consumption	mt	23.1	21.5	23.3	24.1	25.0	25.1	25.3	25.5	25.8	26.0	26.1	25.8
Exports	mt	3.8	6.1	5.9	3.4	4.2	5.2	5.2	5.5	5.6	5.7	5.8	6.4
Closing stocks	mt	4.3	3.1	3.3	4.5	5.0	5.0	5.1	5.2	5.3	5.5	5.6	5.7
Price[c]	CAD/t	146	127	117	127	128	127	125	125	125	123	121	121
EU-25													
Production	mt	135.2	117.1	136.4	136.7	137.1	138.0	138.2	138.5	139.2	139.4	139.5	139.7
EU-15	mt	106.4	91.7	106.6	107.2	107.1	108.3	108.6	109.1	109.7	110.0	110.1	110.6
EU-10	mt	..	25.4	29.8	29.5	30.0	29.8	29.6	29.4	29.5	29.5	29.4	29.1
Consumption	mt	123.9	124.0	123.3	125.5	127.1	128.4	129.5	130.2	130.6	131.5	132.4	133.2
EU-15	mt	95.3	96.4	95.4	96.9	97.8	98.6	99.1	99.2	99.1	99.5	99.8	100.1
EU-10	mt	..	27.6	27.9	28.6	29.3	29.8	30.4	31.0	31.5	32.0	32.6	33.1
Exports[d]	mt	17.7	3.2	11.6	11.4	11.4	11.8	12.0	11.8	11.7	12.4	12.5	12.7
Closing stocks	mt	30.3	28.8	32.9	34.6	35.5	35.8	35.2	35.4	36.1	35.4	33.8	31.5
Price[e]	EUR/t	109	106	103	103	102	101	101	101	101	100	100	100
JAPAN													
Production	mt	0.2	0.2	0.2	0.2	0.2	0.2	0.2	0.2	0.2	0.2	0.2	0.2
Consumption	mt	21.3	21.1	21.1	20.8	20.6	20.4	20.3	20.2	20.1	20.0	19.8	19.8
Imports	mt	21.5	20.8	20.9	20.5	20.2	20.1	20.0	20.0	19.9	19.8	19.7	19.6
Closing stocks	mt	9.5	10.1	10.1	10.0	9.8	9.8	9.8	9.9	10.0	10.1	10.2	10.3
Price[f]	'000 JPY/t	14.5	15.4	14.7	15.2	15.6	15.6	15.6	15.3	15.1	14.9	14.6	14.4
KOREA													
Production	mt	0.3	0.3	0.3	0.3	0.3	0.3	0.3	0.3	0.3	0.3	0.3	0.3
Consumption	mt	9.2	9.9	10.1	10.0	10.1	10.1	10.1	10.1	10.1	10.1	10.1	10.0
Imports	mt	9.0	9.8	9.8	9.7	9.8	9.8	9.9	9.8	9.8	9.8	9.8	9.8
Closing stocks	mt	1.2	1.6	1.5	1.5	1.5	1.5	1.5	1.5	1.5	1.5	1.6	1.6
Price[g]	'000 KRW/t	154.5	159.5	163.4	170.9	175.0	177.2	178.9	180.3	181.4	181.6	182.7	183.6
MEXICO													
Production	mt	25.3	25.3	25.4	26.1	26.7	27.2	28.0	28.8	29.5	30.4	31.2	32.0
Consumption	mt	35.6	36.3	37.1	37.7	38.5	39.4	40.5	41.4	42.1	42.8	43.5	44.1
Imports	mt	10.4	11.1	11.7	11.6	11.8	12.2	12.4	12.7	12.7	12.5	12.3	12.2
Price[h]	MXN/t	1 486	1 715	1 753	1 793	1 827	1 839	1 884	1 881	1 894	1 899	1 902	1 910
NEW ZEALAND													
Production	kt	574.5	573.2	563.3	553.7	554.1	554.4	554.5	554.8	554.8	554.7	554.7	555.0
Consumption	kt	621.3	692.3	707.8	739.5	767.1	789.2	814.2	838.7	862.0	887.5	913.7	943.8
Imports	kt	44.0	119.8	148.4	184.5	212.0	234.0	260.1	284.1	307.6	333.1	359.2	389.0
Price[i]	NZD/t	233	235	198	204	205	206	206	206	206	204	204	204
UNITED STATES													
Production	mt	262.9	275.7	277.1	284.9	292.8	297.0	300.6	303.8	307.2	309.5	310.8	312.4
Consumption	mt	212.8	224.7	222.0	229.4	233.4	235.7	235.6	236.2	238.2	238.8	240.5	242.5
Exports	mt	54.0	54.4	57.7	60.0	63.0	64.2	67.0	70.4	71.5	71.8	72.6	72.8
Closing stocks	mt	45.8	29.8	29.6	27.5	26.3	26.0	26.4	26.0	25.9	27.4	27.8	27.7
Price[j]	USD/t	78	90	91	95	96	97	97	97	97	97	97	97

For notes, see end of the table.

Annex Table 14. **COARSE GRAINS PROJECTIONS** (cont.)

Crop year[a]		Average 98/99-02/03	03/04 est.	04/05	05/06	06/07	07/08	08/09	09/10	10/11	11/12	12/13	13/14
OTHER OECD[k]													
Production	mt	11.7	11.6	11.9	12.1	12.2	12.3	12.5	12.6	12.7	12.8	13.0	13.1
Consumption	mt	12.2	12.7	12.7	12.8	13.0	13.1	13.2	13.3	13.4	13.5	13.6	13.8
Net trade	mt	−0.6	−1.0	−0.8	−0.8	−0.8	−0.7	−0.7	−0.7	−0.7	−0.7	−0.7	−0.7
ARGENTINA													
Production	mt	19.3	19.3	19.1	20.7	22.2	23.0	23.7	24.1	24.4	24.7	24.7	24.9
Consumption	mt	8.7	8.2	8.2	8.2	8.6	8.9	9.0	9.2	9.3	9.6	9.9	10.2
Exports	mt	10.6	11.2	10.9	12.5	13.6	14.1	14.7	14.9	15.1	15.1	14.8	14.7
Closing stocks	mt	0.5	0.4	0.4	0.4	0.4	0.4	0.4	0.4	0.4	0.4	0.4	0.4
Price[l]	ARS/t	135	324	337	367	377	383	379	373	380	382	380	389
BRAZIL													
Production	mt	35.7	43.8	45.2	46.3	47.9	50.0	51.6	53.0	54.3	55.5	56.8	57.9
Consumption	mt	38.5	43.0	46.2	47.7	49.1	50.5	52.0	53.8	55.5	57.3	58.8	59.8
Net trade	mt	−1.2	−0.7	−1.0	−1.5	−1.3	−0.6	−0.6	−0.9	−1.4	−1.9	−2.1	−2.0
Closing stocks	mt	4.1	2.9	3.0	3.1	3.2	3.3	3.4	3.5	3.6	3.7	3.8	3.9
CHINA													
Production	mt	124.2	124.0	130.0	132.3	132.6	133.6	135.2	136.7	138.3	140.3	143.3	144.4
Consumption	mt	131.8	137.5	139.1	142.0	143.3	144.2	146.8	148.8	150.3	152.0	154.7	156.5
Imports	mt	3.1	2.3	8.2	10.5	12.2	13.2	14.3	14.5	14.2	13.7	13.2	13.5
Closing stocks	mt	67.0	23.1	17.2	13.6	11.3	10.7	10.6	10.7	10.8	10.9	11.1	11.2
Price[m]	CNY/t	1 129	1 267	1 260	1 285	1 320	1 396	1 453	1 515	1 550	1 565	1 573	1 619
RUSSIA													
Production	mt	26.7	26.6	28.0	29.0	29.1	29.6	30.3	31.1	31.8	32.8	33.5	34.3
Consumption	mt	27.1	31.8	31.8	31.1	31.6	32.3	32.5	32.9	33.0	33.5	34.0	34.3
Net trade	mt	−0.2	−0.3	−2.9	−2.3	−2.6	−2.7	−2.3	−1.7	−1.3	−0.8	−0.5	0.0
Price[n]	RUR/t	1 205	1 472	1 700	1 647	1 747	1 799	1 858	1 908	1 963	1 974	1 994	2 030
OIS													
Production	mt	22.6	25.5	26.8	28.2	28.6	29.5	29.9	30.0	30.2	30.4	30.5	30.7
Consumption	mt	20.0	23.4	23.5	23.7	23.9	24.2	24.5	24.8	25.1	25.3	25.6	25.9
Net trade	mt	2.6	3.7	2.8	4.5	4.7	5.3	5.4	5.2	5.1	5.0	4.9	4.8
Closing stocks	mt	3.5	2.9	3.5	3.5	3.5	3.5	3.4	3.4	3.4	3.4	3.4	3.4
REST OF WORLD													
Production	mt	162.2	166.1	169.8	172.1	175.6	177.3	179.4	181.5	184.5	187.3	189.7	192.5
Consumption	mt	203.4	201.8	206.0	209.4	215.0	218.8	223.2	227.2	231.7	236.2	240.0	243.8
Net trade	mt	−40.9	−36.3	−36.1	−37.2	−39.7	−41.7	−44.0	−45.9	−47.4	−49.0	−50.3	−51.3
Closing stocks	mt	13.9	12.8	12.7	12.7	12.9	13.1	13.2	13.4	13.5	13.7	13.7	13.7

a) Beginning crop marketing year – see the Glossary of Terms for definitions.
b) Cash price, bulk feed barley, Sydney.
c) CWB final price, No. 1 CW barley, St Lawrence since 1995, Thunder Bay before.
d) Excludes intra-EU-25 trade.
e) Weighted average producer price, barley, year ended 31 December.
f) Farm gate price.
g) Average import price c.i.f., maize, year ended 31 December.
h) Average producer price, maize.
i) Indicative price, feed barley.
j) Maize average producer price.
k) Includes Norway, Switzerland and Turkey. Excludes Iceland.
l) Export price, f.o.b., Argentinean Ports.
m) Maize free market price.
n) Barley average producer price.
est.: estimate.

Source: OECD Secretariat.

Annex Table 15. **RICE PROJECTIONS**

Crop year[a]		Average 98/99-02/03	03/04 est.	04/05	05/06	06/07	07/08	08/09	09/10	10/11	11/12	12/13	13/14
OECD													
AUSTRALIA													
Production	mt	0.8	0.5	0.8	0.9	0.9	0.9	0.9	1.0	1.0	1.0	1.0	1.1
Consumption	mt	0.4	0.4	0.4	0.4	0.4	0.4	0.4	0.4	0.4	0.4	0.4	0.4
Exports	mt	0.5	0.3	0.6	0.6	0.6	0.6	0.6	0.6	0.6	0.7	0.7	0.7
Price[b]	AUD/t	243	219	225	235	246	253	253	253	254	253	254	254
EU-25													
Production	mt	1.5	1.6	1.5	1.5	1.5	1.5	1.5	1.5	1.6	1.6	1.6	1.6
EU-15	mt	1.5	1.6	1.5	1.5	1.5	1.5	1.5	1.5	1.6	1.6	1.6	1.6
EU-10	mt	..	0.0	0.0	0.0	0.0	0.0	0.0	0.0	0.0	0.0	0.0	0.0
Consumption	mt	2.1	2.2	2.3	2.4	2.4	2.4	2.6	2.6	2.7	2.7	2.8	2.8
EU-15	mt	1.8	1.9	2.1	2.1	2.1	2.1	2.3	2.3	2.3	2.4	2.5	2.5
EU-10	mt	..	0.3	0.3	0.3	0.3	0.3	0.3	0.3	0.3	0.3	0.3	0.3
Net trade	mt	−0.6	−0.4	−0.6	−0.8	−0.9	−1.0	−1.0	−1.1	−1.1	−1.2	−1.2	−1.2
Closing stocks	mt	0.8	0.6	0.4	0.3	0.3	0.3	0.3	0.3	0.3	0.3	0.4	0.4
Price[c]	EUR/t	283	283	143	143	143	143	143	143	143	143	144	147
Price[d]	EUR/t	..	255	216	216	216	214	176	149	153	155	158	161
JAPAN													
Production	mt	9.1	7.8	8.4	8.4	8.3	8.2	8.2	8.2	8.2	8.2	8.2	8.2
Consumption	mt	9.8	9.5	9.2	9.0	8.9	8.9	8.8	8.8	8.8	8.7	8.7	8.7
Imports	mt	0.8	0.9	0.9	0.8	0.8	0.8	0.8	0.8	0.8	0.8	0.8	0.8
Closing stocks	mt	4.2	2.8	2.6	2.5	2.5	2.5	2.5	2.5	2.5	2.5	2.6	2.6
Price[e]	'000 JPY/t	250	247	248	257	255	249	244	241	239	236	234	232
KOREA													
Production	mt	5.2	4.5	5.1	4.9	4.8	4.8	4.8	4.8	4.9	4.9	4.9	4.9
Consumption	mt	5.2	4.6	5.2	5.0	4.9	5.0	5.0	5.0	5.0	5.0	5.0	5.0
Imports	mt	0.2	0.2	0.2	0.2	0.2	0.2	0.2	0.2	0.2	0.2	0.2	0.2
Closing stocks	mt	1.2	1.7	1.7	1.8	1.9	2.0	2.1	2.2	2.3	2.3	2.4	2.6
Price[f]	'000 KRW/t	1 984	2 170	2 220	2 269	2 317	2 379	2 442	2 503	2 563	2 623	2 683	2 742
MEXICO													
Production	mt	0.2	0.2	0.2	0.2	0.2	0.2	0.2	0.2	0.2	0.2	0.2	0.2
Consumption	mt	0.7	0.7	0.7	0.7	0.7	0.8	0.8	0.8	0.9	0.9	0.9	0.9
Imports	mt	0.4	0.5	0.5	0.5	0.6	0.6	0.6	0.6	0.7	0.7	0.7	0.7
Price[b]	MXN/t	1 601	2 260	2 324	2 478	2 742	2 844	2 935	3 018	3 069	3 151	3 196	3 276
UNITED STATES													
Production	mt	6.4	6.3	6.9	6.9	7.0	7.0	7.1	7.2	7.2	7.3	7.3	7.4
Consumption	mt	3.7	3.9	4.0	4.0	4.1	4.2	4.3	4.3	4.4	4.5	4.6	4.7
Exports	mt	3.0	3.1	3.3	3.5	3.5	3.4	3.4	3.4	3.4	3.4	3.4	3.3
Closing stocks	mt	0.9	0.7	0.8	0.8	0.8	0.8	0.8	0.8	0.8	0.8	0.8	0.8
Price[b]	USD/t	127	161	136	146	152	157	162	163	166	167	170	172
OTHER OECD[g]													
Production	mt	0.2	0.2	0.2	0.2	0.2	0.2	0.2	0.2	0.2	0.2	0.2	0.2
Consumption	mt	0.9	1.0	1.0	1.1	1.1	1.1	1.2	1.2	1.3	1.3	1.4	1.4
Net trade	mt	−0.7	−0.8	−0.8	−0.9	−0.9	−0.9	−1.0	−1.0	−1.1	−1.1	−1.1	−1.2

For notes, see end of the table.

Annex Table 15. **RICE PROJECTIONS** (cont.)

Crop year[a]		Average 98/99-02/03	03/04 est.	04/05	05/06	06/07	07/08	08/09	09/10	10/11	11/12	12/13	13/14
OECD NON-MEMBER COUNTRIES													
ARGENTINA													
Production	mt	0.7	0.7	0.9	1.1	1.2	1.2	1.3	1.3	1.3	1.3	1.4	1.4
Consumption	mt	0.4	0.5	0.5	0.5	0.5	0.5	0.5	0.5	0.5	0.5	0.5	0.5
Exports	mt	0.3	0.3	0.4	0.6	0.7	0.8	0.8	0.8	0.8	0.8	0.8	0.9
Closing stocks	mt	0.1	0.1	0.1	0.1	0.1	0.1	0.1	0.1	0.1	0.1	0.1	0.1
Price[h]	ARS/t	316	488	562	614	653	680	685	682	680	689	707	723
BRAZIL													
Production	mt	6.8	7.0	7.1	7.4	7.5	7.8	8.0	8.1	8.3	8.4	8.6	8.7
Consumption	mt	7.8	7.7	7.7	7.9	8.0	8.2	8.3	8.5	8.6	8.8	9.0	9.2
Net trade	mt	−0.6	−0.6	−0.6	−0.5	−0.5	−0.4	−0.4	−0.4	−0.4	−0.4	−0.5	−0.5
Closing stocks	mt	1	1	1	1	1	1	1	1	1	1	1	1
CHINA													
Production	mt	131.3	119.2	127.5	128.7	129.6	129.3	128.9	129.8	129.5	129.0	128.7	128.5
Consumption	mt	134.1	135.2	133.5	132.9	132.3	132.6	132.4	132.8	132.8	132.1	131.6	131.5
of which feed	mt	39.1	39.0	39.8	40.2	40.2	40.9	41.4	42.1	42.3	42.3	42.7	43.0
Imports	mt	0.3	0.3	2.3	3.4	4.4	5.3	5.3	5.3	5.3	4.9	4.6	4.5
Closing stocks	mt	87.9	50.3	44.1	40.8	39.9	39.5	39.3	39.5	39.4	39.2	39.1	39.1
Price[i]	CNY/t	2 099	2 265	2 526	2 723	2 887	3 021	3 259	3 374	3 412	3 523	3 676	3 818
INDIA													
Production	mt	85.9	89.0	90.6	92.3	93.4	94.4	95.5	96.6	97.7	98.7	99.8	100.8
Consumption	mt	81.9	85.1	84.1	84.9	85.7	86.8	87.9	89.0	90.0	91.0	92.0	93.0
Closing stocks	mt	18.5	14.4	16.6	18.8	20.2	20.9	21.1	21.2	21.2	21.2	21.1	21.2
Price[j]	INR/t	5 000	5 243	5 893	6 390	6 705	7 063	7 501	7 937	8 296	8 552	8 862	9 152
INDONESIA													
Production	mt	32.8	33.5	33.7	34.2	34.7	35.1	35.5	35.9	36.2	36.6	37.0	37.4
Consumption	mt	35.9	37.5	38.1	38.6	39.0	39.4	39.8	40.3	40.7	41.1	41.4	41.8
Imports	mt	2.7	3.5	4.1	4.3	4.3	4.3	4.3	4.4	4.5	4.5	4.5	4.4
Closing stocks	mt	5.5	4.3	4.0	3.8	3.8	3.8	3.7	3.7	3.7	3.7	3.7	3.7
Price[k]	'000 IDR/t	1 150	1 334	1 475	1 758	2 034	2 291	2 573	2 858	3 202	3 549	3 981	4 454
OIS													
Production	mt	0.4	0.4	0.4	0.4	0.4	0.4	0.4	0.4	0.4	0.4	0.4	0.4
Consumption	mt	0.6	0.6	0.7	0.7	0.7	0.7	0.7	0.7	0.7	0.7	0.7	0.7
Net trade	mt	−0.2	−0.3	−0.3	−0.3	−0.3	−0.3	−0.3	−0.3	−0.3	−0.3	−0.3	−0.3
RUSSIA													
Production	mt	0.3	0.3	0.2	0.2	0.3	0.3	0.3	0.3	0.2	0.2	0.2	0.2
Consumption	mt	0.6	0.8	0.8	0.8	0.8	0.8	0.8	0.8	0.8	0.9	0.9	0.9
Net trade	mt	−0.4	−0.5	−0.5	−0.6	−0.6	−0.6	−0.6	−0.6	−0.6	−0.6	−0.6	−0.7
THAILAND													
Production	mt	16.8	17.8	18.2	18.8	19.1	19.5	19.8	20.0	20.2	20.4	20.5	20.7
Consumption	mt	9.4	9.6	9.5	9.5	9.5	9.5	9.5	9.4	9.4	9.4	9.3	9.3
Exports	mt	7.1	8.6	8.8	9.2	9.6	9.9	10.3	10.6	10.8	11.0	11.2	11.4
Closing stocks	mt	1.8	1.5	1.4	1.5	1.5	1.5	1.5	1.5	1.5	1.5	1.5	1.5
Price[k]	THB/t	5 313	5 634	6 950	8 050	8 682	8 980	9 121	9 222	9 412	9 244	9 449	9 629

For notes, see end of the table.

Annex Table 15. **RICE PROJECTIONS** (cont.)

Crop year[a]		Average 98/99-02/03	03/04 est.	04/05	05/06	06/07	07/08	08/09	09/10	10/11	11/12	12/13	13/14
REST OF WORLD													
Production	mt	96.5	103.1	106.5	109.7	111.5	113.3	115.2	117.4	119.5	121.4	123.2	125.3
Consumption	mt	105.7	114.2	115.8	118.8	121.0	122.9	125.2	127.4	129.5	131.6	133.5	135.6
Net trade	mt	−10.0	−9.3	−8.7	−9.0	−9.2	−9.2	−9.5	−9.8	−10.0	−10.3	−10.5	−10.5
Closing stocks	mt	13.7	12.6	12.1	12.0	11.8	11.3	10.8	10.6	10.5	10.7	10.9	11.1

a) Beginning crop marketing year – see the Glossary of Terms for definitions.
b) Producer price.
c) Producer price, paddy rice.
d) Market price (consumers), paddy equivalent.
e) Market price, husked rice.
f) Producer price, native king, polished grade b.
g) Includes Norway, Switzerland and Turkey. Excludes Iceland.
h) Export price.
i) Free market price, weighted average of japonica and indica.
j) Farm harvest price, rough basis.
k) Paddy, farm harvest price.
est.: estimate.
Source: OECD secretariat.

Annex Table 16. **OILSEED PROJECTIONS**

Crop year[a]		Average 98/99-02/03	03/04 est.	04/05	05/06	06/07	07/08	08/09	09/10	10/11	11/12	12/13	13/14
AUSTRALIA													
Production	mt	1.9	1.5	2.5	2.5	2.4	2.4	2.3	2.4	2.4	2.4	2.4	2.5
Consumption	mt	0.6	0.5	0.6	0.7	0.7	0.7	0.7	0.7	0.7	0.7	0.7	0.7
crush	mt	0.5	0.5	0.6	0.7	0.7	0.7	0.7	0.7	0.7	0.6	0.6	0.6
Exports	mt	1.3	1.0	1.9	1.8	1.7	1.7	1.7	1.7	1.7	1.8	1.8	1.9
Price[b]	AUD/t	362	350	349	334	342	345	343	341	338	344	345	347
CANADA													
Production	mt	9.1	9.1	10.5	11.8	11.0	11.5	11.8	12.0	12.0	12.2	12.6	12.9
Consumption	mt	5.5	5.7	5.7	5.9	6.1	6.1	6.2	6.3	6.4	6.5	6.6	6.7
crush	mt	4.5	4.9	4.9	5.0	5.1	5.1	5.2	5.2	5.2	5.3	5.4	5.4
Exports	mt	4.3	4.2	5.1	6.7	5.9	6.0	6.3	6.3	6.3	6.4	6.7	6.8
Price[c]	CAD/t	345	397	399	378	387	390	387	382	378	384	385	387
EU-25													
Production	mt	15.2	12.9	16.3	15.5	15.2	14.6	14.3	14.0	14.3	14.5	14.5	14.5
EU-15	mt	11.9	10.4	13.3	12.6	12.4	11.7	11.4	11.1	11.3	11.4	11.4	11.4
EU-10	mt	..	2.5	3.0	2.9	2.8	2.8	2.9	3.0	3.0	3.1	3.1	3.1
Consumption	mt	33.0	31.3	33.1	32.9	32.7	33.0	33.2	33.5	33.7	33.7	33.9	34.1
EU-15	mt	30.4	29.2	30.8	30.6	30.4	30.6	30.9	31.1	31.3	31.3	31.5	31.7
EU-10	mt	..	2.1	2.3	2.3	2.3	2.3	2.4	2.4	2.4	2.4	2.4	2.4
crush	mt	30.5	29.8	31.0	30.8	30.6	30.8	31.1	31.3	31.5	31.6	31.8	31.9
Net trade[d]	mt	−17.5	−18.7	−17.2	−17.4	−17.5	−18.4	−19.0	−19.5	−19.4	−19.3	−19.5	−19.6
Closing stocks	mt	2.6	2.3	2.7	2.7	2.7	2.7	2.7	2.8	2.8	2.8	2.9	2.9
Price[e]	EUR/t	227	306	251	247	257	257	256	254	251	254	255	255
JAPAN[f]													
Production	mt	0.2	0.3	0.3	0.3	0.3	0.3	0.3	0.3	0.3	0.3	0.3	0.3
Consumption	mt	7.3	7.6	7.7	7.8	7.9	7.9	7.9	7.9	7.9	7.9	7.9	7.8
crush	mt	6.4	6.5	6.6	6.7	6.7	6.8	6.8	6.8	6.8	6.7	6.7	6.7
Imports	mt	7.1	7.4	7.4	7.6	7.6	7.6	7.6	7.6	7.6	7.6	7.5	7.5
Closing stocks	mt	1.8	1.9	1.9	2.0	2.0	2.0	2.0	2.0	2.0	2.0	2.0	2.0
Price[g]	'000 JPY/t	31	33	30	26	25	26	25	24	23	23	22	22
KOREA													
Production	mt	0.1	0.1	0.1	0.1	0.1	0.1	0.1	0.1	0.1	0.1	0.1	0.1
Consumption	mt	1.6	1.7	1.9	1.9	1.9	1.9	1.9	2.0	2.0	2.0	2.0	2.1
crush	mt	1.2	1.3	1.4	1.4	1.4	1.4	1.4	1.5	1.5	1.5	1.5	1.5
Imports	mt	1.5	1.6	1.8	1.8	1.8	1.8	1.8	1.9	1.9	1.9	2.0	2.0
Price[h]	'000 KRW/t	2 876	2 662	2 734	2 808	2 883	2 970	3 060	3 153	3 249	3 348	3 451	3 558
MEXICO													
Production	mt	0.1	0.1	0.1	0.1	0.1	0.1	0.1	0.1	0.1	0.1	0.0	0.0
Consumption	mt	5.1	5.3	5.5	5.6	5.8	5.9	6.0	6.2	6.4	6.6	6.7	6.9
crush	mt	4.4	4.9	5.0	5.1	5.2	5.3	5.5	5.6	5.7	5.9	6.0	6.1
Imports	mt	5.0	5.2	5.4	5.5	5.7	5.8	6.0	6.1	6.3	6.5	6.6	6.8
Price[i]	MXN/t	2 087	2 913	2 601	2 531	2 526	2 602	2 588	2 572	2 547	2 540	2 563	2 566

For notes, see end of the table.

Annex Table 16. **OILSEED PROJECTIONS** (cont.)

Crop year[a]		Average 98/99-02/03	03/04 est.	04/05	05/06	06/07	07/08	08/09	09/10	10/11	11/12	12/13	13/14
UNITED STATES													
Production	mt	77.6	67.7	81.0	80.5	80.5	82.4	83.0	83.6	84.3	85.1	86.2	86.8
Consumption	mt	51.0	44.3	48.8	51.7	53.3	54.8	56.2	57.5	58.6	59.4	60.1	60.9
crush	mt	45.8	40.3	44.2	47.1	48.8	50.3	51.7	52.9	54.0	54.8	55.6	56.3
Exports	mt	27.0	25.6	29.9	28.5	28.0	27.7	27.0	26.3	25.9	26.3	26.4	26.3
Closing stocks	mt	7.2	3.5	6.2	6.9	6.6	6.9	7.1	7.3	7.4	7.2	7.2	7.2
Price[j]	USD/t	176	270	210	203	214	213	212	210	207	210	210	211
OTHER OECD[k]													
Production	mt	0.9	1.0	1.0	1.0	1.0	1.1	1.1	1.1	1.1	1.1	1.1	1.2
Consumption	mt	2.2	2.5	2.6	2.6	2.6	2.7	2.7	2.7	2.8	2.8	2.8	2.9
crush	mt	2.2	2.2	2.2	2.2	2.3	2.3	2.3	2.4	2.4	2.4	2.4	2.5
Net trade	mt	−1.3	−1.5	−1.5	−1.6	−1.6	−1.6	−1.6	−1.6	−1.7	−1.7	−1.7	−1.7
ARGENTINA													
Production	mt	31.1	40.0	40.5	42.3	43.6	45.4	47.1	48.6	50.1	51.6	53.5	55.5
Consumption	mt	24.5	30.3	31.9	32.3	32.4	33.3	34.2	34.9	35.7	36.3	37.2	38.1
crush	mt	23.6	29.4	30.9	31.3	31.4	32.2	33.1	33.8	34.6	35.3	36.1	37.0
Exports	mt	6.8	10.4	8.5	10.2	11.5	12.4	13.3	14.1	14.8	15.7	16.7	17.9
Closing stocks	mt	0.5	0.6	1.0	1.1	1.2	1.2	1.3	1.3	1.4	1.4	1.4	1.5
Price, (soyabeans)[l]	ARS/t	277	601	591	595	649	656	658	652	664	697	719	744
Price, (sunflower)[l]	ARS/t	307	693	678	683	740	748	750	744	757	792	816	843
BRAZIL													
Production	mt	35.1	49.1	52.0	52.0	54.1	56.4	59.2	62.5	65.1	66.4	68.2	70.4
Consumption	mt	23.6	28.2	30.6	31.9	32.5	33.4	34.6	35.8	37.1	38.7	40.4	42.2
crush	mt	23.0	27.6	30.0	31.3	31.9	32.7	33.9	35.1	36.4	38.0	39.6	41.4
Net trade	mt	11.5	20.8	20.0	20.1	21.1	22.5	24.4	26.5	27.7	27.4	27.5	27.9
Closing stocks	mt	2.3	2.2	3.5	3.5	4.0	4.5	4.7	4.9	5.1	5.3	5.5	5.8
CHINA													
Production	mt	27.5	29.7	30.7	31.4	31.7	32.2	32.8	33.2	33.6	34.2	34.7	35.3
Consumption	mt	40.7	51.7	54.5	55.8	56.8	58.4	60.0	61.6	63.2	64.7	66.4	68.1
crush	mt	31.1	40.7	42.8	43.8	44.7	46.0	47.3	48.6	50.0	51.2	52.6	54.1
Exports	mt	0.3	0.4	0.4	0.4	0.4	0.4	0.4	0.4	0.4	0.4	0.4	0.4
Imports	mt	13.7	21.3	24.9	24.9	25.5	26.6	27.6	28.8	30.1	31.0	32.1	33.3
Price[m]	CNY/t	1 833	2 674	2 263	2 205	2 299	2 295	2 282	2 310	2 317	2 368	2 398	2 433
RUSSIA													
Production	mt	4.0	5.1	4.8	5.1	5.0	5.2	5.3	5.3	5.4	5.5	5.6	5.7
Consumption	mt	3.4	4.2	4.4	4.6	4.7	4.7	4.7	4.7	4.7	4.8	4.8	4.9
crush	mt	3.2	4.1	4.3	4.5	4.6	4.6	4.7	4.7	4.6	4.7	4.7	4.8
Net trade	mt	0.6	1.0	0.5	0.5	0.4	0.5	0.5	0.6	0.7	0.8	0.8	0.9
Price[b]	RUR/t	6 738	8 495	9 133	9 494	10 021	10 518	10 643	10 997	11 146	11 381	11 594	11 702
OIS													
Production	mt	3.5	4.0	4.1	4.2	4.4	4.5	4.6	4.8	4.9	5.1	5.2	5.4
Consumption	mt	2.8	3.5	3.5	3.6	3.7	3.7	3.8	3.9	4.0	4.0	4.1	4.2
crush	mt	2.6	3.3	3.4	3.4	3.5	3.6	3.7	3.7	3.8	3.9	4.0	4.0
Net trade	mt	0.6	0.5	0.6	0.6	0.7	0.8	0.8	0.9	1.0	1.0	1.1	1.2
Closing stocks	mt	0.0	0.1	0.1	0.1	0.1	0.1	0.1	0.1	0.1	0.1	0.1	0.2

For notes, see end of the table.

Annex Table 16. **OILSEED PROJECTIONS** (cont.)

Crop year[a]		Average 98/99-02/03	03/04 est.	04/05	05/06	06/07	07/08	08/09	09/10	10/11	11/12	12/13	13/14
REST OF WORLD													
Production	mt	22.2	23.4	25.4	24.4	24.5	25.2	25.6	25.9	26.3	26.6	27.2	27.5
Consumption	mt	27.1	29.8	32.2	32.8	33.0	33.9	34.8	35.5	36.3	36.8	37.6	38.4
crush	mt	21.4	23.8	25.5	25.8	26.0	26.7	27.4	28.0	28.6	29.1	29.8	30.5
Net trade	mt	−4.9	−6.2	−7.4	−8.6	−8.5	−8.7	−9.2	−9.6	−10.0	−10.2	−10.5	−10.8
Closing stocks	mt	1.4	1.0	1.5	1.7	1.7	1.7	1.7	1.7	1.7	1.7	1.7	1.7

a) Beginning crop marketing year – see the Glossary of Terms for definitions.
b) Producer price, rapeseed.
c) Canola price, stored in Vancouver.
d) Excludes intra-EU-25 trade.
e) Import price, rapeseed c.i.f. Hamburg.
f) Excludes sunflower seed.
g) Import price c.i.f., soyabeans, year ended 31 December.
h) Producer price, soyabeans.
i) Average producer price, soyabeans.
j) Average price received by farmers, soyabeans.
k) Includes Norway, New Zealand, Switzerland and Turkey.
l) Export price, f.o.b., Argentinean Ports.
m) Soyabeans free market price.
est.: estimate.

Source: OECD Secretariat.

Annex Table 17. **OILSEED MEALS PROJECTIONS**

Marketing year[a]		Average 98/99-02/03	03/04 est.	04/05	05/06	06/07	07/08	08/09	09/10	10/11	11/12	12/13	13/14
AUSTRALIA													
Production	mt	0.3	0.3	0.4	0.4	0.4	0.4	0.4	0.4	0.4	0.4	0.4	0.4
Consumption	mt	0.6	0.6	0.7	0.7	0.7	0.6	0.7	0.7	0.7	0.7	0.7	0.7
Imports	mt	0.2	0.3	0.3	0.2	0.2	0.2	0.3	0.3	0.3	0.3	0.3	0.3
Price[b]	AUD/t	305	272	259	250	251	252	253	254	255	256	258	259
CANADA													
Production	mt	3.0	3.1	3.0	3.1	3.2	3.2	3.3	3.3	3.3	3.3	3.4	3.4
Consumption	mt	2.8	2.3	2.5	3.0	3.0	3.0	3.1	3.1	3.2	3.3	3.3	3.3
Imports	mt	0.9	1.0	1.1	1.5	1.4	1.5	1.4	1.4	1.5	1.5	1.5	1.5
Exports	mt	1.1	1.7	1.6	1.6	1.6	1.7	1.5	1.6	1.6	1.5	1.5	1.6
Price[c]	CAD/t	195	273	220	214	216	217	219	220	218	218	218	218
EU-25													
Production	mt	21.3	21.2	22.1	21.9	21.8	22.0	22.2	22.3	22.5	22.5	22.7	22.8
EU-15	mt	19.7	19.8	20.5	20.3	20.2	20.4	20.5	20.7	20.8	20.8	21.0	21.1
EU-10	mt	. .	1.4	1.6	1.6	1.6	1.6	1.6	1.6	1.7	1.7	1.7	1.7
Consumption	mt	41.6	44.6	46.5	46.9	46.8	47.0	47.3	47.7	47.9	48.1	48.4	48.8
EU-15	mt	37.8	40.1	41.9	42.2	42.1	42.3	42.5	42.8	42.9	43.0	43.3	43.5
EU-10	mt	. .	4.5	4.6	4.7	4.7	4.7	4.8	4.9	5.0	5.1	5.2	5.3
Net trade[d]	mt	−20.6	−23.4	−24.6	−25.0	−25.0	−25.0	−25.2	−25.4	−25.5	−25.6	−25.8	−26.0
Closing stocks	mt	1.4	1.4	1.5	1.5	1.5	1.5	1.5	1.5	1.5	1.5	1.6	1.6
Price[e]	EUR/t	195	233	189	178	180	179	178	178	177	177	177	177
JAPAN[f]													
Production	mt	4.5	4.6	4.7	4.7	4.8	4.8	4.8	4.8	4.8	4.8	4.7	4.7
Consumption	mt	5.4	5.8	5.8	5.9	6.0	5.9	5.9	5.9	5.9	5.9	5.9	5.9
Imports	mt	0.9	1.2	1.1	1.2	1.2	1.1	1.1	1.1	1.1	1.2	1.2	1.2
Price[g]	'000 JPY/t	26	28	28	28	27	27	27	26	26	26	25	25
KOREA													
Production	mt	0.9	1.0	1.1	1.1	1.1	1.1	1.1	1.2	1.2	1.2	1.2	1.2
Consumption	mt	2.7	3.0	3.3	3.3	3.4	3.5	3.5	3.6	3.6	3.7	3.7	3.7
Imports	mt	1.7	2.0	2.3	2.2	2.3	2.4	2.4	2.4	2.5	2.5	2.5	2.5
MEXICO													
Production	mt	3.3	3.6	3.7	3.8	3.9	4.0	4.1	4.2	4.2	4.3	4.4	4.5
Consumption	mt	3.5	4.1	4.3	4.6	4.8	5.0	5.3	5.5	5.7	5.8	6.0	6.2
Imports	mt	0.2	0.5	0.6	0.8	0.9	1.1	1.2	1.3	1.4	1.5	1.6	1.7
Price[h]	MXN/t	1 726	2 842	2 371	2 222	2 243	2 212	2 195	2 187	2 156	2 149	2 137	2 127
UNITED STATES													
Production	mt	35.9	31.8	35.1	37.4	38.8	39.8	41.0	41.9	42.8	43.4	44.1	44.6
Consumption	mt	30.5	30.1	31.3	32.1	32.6	33.4	34.1	34.7	35.3	35.9	36.5	37.0
Imports	mt	1.1	1.7	1.0	1.3	1.3	1.4	1.4	1.4	1.4	1.4	1.4	1.4
Exports	mt	6.5	3.4	4.8	6.6	7.5	7.8	8.3	8.7	8.9	8.9	8.9	9.1
Price[i]	USD/t	183	264	216	202	204	201	200	199	196	196	194	194

For notes, see end of the table.

Annex Table 17. **OILSEED MEALS PROJECTIONS** (cont.)

Marketing year[a]		Average 98/99-02/03	03/04 est.	04/05	05/06	06/07	07/08	08/09	09/10	10/11	11/12	12/13	13/14
OTHER OECD[j]													
Production	mt	1.3	1.3	1.4	1.4	1.4	1.4	1.4	1.5	1.5	1.5	1.5	1.5
Consumption	mt	2.0	2.5	2.5	2.5	2.6	2.6	2.7	2.7	2.7	2.8	2.8	2.9
Net trade	mt	−0.5	−1.1	−1.1	−1.2	−1.2	−1.2	−1.2	−1.3	−1.3	−1.3	−1.3	−1.4
ARGENTINA													
Production	mt	17.1	21.8	22.9	23.2	23.2	23.8	24.4	25.0	25.5	25.9	26.6	27.1
Consumption	mt	0.7	0.7	0.8	1.0	1.1	1.1	1.2	1.2	1.3	1.3	1.3	1.4
Exports	mt	16.4	21.1	22.1	22.2	22.1	22.7	23.2	23.7	24.2	24.6	25.2	25.7
Closing stocks	mt	0.8	0.9	0.9	1.0	1.0	1.0	1.0	1.0	1.0	1.0	1.0	1.0
Price, (soya meal)[k]	ARS/t	215	496	506	488	513	518	522	525	537	557	577	597
Price, (sun meal)[k]	ARS/t	111	292	315	299	320	323	325	327	336	352	367	383
BRAZIL													
Production	mt	17.4	21.3	23.0	24.0	24.5	25.2	26.1	27.0	28.0	29.2	30.5	31.8
Consumption	mt	6.8	8.1	8.3	8.6	8.9	9.3	9.6	10.1	10.5	10.9	11.4	11.9
Net trade	mt	10.5	13.3	14.7	15.4	15.6	15.9	16.4	16.9	17.5	18.2	19.0	19.9
CHINA													
Production	mt	22.6	30.1	31.7	32.5	33.1	34.1	35.0	36.0	37.0	37.9	39.0	40.1
Consumption	mt	22.1	29.9	32.1	33.5	34.4	35.5	37.0	38.2	39.2	40.2	41.5	42.9
Imports	mt	0.4	1.3	1.9	2.5	2.8	3.0	3.5	3.6	3.7	3.8	4.0	4.4
Price[l]	CNY/t	1 546	2 064	1 641	1 524	1 568	1 559	1 548	1 593	1 601	1 634	1 664	1 696
RUSSIA													
Production	mt	1.4	1.8	1.9	2.0	2.1	2.1	2.1	2.1	2.1	2.1	2.2	2.2
Consumption	mt	1.6	1.8	1.6	1.7	1.7	1.8	1.9	1.9	2.0	2.0	2.1	2.2
Net trade	mt	−0.2	0.1	0.3	0.4	0.3	0.3	0.2	0.2	0.1	0.1	0.1	0.0
Price	RUR/t	3 818	6 701	7 672	8 269	8 876	9 303	9 287	9 596	9 644	9 839	9 967	10 105
OIS													
Production	mt	1.1	1.5	1.5	1.5	1.6	1.6	1.6	1.7	1.7	1.8	1.8	1.8
Consumption	mt	0.9	1.1	1.1	1.2	1.2	1.3	1.3	1.4	1.4	1.5	1.5	1.6
Net trade	mt	0.3	0.4	0.4	0.4	0.4	0.4	0.3	0.3	0.3	0.3	0.3	0.2
REST OF WORLD													
Production	mt	16.2	17.0	18.0	18.1	18.2	18.7	19.2	19.6	20.0	20.4	20.8	21.3
Consumption	mt	25.0	26.7	28.9	30.0	30.9	32.0	33.2	34.3	35.5	36.6	37.7	38.9
Net trade	mt	−8.8	−9.1	−11.4	−12.1	−12.7	−13.3	−14.0	−14.7	−15.5	−16.2	−16.9	−17.6
Closing stocks	mt	1.7	1.5	1.9	2.1	2.1	2.1	2.2	2.2	2.2	2.3	2.3	2.3

a) Beginning crop marketing year – see the Glossary of Terms for definitions.
b) Average import price c.i.f., soyabean and other oilseed meals, year beginning 1 July.
c) Canola meal price, f.o.b. Vancouver (prior 2002, f.o.b plant).
d) Excludes intra-EU-25 trade.
e) Soyabean meal price, 44/45%, f.o.b. ex-mill Hamburg.
f) Excludes sunflower seed.
g) Average import price c.i.f., soyabean cake, year ended 31 December.
h) Calculated import price of soyabean meal.
i) Wholesale price, soyabean meal, 48% solvent, Decatur.
j) Includes Norway, New Zealand, Switzerland and Turkey. Excludes Iceland.
k) Export price, f.o.b., Argentinean Ports.
l) Calculated import price.
est.: estimate.

Source: OECD Secretariat.

Annex Table 18. **VEGETABLE OILS PROJECTIONS**

Marketing year[a]		Average 98/99-02/03	03/04 est.	04/05	05/06	06/07	07/08	08/09	09/10	10/11	11/12	12/13	13/14
AUSTRALIA													
Production	kt	198	187	233	263	261	250	248	245	242	240	240	240
Consumption	kt	285	266	342	339	348	358	372	384	395	405	414	424
Imports	kt	132	133	131	116	128	150	167	183	199	211	221	232
Price[b]	AUD/t	352	344	346	348	358	370	379	383	386	394	403	411
CANADA													
Production	kt	1 418	1 443	1 474	1 510	1 534	1 530	1 534	1 540	1 561	1 579	1 600	1 617
Consumption	kt	885	759	764	720	724	727	726	727	729	729	731	735
Imports	kt	136	47	47	75	75	75	75	75	75	75	75	75
Exports	kt	670	731	757	864	885	878	882	886	907	925	943	956
Price[c]	CAD/t	601	595	578	579	592	604	611	606	601	609	615	619
EU-25													
Production	kt	8 739	8 191	8 533	8 486	8 436	8 503	8 573	8 640	8 682	8 698	8 755	8 800
EU-15	kt	7 858	7 388	7 655	7 594	7 540	7 606	7 672	7 736	7 776	7 787	7 841	7 882
EU-10	kt	. .	802	878	892	896	897	901	904	906	911	914	917
Consumption	kt	10 332	11 374	11 464	11 762	11 894	12 024	12 170	12 380	12 587	12 751	12 924	13 100
EU-15	kt	9 254	10 231	10 284	10 539	10 643	10 749	10 866	11 041	11 210	11 342	11 484	11 627
EU-10	kt	. .	1 143	1 180	1 223	1 250	1 274	1 304	1 339	1 377	1 408	1 440	1 473
Net trade[d]	kt	−1 599	−3 197	−2 945	−3 278	−3 455	−3 523	−3 602	−3 747	−3 908	−4 053	−4 173	−4 304
Closing stocks	kt	783	749	763	765	762	765	770	777	780	781	784	788
Price[e]	EUR/t	456	566	553	551	563	573	578	573	569	574	578	581
JAPAN[f]													
Production	kt	1 598	1 646	1 657	1 676	1 696	1 702	1 703	1 703	1 701	1 697	1 689	1 679
Consumption	kt	1 949	2 021	2 032	2 048	2 059	2 071	2 082	2 098	2 114	2 126	2 138	2 150
Imports	kt	355	373	379	376	363	367	378	395	412	428	447	469
Closing stocks	kt	203	201	204	208	208	207	206	206	205	204	202	201
Price[g]	'000 JPY/t	64	76	78	77	76	76	75	73	71	70	69	68
KOREA													
Production	kt	224	238	259	261	260	265	270	275	279	281	285	289
Consumption	kt	439	462	478	491	498	506	516	530	543	554	565	577
Imports	kt	221	232	222	233	242	250	257	267	278	287	295	305
Price[g]	'000 KRW/t	537	712	696	694	711	725	731	724	719	727	732	736
MEXICO													
Production	kt	896	1 094	1 066	1 039	1 060	1 091	1 140	1 185	1 228	1 276	1 315	1 358
Consumption	kt	1 387	1 555	1 588	1 679	1 718	1 763	1 823	1 898	1 962	2 020	2 094	2 172
Imports	kt	522	448	543	662	679	693	705	735	755	765	800	835
UNITED STATES													
Production	kt	8 955	7 867	8 705	9 270	9 608	9 904	10 188	10 439	10 662	10 820	10 981	11 132
Consumption	kt	8 434	8 685	8 770	8 929	9 056	9 226	9 345	9 501	9 638	9 760	9 940	10 122
Imports	kt	724	946	1 073	1 130	1 192	1 257	1 324	1 393	1 466	1 539	1 615	1 693
Exports	kt	1 241	360	798	1 348	1 715	1 913	2 145	2 311	2 474	2 588	2 645	2 692
Closing stocks	kt	1 042	515	725	849	878	900	921	941	957	967	978	989
Price[h]	USD/t	389	614	600	598	610	620	625	620	616	621	625	628

For notes, see end of the table.

Annex Table 18. **VEGETABLE OILS PROJECTIONS** (cont.)

Marketing year[a]		Average 98/99-02/03	03/04 est.	04/05	05/06	06/07	07/08	08/09	09/10	10/11	11/12	12/13	13/14
OTHER OECD[i]													
Production	kt	679	624	637	646	655	664	673	683	692	702	711	721
Consumption	kt	1 237	1 299	1 317	1 334	1 352	1 370	1 389	1 407	1 426	1 445	1 465	1 485
Net trade	kt	−547	−665	−670	−679	−687	−697	−706	−715	−725	−734	−744	−754
ARGENTINA													
Production	kt	5 378	6 252	6 609	6 718	6 767	6 989	7 197	7 383	7 594	7 773	8 009	8 239
Consumption	kt	672	656	708	746	761	781	810	862	899	918	941	966
Exports	kt	4 716	5 606	5 892	5 968	6 003	6 203	6 382	6 515	6 690	6 850	7 063	7 267
Closing stocks	kt	259	222	231	234	237	242	248	253	257	261	267	272
Price, (soya oil)[j]	ARS/t	501	1 278	1 298	1 337	1 408	1 456	1 472	1 442	1 463	1 517	1 566	1 613
Price, (sunflower oil)[j]	ARS/t	654	1 591	1 697	1 750	1 840	1 905	1 935	1 912	1 948	2 022	2 092	2 161
BRAZIL													
Production	kt	4 307	5 361	5 802	6 054	6 172	6 332	6 566	6 791	7 039	7 342	7 657	7 998
Consumption	kt	3 004	3 154	3 367	3 570	3 735	3 883	4 053	4 249	4 463	4 684	4 886	5 104
Net trade	kt	1 279	2 229	2 421	2 472	2 426	2 437	2 501	2 530	2 564	2 644	2 756	2 879
CHINA													
Production	kt	7 222	8 942	9 458	9 690	9 844	10 125	10 413	10 706	10 992	11 257	11 567	11 880
Consumption	kt	9 028	12 479	13 266	13 807	13 984	14 491	15 205	15 687	16 236	16 782	17 343	17 927
Imports	kt	1 933	3 487	3 936	4 228	4 239	4 463	4 884	5 069	5 327	5 603	5 852	6 119
Closing stocks	kt	273	243	255	258	260	265	269	274	279	283	288	292
Price[k]	CNY/t	..	5 870	5 634	5 564	5 798	5 918	5 950	6 044	6 093	6 267	6 443	6 619
RUSSIA													
Production	kt	1 218	1 550	1 620	1 704	1 733	1 743	1 766	1 765	1 761	1 779	1 788	1 809
Consumption	kt	1 845	2 127	2 177	2 203	2 266	2 347	2 417	2 506	2 586	2 648	2 709	2 771
Net trade	kt	−616	−595	−564	−499	−534	−604	−651	−742	−825	−869	−921	−962
OIS													
Production	kt	1 057	1 344	1 374	1 405	1 437	1 469	1 502	1 536	1 571	1 606	1 642	1 679
Consumption	kt	749	807	832	856	882	909	936	964	993	1 023	1 053	1 085
Net trade	kt	304	526	541	547	553	559	565	570	576	582	587	593
Closing stocks	kt	20	40	41	43	44	46	47	49	51	52	54	56
REST OF WORLD													
Production	kt	25 524	27 999	29 069	29 728	30 549	31 808	33 250	34 720	36 133	37 479	38 874	40 258
Consumption	kt	26 989	27 360	28 588	29 751	30 675	31 775	33 011	34 229	35 436	36 599	37 779	38 955
Net trade	kt	−1 612	503	−72	−91	−145	−65	123	359	612	830	1 005	1 218
Closing stocks	kt	4 604	4 158	4 710	4 778	4 797	4 895	5 012	5 144	5 230	5 279	5 369	5 454

a) Beginning crop marketing year – see the Glossary of Terms for definitions.
b) Average import price c.i.f., soyabean, sunflower & other oilseed oils, year beginning 1 July.
c) Canola oil price, f.o.b Vancouver (prior 2002, f.o.b. plant).
d) Excludes intra-EU-25 trade.
e) Rapeseed oil price, f.o.b. ex-mill Hamburg.
f) Excludes sunflower seeds.
g) Calculated import price.
h) Wholesale price, crude soyabean oil, Decatur.
i) Includes Norway, New Zealand, Switzerland and Turkey. Excludes Iceland.
j) Export price, f.o.b, Argentinean Ports.
k) Calculated import price.
est.: estimate.

Source: OECD Secretariat.

Annex Table 19. **BEEF AND VEAL PROJECTIONS**[a]

Calendar year[b]		Average 1998-02	2003 est.	2004	2005	2006	2007	2008	2009	2010	2011	2012	2013
PACIFIC MARKET													
AUSTRALIA													
Production	kt cwe	2 033	1 938	2 009	1 837	1 871	1 965	2 012	2 082	2 155	2 224	2 279	2 324
Consumption	kt cwe	706	748	660	714	709	720	727	742	755	766	770	772
Exports	kt cwe	1 330	1 192	1 351	1 125	1 164	1 246	1 287	1 342	1 401	1 460	1 511	1 553
Price[c]	AUD/100 kg dw	238	272	330	272	277	267	260	248	241	235	233	235
CANADA													
Production	kt cwe	1 248	1 160	1 448	1 377	1 306	1 414	1 345	1 334	1 374	1 397	1 432	1 454
Consumption	kt cwe	1 005	1 028	1 024	998	973	969	969	975	977	976	970	969
Imports	kt cwe	280	283	184	192	199	208	217	227	246	263	277	288
Exports	kt cwe	521	415	609	572	532	652	593	586	643	685	739	773
Price[d]	CAD/100 kg dw	366	314	298	426	432	417	404	385	374	368	365	367
JAPAN													
Production	kt cwe	519	460	476	482	494	505	519	536	554	572	591	609
Consumption	kt cwe	1 437	1 442	1 443	1 570	1 568	1 607	1 634	1 657	1 693	1 729	1 765	1 800
Imports	kt cwe	921	968	967	1 088	1 074	1 102	1 115	1 122	1 139	1 156	1 173	1 191
Price[e]	'000 JPY/100 kg dw	95	87	105	91	91	88	84	80	77	74	72	71
KOREA													
Production	kt cwe	290	162	200	171	186	190	196	202	201	197	190	182
Consumption	kt cwe	551	591	561	616	623	641	654	675	700	714	734	753
Imports	kt cwe	263	418	360	443	436	449	456	471	497	516	542	570
Price[f]	'000 KRW/100 kg dw	577	813	807	642	551	540	532	518	509	503	501	502
MEXICO													
Production	kt cwe	1 320	1 326	1 331	1 284	1 322	1 355	1 357	1 359	1 359	1 348	1 345	1 349
Consumption	kt cwe	1 608	1 701	1 728	1 628	1 661	1 705	1 727	1 771	1 818	1 837	1 862	1 912
Imports	kt cwe	289	378	399	346	341	352	373	415	462	493	519	566
Price[g]	MXN/100 kg dw	2 137	2 356	1 778	2 106	2 159	2 077	2 022	1 925	1 861	1 818	1 803	1 816
NEW ZEALAND													
Production	kt cwe	587	681	685	689	669	666	669	672	680	684	688	692
Consumption	kt cwe	115	107	98	103	106	107	108	109	110	112	113	115
Exports	kt cwe	472	565	586	585	563	558	560	562	569	575	582	590
Price[h]	NZD/100 kg dw	254	203	283	270	258	248	231	214	205	198	199	202
UNITED STATES													
Production	kt cwe	12 127	12 040	11 866	11 439	11 471	11 637	11 791	12 002	12 215	12 421	12 599	12 752
Consumption	kt cwe	12 394	12 229	12 603	11 826	11 743	11 844	11 856	11 942	12 099	12 278	12 436	12 518
Imports	kt cwe	1 355	1 287	1 366	1 581	1 502	1 456	1 355	1 312	1 317	1 315	1 303	1 233
Exports	kt cwe	1 067	1 186	606	1 198	1 230	1 249	1 291	1 371	1 433	1 457	1 466	1 468
Price[i]	USD/100 kg dw	239	302	261	329	336	324	316	301	291	285	283	285
OTHERS													
Chinese Taipei: imports	kt cwe	85	95	91	99	103	108	112	117	123	128	134	140
Singapore: imports	kt cwe	25	28	25	27	27	28	28	28	29	29	29	30
Hong Kong (China): imports	kt cwe	68	72	68	71	71	72	73	74	75	77	78	79

For notes, see end of the table.

Annex Table 19. **BEEF AND VEAL PROJECTIONS**[a] (cont.)

Calendar year[b]		Average 1998-02	2003 est.	2004	2005	2006	2007	2008	2009	2010	2011	2012	2013
MERCOSUR MARKET													
ARGENTINA													
Production	kt cwe	2 566	2 526	2 620	2 775	2 891	3 043	3 096	3 159	3 192	3 211	3 208	3 209
Consumption	kt cwe	2 261	2 133	2 187	2 223	2 282	2 329	2 375	2 449	2 535	2 559	2 604	2 642
Exports	kt cwe	298	394	467	569	626	731	738	726	673	669	621	595
Price[j]	ARS/100 kg dw	182	360	379	411	408	409	407	395	378	381	385	391
BRAZIL													
Production	kt cwe	6 408	7 244	7 549	7 732	7 958	8 195	8 464	8 760	9 057	9 352	9 665	9 980
Consumption	kt cwe	5 821	6 011	6 346	6 622	6 813	7 082	7 268	7 527	7 743	7 994	8 224	8 482
Exports	kt cwe	661	1 299	1 271	1 175	1 210	1 181	1 264	1 301	1 381	1 426	1 508	1 566
Price	BRL/100 kg dw	239	333	314	334	335	341	344	341	338	343	349	357
CHILE													
Production[k]	kt cwe	226	201	205	204	204	203	202	202	200	198	196	195
Consumption	kt cwe	328	309	335	339	355	366	378	393	410	427	444	460
Net trade	kt cwe	−102	−107	−130	−135	−151	−163	−175	−191	−209	−229	−247	−265
PARAGUAY													
Production[k]	kt cwe	245	255	264	267	272	275	279	283	286	287	289	292
Consumption	kt cwe	198	193	205	206	209	213	208	214	222	225	228	231
Net trade	kt cwe	47	62	59	61	63	63	71	69	64	62	61	60
URUGUAY													
Production[k]	kt cwe	431	442	455	460	468	473	478	483	487	489	491	494
Consumption	kt cwe	222	210	220	223	230	236	242	249	257	263	269	275
Net trade	kt cwe	208	232	235	237	237	237	236	234	230	226	222	219
OTHER MARKETS													
EU-25													
Production	kt cwe	8 379	8 153	8 168	8 154	8 134	8 153	8 173	8 190	8 188	8 203	8 211	8 222
EU-15	kt cwe	..	7 468	7 486	7 448	7 438	7 465	7 491	7 515	7 520	7 543	7 557	7 574
EU-10	kt cwe	..	685	682	706	696	688	682	675	668	661	654	648
Consumption	kt cwe	7 948	8 220	8 151	8 131	8 181	8 207	8 217	8 220	8 197	8 217	8 219	8 227
EU-15	kt cwe	7 270	7 606	7 578	7 562	7 601	7 619	7 623	7 621	7 590	7 600	7 594	7 594
EU-10	kt cwe	..	613	573	569	580	587	593	600	607	616	625	633
Imports[l]	kt cwe	363	414	396	406	422	418	415	412	407	404	400	396
Exports[l]	kt cwe	649	386	469	506	418	342	348	358	375	367	368	368
Closing stocks	kt cwe	409	564	484	384	318	318	318	318	318	318	318	318
Price[m]	EUR/100 kg cwe	245	245	243	241	245	249	249	248	247	248	248	248
Other OECD[n]													
Production	kt cwe	710	730	734	738	742	746	750	754	759	763	767	772
Consumption	kt cwe	715	734	738	742	746	750	754	758	763	767	771	776
Net trade	kt cwe	−4	−4	−4	−4	−4	−4	−4	−4	−4	−4	−4	−4
CHINA													
Production	kt cwe	5 303	6 019	6 270	6 506	6 605	6 711	6 908	7 084	7 222	7 510	7 860	8 182
Consumption	kt cwe	5 260	6 007	6 262	6 503	6 609	6 719	6 922	7 106	7 252	7 551	7 911	8 243
Price[o]	CNY/100 kg	1 229	1 391	1 293	1 338	1 409	1 505	1 591	1 677	1 770	1 856	1 928	2 010

For notes, see end of the table.

Annex Table 19. **BEEF AND VEAL PROJECTIONS**[a] (cont.)

Calendar year[b]		Average 1998-02	2003 est.	2004	2005	2006	2007	2008	2009	2010	2011	2012	2013
RUSSIA													
Production	kt cwe	1 948	1 790	1 736	1 689	1 634	1 791	1 885	1 895	1 935	1 963	1 990	2 019
Consumption	kt cwe	2 488	2 346	2 315	2 324	2 350	2 463	2 510	2 534	2 593	2 649	2 705	2 761
Imports[p]	kt cwe	546	563	585	641	722	678	631	644	664	692	721	747
Price[o]	RUR/100 kg	251	380	479	556	586	609	618	630	633	643	652	660

a) Excludes trade of live animals.
b) Year ended 30 September for New Zealand.
c) Weighted average price of cows 201-260 kg, steers 301-400 kg, yearling < 200 kg dw.
d) Grade A slaughter steers > 1251 lb lw, Ontario – lw to dw conversion factor 0.6.
e) Wholesale carcass price B2-B3 steers, Tokyo.
f) Farm price of native cattle male 500 kg.
g) Huasteco steers grade 1A, 400 kg lw.
h) Schedule price M grade cow, 145.5-170 kg dw.
i) Choice steers, 1100-1300 lb lw, Nebraska-lw to dw conversion factor 0.63.
j) Buenos Aires wholesale liner, young bull, lw to dw conversion factor 0.55.
k) Indigenous basis, including live exports but excluding live imports.
l) Excludes intra-EU-25 trade.
m) Producer price.
n) Includes Norway, Switzerland and Turkey. Excludes Island.
o) Producer price.
p) Includes trade of live animals.
est.: estimate.

Source: OECD Secretariat.

Annex Table 20. **PIG MEAT PROJECTIONS**[a]

Calendar year[b]		Average 1998-02	2003 est.	2004	2005	2006	2007	2008	2009	2010	2011	2012	2013
PACIFIC MARKET													
CANADA													
Production	kt cwe	1 637	1 954	2 028	2 075	2 165	2 175	2 211	2 222	2 258	2 253	2 247	2 233
Consumption[c]	kt cwe	893	827	844	876	886	900	907	906	907	914	905	889
Exports	kt cwe	647	1 018	1 061	1 099	1 147	1 155	1 190	1 204	1 237	1 226	1 232	1 245
Price[d]	CAD/100 kg dw	143	136	126	135	132	126	124	123	124	125	125	131
JAPAN													
Production	kt cwe	1 262	1 264	1 274	1 249	1 216	1 199	1 190	1 175	1 165	1 149	1 126	1 118
Consumption	kt cwe	2 189	2 332	2 408	2 389	2 414	2 395	2 364	2 361	2 353	2 347	2 323	2 288
Imports	kt cwe	926	1 111	1 133	1 140	1 198	1 197	1 174	1 187	1 189	1 198	1 198	1 172
Price[e]	'000 JPY/100 kg dw	46	41	39	40	39	38	37	37	36	36	36	36
KOREA													
Production	kt cwe	940	1 022	1 037	1 053	1 069	1 087	1 107	1 127	1 148	1 169	1 192	1 217
Consumption	kt cwe	988	1 053	1 082	1 109	1 136	1 160	1 188	1 219	1 254	1 290	1 323	1 358
Imports	kt cwe	120	84	88	104	93	106	127	142	157	171	181	191
Exports	kt cwe	62	50	50	50	50	50	50	50	50	50	50	50
Price[f]	'000 KRW/100 kg dw	179	184	189	195	201	207	213	219	225	232	240	248
MEXICO													
Production	kt cwe	1 025	1 097	1 180	1 239	1 273	1 294	1 313	1 341	1 366	1 388	1 404	1 426
Consumption	kt cwe	1 143	1 282	1 310	1 392	1 427	1 456	1 493	1 545	1 580	1 614	1 638	1 656
Imports	kt cwe	146	209	158	187	192	207	230	258	272	289	301	301
Price[g]	MXN/100 kg dw	1 917	2 035	2 247	2 390	2 197	2 238	2 435	2 525	2 301	2 293	2 496	2 688
UNITED STATES													
Production	kt cwe	8 719	9 045	8 965	9 152	9 349	9 531	9 647	9 728	9 770	9 864	9 911	9 833
Consumption	kt cwe	8 486	8 865	8 796	8 936	9 219	9 401	9 562	9 647	9 716	9 741	9 833	9 821
Imports	kt cwe	410	486	563	614	669	679	718	750	801	776	806	840
Exports	kt cwe	632	673	725	829	791	809	808	832	853	899	882	859
Price[h]	USD/100 kg dw	119	120	120	126	124	119	118	117	118	118	119	123
CHINESE TAIPEI													
Production	kt cwe	906	902	919	914	954	963	982	1 000	1 016	1 024	1 034	1 059
Consumption	kt cwe	968	947	978	989	1 018	1 027	1 051	1 068	1 087	1 107	1 121	1 132
Imports	kt cwe	42	45	40	40	40	40	40	40	40	40	40	40
OCEANIA													
AUSTRALIA													
Production	kt cwe	376	423	415	410	412	414	417	419	422	424	425	427
Consumption	kt cwe	376	406	411	410	407	416	423	432	439	446	452	458
Exports	kt cwe	59	57	54	62	67	67	68	68	68	69	70	72
Price[i]	AUD/100 kg dw	236	276	273	231	250	251	255	254	253	253	248	251
NEW ZEALAND													
Production	kt cwe	48	49	51	53	54	55	56	55	55	54	54	53
Consumption	kt cwe	66	74	76	86	94	97	100	100	101	102	103	103
Imports	kt cwe	18	26	28	30	31	33	35	36	37	38	39	39
Price[j]	NZD/100 kg dw	299	338	331	279	261	257	258	257	254	252	247	246

For notes, see end of the table.

Annex Table 20. **PIG MEAT PROJECTIONS**[a] (cont.)

Calendar year[b]		Average 1998-02	2003 est.	2004	2005	2006	2007	2008	2009	2010	2011	2012	2013
OTHER MARKETS													
EU-25													
Production	kt cwe	..	21 913	22 003	22 200	22 331	22 496	22 750	22 915	22 962	23 082	23 282	23 431
EU-15	kt cwe	..	18 070	18 287	18 383	18 444	18 541	18 704	18 810	18 798	18 877	18 999	19 101
EU-10	kt cwe	..	3 845	3 718	3 818	3 889	3 956	4 048	4 107	4 166	4 207	4 285	4 332
Consumption	kt cwe	19 593	20 645	20 597	20 785	20 843	20 975	21 205	21 375	21 399	21 497	21 650	21 789
EU-15	kt cwe	16 501	17 207	17 069	17 234	17 240	17 321	17 491	17 601	17 549	17 611	17 720	17 786
EU-10	kt cwe	..	3 438	3 528	3 551	3 604	3 654	3 714	3 774	3 850	3 886	3 930	4 002
Imports[k]	kt cwe	..	23	4	4	4	4	4	4	4	4	4	4
Exports to Pacific markets	kt cwe	..	581	592	591	658	682	696	691	704	716	752	760
Exports to other markets[k]	kt cwe	..	736	800	825	830	840	850	860	870	880	890	900
Price[l]	EUR/100 kg dw	127	125	123	129	133	135	133	133	130	134	134	135
OTHER OECD[n]													
Production	kt cwe	337	339	338	338	337	337	337	337	336	336	336	336
Consumption	kt cwe	346	345	344	344	343	343	343	343	342	342	342	342
Net trade	kt cwe	−9	−6	−6	−6	−6	−6	−6	−6	−6	−6	−6	−6
ARGENTINA													
Production	kt cwe	203	151	160	177	196	199	207	214	223	226	234	238
Consumption	kt cwe	264	178	177	186	203	216	233	252	275	292	317	337
Net trade	kt cwe	−60	−27	−18	−8	−7	−16	−26	−38	−52	−65	−84	−99
BRAZIL													
Production	kt cwe	2 311	2 710	2 398	2 480	2 561	2 648	2 725	2 820	2 909	3 013	3 125	3 244
Consumption	kt cwe	2 084	2 235	1 939	2 034	2 119	2 201	2 271	2 359	2 435	2 526	2 619	2 724
Net trade	kt cwe	227	475	459	446	442	447	454	461	475	488	506	520
CHINA													
Production	kt cwe	40 694	43 629	44 132	45 029	45 532	46 086	47 462	48 675	49 367	49 725	51 369	53 210
Consumption	kt cwe	40 614	43 378	43 887	44 795	45 315	45 882	47 277	48 499	49 199	49 569	51 213	53 047
Exports	kt cwe	131	304	310	309	301	294	289	287	285	282	285	291
Price[o]	CNY/100 kg	914	983	984	1 033	1 029	1 047	1 072	1 108	1 143	1 195	1 203	1 230
RUSSIA													
Production	kt cwe	1 530	1 749	1 794	1 808	1 882	1 940	1 982	1 987	2 013	2 022	2 063	2 094
Consumption	kt cwe	2 123	2 209	2 207	2 226	2 319	2 371	2 418	2 428	2 459	2 472	2 517	2 550
Imports	kt cwe	514	472	425	429	448	443	448	453	458	462	466	468
Price[m]	RUR/100 kg	329	534	594	667	733	776	797	828	841	893	930	967

a) Excludes trade of live animals.
b) Year ended 30 September for New Zealand.
c) Excluding non-food parts.
d) Carcass price, index 100, Ontario.
e) Wholesale carcass price, excellent grade, Tokyo.
f) Farm price of pigs 100 kg.
g) Supreme grade.
h) Barrows and gilts, No. 1-3, 230-250 lb lw, Iowa/South Minnesota – lw to dw conversion factor 0.72.
i) Weighted average price, pigs 60-73 kg dw.
j) Schedule price, pigs > 50 kg dw, Canterbury.
k) Excludes intra-EU-25 trade.
l) Pig producer price.
m) Producer price.
n) Includes Norway, Switzerland and Turkey. Excludes Iceland.
o) Pig meat reference price.
est.: estimate.

Source: OECD Secretariat.

Annex Table 21. **POULTRY MEAT PROJECTIONS**

Calendar year[a]		Average 1998-02	2003 est.	2004	2005	2006	2007	2008	2009	2010	2011	2012	2013
AUSTRALIA													
Production	kt rtc	654	715	736	713	705	699	701	712	721	730	740	749
Consumption	kt rtc	628	692	712	689	681	675	677	688	697	706	716	725
Exports	kt rtc	26	23	24	24	24	24	24	24	24	24	24	24
Price[b]	AUD/100 kg rtc	361	411	362	381	401	419	435	443	452	461	469	481
CANADA													
Production	kt rtc	1 046	1 070	1 066	1 128	1 155	1 167	1 182	1 204	1 228	1 248	1 271	1 299
Consumption	kt rtc	1 074	1 139	1 124	1 184	1 216	1 230	1 245	1 267	1 292	1 313	1 337	1 366
Imports	kt rtc	150	165	163	164	170	174	176	179	183	186	190	193
Price[c]	CAD/100 kg rtc	155	154	152	148	147	148	151	152	153	154	154	154
EU-25													
Production	kt rtc	10 410	10 209	10 674	11 357	11 445	11 548	11 590	11 638	11 669	11 744	11 829	11 926
EU-15	kt rtc	8 895	8 390	8 844	9 327	9 287	9 275	9 216	9 149	9 118	9 106	9 104	9 158
EU-10	kt rtc	..	1 819	1 832	2 024	2 155	2 275	2 389	2 504	2 565	2 650	2 736	2 777
Consumption	kt rtc	9 841	9 863	10 264	11 004	11 120	11 215	11 249	11 289	11 308	11 373	11 448	11 534
EU-15	kt rtc	8 402	8 404	8 751	9 458	9 536	9 561	9 554	9 542	9 516	9 533	9 570	9 618
EU-10	kt rtc	..	1 459	1 513	1 546	1 585	1 654	1 695	1 747	1 792	1 840	1 879	1 916
Imports[d]	kt rtc	398	635	700	767	822	861	900	939	976	1 015	1 053	1 091
Exports[d]	kt rtc	1 052	981	1 110	1 120	1 147	1 194	1 240	1 288	1 338	1 386	1 434	1 483
Price[e]	EUR/100 kg rtc	99	104	103	98	98	98	97	97	97	97	97	97
JAPAN													
Production	kt rtc	1 209	1 256	1 246	1 209	1 172	1 137	1 111	1 093	1 075	1 062	1 051	1 040
Consumption	kt rtc	1 748	1 742	1 773	1 765	1 767	1 770	1 774	1 773	1 771	1 769	1 767	1 769
Imports	kt rtc	547	459	531	560	598	636	666	684	700	711	719	733
Price[f]	'000 JPY/100 kg rwt	118	126	100	100	100	99	98	97	96	95	94	93
KOREA													
Production	kt rtc	469	520	561	559	575	582	594	603	612	621	630	640
Consumption	kt rtc	555	687	718	701	702	707	717	724	732	740	748	756
Imports	kt rtc	87	173	158	143	128	126	125	123	122	121	119	118
Price[g]	'000 KRW/100 kg rtc	179	173	183	194	203	205	206	207	207	208	209	210
MEXICO													
Production	kt rtc	1 832	2 149	2 181	2 373	2 481	2 588	2 714	2 818	2 925	3 028	3 134	3 255
Consumption	kt rtc	2 084	2 457	2 492	2 690	2 819	2 903	3 030	3 185	3 305	3 418	3 547	3 679
Imports	kt rtc	255	310	315	325	340	320	325	370	386	401	416	431
Price[h]	MXN/100 kg rtc	1 411	1 593	1 601	1 623	1 666	1 689	1 704	1 726	1 783	1 841	1 905	1 956

For notes, see end of the table.

Annex Table 21. **POULTRY MEAT PROJECTIONS** (cont.)

Calendar year[a]		Average 1998-02	2003 est.	2004	2005	2006	2007	2008	2009	2010	2011	2012	2013
NEW ZEALAND													
Production	kt rtc	113	144	165	171	173	175	176	177	179	180	183	187
Consumption	kt rtc	113	144	165	171	173	175	176	177	179	180	183	187
UNITED STATES													
Production	kt rtc	16 302	17 422	17 383	18 336	18 623	18 834	19 184	19 545	19 848	20 131	20 293	20 636
Consumption	kt rtc	13 752	15 037	14 807	15 745	15 976	16 129	16 421	16 722	16 981	17 218	17 341	17 644
Exports	kt rtc	2 548	2 482	2 549	2 597	2 654	2 712	2 770	2 829	2 875	2 920	2 959	2 999
Price[i]	USD/100 kg rtc	129	137	139	139	141	143	142	142	141	142	143	143
OTHER OECD[j]													
Production	kt rtc	744	804	825	846	868	891	915	938	961	985	1 008	1 032
Consumption	kt rtc	782	846	867	888	910	933	957	980	1 003	1 027	1 050	1 074
Net trade	kt rtc	−39	−42	−42	−42	−42	−42	−42	−42	−42	−42	−42	−42
ARGENTINA													
Production	kt rtc	853	714	796	836	903	994	1 036	1 082	1 124	1 176	1 241	1 296
Consumption	kt rtc	861	670	727	758	806	873	891	911	929	964	1 002	1 048
Imports	kt rtc	40	16	20	22	24	26	28	30	32	34	36	38
Price[k]	ARS/100 kg rtc	127	240	251	264	279	290	298	301	304	307	311	312
BRAZIL													
Production	kt rtc	6 075	7 532	7 693	7 929	8 126	8 308	8 530	8 791	9 053	9 322	9 631	9 931
Consumption	kt rtc	5 036	5 761	5 933	6 122	6 369	6 681	6 995	7 265	7 529	7 895	8 160	8 504
Net trade	kt rtc	1 039	1 771	1 760	1 807	1 757	1 627	1 535	1 526	1 523	1 427	1 472	1 427
CHINA													
Production	kt rtc	9 039	10 141	10 172	10 236	10 388	10 494	10 703	10 911	11 013	11 084	11 121	11 068
Consumption	kt rtc	9 124	10 140	10 164	10 258	10 450	10 607	10 890	11 149	11 314	11 456	11 560	11 575
Imports	kt rtc	504	406	393	408	430	454	490	522	562	610	661	716
Exports	kt rtc	419	407	401	386	368	341	303	284	261	239	222	209
Price[l]	CNY/100 kg rtc	1 021	1 214	1 265	1 275	1 333	1 423	1 502	1 588	1 679	1 810	1 943	2 095
RUSSIA													
Production	kt rtc	389	566	533	690	728	766	808	850	896	944	997	1 054
Consumption	kt rtc	1 606	1 661	1 658	1 735	1 773	1 811	1 853	1 895	1 941	1 989	2 042	2 099
Imports	kt rtc	1 246	1 050	1 050	1 050	1 050	1 050	1 050	1 050	1 050	1 050	1 050	1 050
Price[l]	RUR/100 kg rtc	331	663	624	603	639	666	690	719	739	764	787	811

a) Year ended 30 September for New Zealand.
b) Average retail price of chicken.
c) Weighted average producer price of broilers < 2kg, Ontario – lw to rtc conversion factor 0.75.
d) Excludes intra-EU-25 trade.
e) Weighted average farmgate live fowls, top quality, lw to rtc conversion of 0.75, EU-15 starting in 1995.
f) Consumer price. Young boneless broilers.
g) Farm price of hi-broiler 1 kg.
h) Average producer price, chicken.
i) Wholesale weighted average broiler price 12 cities.
j) Includes Norway, Switzerland and Turkey. Excludes Iceland.
k) Brazil export price.
l) Producer price.
est.: estimate.

Source: OECD Secretariat.

Annex Table 22. **SHEEP MEAT PROJECTIONS**[a]

Calendar year[b]		Average 1998-02	2003 est.	2004	2005	2006	2007	2008	2009	2010	2011	2012	2013
AUSTRALIA													
Production	kt cwe	650	579	591	599	600	606	611	616	620	625	629	634
Consumption	kt cwe	307	295	311	302	311	313	316	314	314	315	317	321
Exports	kt cwe	342	284	278	296	289	294	295	302	306	310	312	312
Price[c] (lamb)	AUD/100 kg dw	232	388	250	252	265	273	282	290	299	308	318	327
Price[d] (mutton)	AUD/100 kg dw	106	193	125	120	123	127	131	135	139	143	147	152
CANADA													
Production	kt cwe	13	15	15	16	16	17	17	17	18	18	19	19
Consumption	kt cwe	29	33	35	36	39	42	46	49	53	57	62	68
Imports	kt cwe	16	19	20	21	23	26	29	32	36	39	43	49
Price[e] (lamb)	CAD/100 kg dw	554	544	478	539	542	527	514	498	488	483	482	486
EU-25													
Production	kt cwe	..	1 050	1 045	1 032	1 027	1 019	1 015	1 003	997	986	975	964
EU-15	kt cwe	..	1 018	1 013	1 000	994	987	983	972	967	955	945	935
EU-10	kt cwe	..	33	33	32	32	32	31	31	31	30	30	29
Consumption	kt cwe	..	1 329	1 332	1 339	1 343	1 342	1 341	1 340	1 336	1 332	1 328	1 324
EU-15	kt cwe	1 329	1 301	1 303	1 311	1 315	1 314	1 314	1 313	1 310	1 307	1 303	1 299
EU-10	kt cwe	..	28	28	28	28	28	27	27	26	26	25	25
Net trade[f]	kt cwe	..	−279	−286	−307	−316	−323	−326	−336	−337	−344	−349	−355
Price[g]	EUR/100 kg dw	368	363	331	336	336	338	340	343	345	348	350	353
JAPAN													
Consumption	kt cwe	50	41	40	40	40	40	40	40	40	40	40	40
Imports	kt cwe	50	41	40	40	40	40	40	40	40	40	40	40
MEXICO													
Production	kt cwe	47	57	58	60	62	63	65	67	68	70	72	74
Consumption	kt cwe	87	105	108	110	113	115	118	120	123	125	128	131
Imports	kt cwe	40	48	49	50	51	52	53	54	55	55	56	57
Price[h]	MXN/100 kg dw	3 144	4 768	4 504	5 083	5 103	4 962	4 844	4 693	4 596	4 547	4 539	4 579
NEW ZEALAND													
Production	kt cwe	536	584	501	515	520	525	531	530	524	517	510	501
Consumption	kt cwe	92	95	104	97	94	97	100	102	104	106	108	110
Exports	kt cwe	429	485	402	423	430	431	433	429	422	413	404	393
Price[i] (lamb)	NZD/100 kg dw	325	379	363	347	346	347	348	346	347	349	352	355
Price[i] (mutton)	NZD/100 kg dw	148	203	182	173	172	171	171	171	170	170	170	170
UNITED STATES													
Production	kt cwe	107	101	96	98	91	102	103	104	104	104	104	104
Consumption	kt cwe	165	177	181	185	188	190	191	192	192	192	192	192
Imports	kt cwe	60	81	87	89	99	90	90	90	90	90	90	90
Price[j] (lamb)	USD/100 kg dw	330	379	350	395	397	386	377	365	357	354	353	356
OTHER OECD[k]													
Production	kt cwe	259	240	232	224	216	208	199	191	183	175	167	158
Consumption	kt cwe	263	246	238	230	221	213	205	197	189	180	172	164
Net trade	kt cwe	−4	−6	−6	−6	−6	−6	−6	−6	−6	−6	−6	−6

a) Excludes trade of live animal.
b) Year ended 30 September for New Zealand.
c) Saleyard price, lamb, 16-20 kg dw.
d) Saleyard price, wethers, < 22kg dw.
e) A/B grade slaughter lambs, 80-94 lb lw, Toronto – lw to dw conversion factor 0.5.
f) Excludes intra-EU-25 trade.
g) Market price for sheep meat, EU-15 starting in 1995.
h) Average producer price, sheep.
i) Schedule price, all grade average.
j) Choice grade slaughter lamb, 95-115 lb. lw, San Angelo – lw to dw conversion factor 0.5.
k) Includes Korea, Norway, Switzerland and Turkey. Excludes Iceland.
est.: estimate.

Source: : OECD Secretariat.

Annex Table 23. **MEAT PER CAPITA CONSUMPTION PROJECTIONS**

Calendar year[a]		Average 1998-02	2003 est.	2004	2005	2006	2007	2008	2009	2010	2011	2012	2013
AUSTRALIA													
Total meat	kg/person	83.8	86.1	84.1	83.8	82.9	82.8	82.9	83.5	84.0	84.5	84.9	85.2
Beef and veal	kg/person	25.7	26.4	23.1	24.7	24.4	24.6	24.6	25.0	25.2	25.4	25.4	25.3
Pig meat	kg/person	15.3	16.0	16.0	15.8	15.6	15.8	16.0	16.2	16.3	16.5	16.6	16.7
Poultry meat	kg/person	28.7	30.7	31.3	30.0	29.4	29.0	28.8	29.1	29.3	29.5	29.7	29.9
Sheep meat	kg/person	14.1	13.1	13.7	13.2	13.5	13.4	13.4	13.3	13.2	13.2	13.2	13.3
CANADA													
Total meat[b]	kg/person	76.9	75.7	75.0	76.4	76.6	76.9	77.1	77.6	78.1	78.6	78.7	79.0
Beef and veal	kg/person	22.8	22.7	22.5	21.7	21.1	20.9	20.8	20.8	20.7	20.6	20.4	20.3
Pig meat	kg/person	22.6	20.4	20.6	21.3	21.4	21.6	21.6	21.5	21.4	21.5	21.2	20.8
Poultry meat	kg/person	30.7	31.7	31.0	32.4	33.1	33.3	33.5	33.9	34.5	34.9	35.4	36.1
EU-25													
Total meat	kg/person	..	69.7	70.1	71.8	72.1	72.4	72.8	73.1	73.0	73.3	73.6	73.9
Beef and veal	kg/person	12.3	12.6	12.5	12.5	12.5	12.5	12.5	12.5	12.5	12.5	12.5	12.5
Pig meat	kg/person	33.8	35.4	35.2	35.5	35.6	35.7	36.1	36.3	36.3	36.5	36.7	36.9
Poultry meat	kg/person	19.1	19.1	19.8	21.2	21.4	21.6	21.6	21.6	21.7	21.8	21.9	22.0
Sheep meat[c]	kg/person	..	2.6	2.6	2.6	2.6	2.6	2.6	2.6	2.6	2.5	2.5	2.5
EU-15													
Total meat	kg/person	70.3	71.5	71.7	73.4	73.4	73.5	73.6	73.6	73.1	73.0	73.0	73.0
Beef and veal	kg/person	13.5	14.0	13.8	13.8	13.8	13.8	13.8	13.7	13.6	13.6	13.5	13.5
Pig meat	kg/person	34.1	35.2	34.7	35.0	34.9	34.9	35.2	35.3	35.1	35.1	35.2	35.2
Poultry meat	kg/person	19.6	19.4	20.1	21.7	21.8	21.8	21.7	21.6	21.5	21.4	21.4	21.5
Sheep meat[c]	kg/person	3.1	3.0	3.0	3.0	3.0	3.0	3.0	3.0	3.0	2.9	2.9	2.9
EU-10													
Total meat	kg/person	..	60.2	62.0	63.1	64.7	66.7	68.4	70.3	72.5	74.6	76.9	79.6
Beef and veal	kg/person	..	5.8	5.5	5.5	5.7	5.8	5.9	6.0	6.1	6.3	6.5	6.7
Pig meat	kg/person	..	36.5	37.8	38.4	39.2	40.1	41.0	42.0	43.3	44.3	45.5	47.1
Poultry meat	kg/person	..	17.5	18.3	18.9	19.5	20.5	21.1	22.0	22.7	23.7	24.5	25.5
Sheep meat[c]	kg/person	..	0.3	0.3	0.3	0.3	0.3	0.3	0.3	0.3	0.3	0.3	0.3
JAPAN													
Total meat[b]	kg/person	33.8	34.5	35.2	35.7	35.8	36.0	36.0	36.2	36.4	36.6	36.8	36.9
Beef and veal	kg/person	7.9	7.9	7.9	8.6	8.6	8.8	9.0	9.1	9.3	9.6	9.8	10.0
Pig meat	kg/person	13.4	14.3	14.7	14.6	14.8	14.7	14.5	14.5	14.5	14.5	14.4	14.2
Poultry meat	kg/person	12.1	12.0	12.2	12.2	12.2	12.2	12.3	12.3	12.3	12.3	12.3	12.4
KOREA													
Total meat[b]	kg/person	35.0	38.4	38.7	39.4	39.7	40.2	40.8	41.5	42.4	43.1	43.9	44.7
Beef and veal	kg/person	8.2	8.6	8.1	8.9	8.9	9.1	9.3	9.5	9.8	10.0	10.2	10.5
Pig meat	kg/person	16.4	17.1	17.5	17.8	18.1	18.4	18.7	19.1	19.6	20.1	20.5	21.0
Poultry meat	kg/person	10.4	12.6	13.1	12.7	12.6	12.6	12.8	12.8	12.9	13.0	13.1	13.2
MEXICO													
Total meat[b]	kg/person	40.0	43.3	43.4	44.4	45.5	46.1	47.0	48.4	49.3	49.9	50.7	51.5
Beef and veal	kg/person	11.5	11.6	11.6	10.8	10.9	11.0	11.0	11.2	11.4	11.3	11.4	11.5
Pig meat	kg/person	9.1	9.7	9.8	10.3	10.4	10.5	10.6	10.9	11.0	11.1	11.1	11.1
Poultry meat	kg/person	18.7	21.1	21.1	22.4	23.2	23.6	24.4	25.3	26.0	26.5	27.2	27.9

For notes, see end of the table.

Annex Table 23. **MEAT PER CAPITA CONSUMPTION PROJECTIONS** (cont.)

Calendar year[a]		Average 1998-02	2003 est.	2004	2005	2006	2007	2008	2009	2010	2011	2012	2013
NEW ZEALAND													
Total meat	kg/person	80.9	86.8	91.7	93.9	95.7	97.1	98.2	98.9	99.7	100.4	101.4	102.8
Beef and veal	kg/person	20.9	18.9	17.3	18.0	18.5	18.7	18.8	18.9	19.0	19.2	19.4	19.5
Pig meat	kg/person	13.2	14.6	14.8	16.9	18.4	18.9	19.3	19.2	19.4	19.4	19.5	19.6
Poultry meat	kg/person	25.7	32.0	36.7	37.7	38.0	38.3	38.4	38.5	38.8	38.9	39.3	40.0
Sheep meat	kg/person	21.1	21.2	23.0	21.3	20.8	21.2	21.8	22.3	22.6	22.9	23.2	23.6
UNITED STATES													
Total meat[b]	kg/person	97.5	99.0	98.3	99.0	99.5	100.0	100.7	101.3	101.9	102.4	102.7	103.0
Beef and veal	kg/person	30.7	29.4	30.0	28.0	27.6	27.7	27.5	27.5	27.7	27.9	28.1	28.1
Pig meat	kg/person	23.4	23.7	23.4	23.6	24.2	24.5	24.7	24.8	24.8	24.7	24.8	24.6
Poultry meat	kg/person	42.8	45.4	44.4	46.9	47.2	47.4	47.9	48.4	48.9	49.2	49.3	49.8
Other OECD[d]													
Total meat[b]	kg/person	21.4	21.2	21.1	21.0	21.0	21.0	21.0	20.9	20.9	20.9	20.9	20.9
Beef and veal	kg/person	6.3	6.2	6.2	6.2	6.1	6.1	6.1	6.1	6.0	6.0	6.0	6.0
Pig meat	kg/person	3.4	3.3	3.2	3.2	3.2	3.1	3.1	3.1	3.0	3.0	3.0	2.9
Poultry meat	kg/person	8.7	9.0	9.2	9.3	9.4	9.6	9.7	9.8	10.0	10.1	10.3	10.4
ARGENTINA													
Total meat[b]	kg/person	68.8	57.8	59.5	60.4	62.4	64.4	65.3	66.9	68.6	69.4	70.9	72.2
Beef and veal	kg/person	42.7	38.9	39.4	39.7	40.4	40.9	41.3	42.3	43.3	43.4	43.8	44.0
Pig meat	kg/person	5.6	3.6	3.6	3.7	4.0	4.2	4.5	4.9	5.2	5.5	5.9	6.3
Poultry meat	kg/person	20.5	15.3	16.5	17.0	17.9	19.3	19.5	19.8	20.0	20.6	21.2	22.0
BRAZIL													
Total meat[b]	kg/person	59.5	62.5	62.6	64.3	65.9	68.0	69.8	71.6	73.2	75.4	77.1	79.2
Beef and veal	kg/person	23.9	23.9	24.9	25.7	26.1	26.9	27.3	28.0	28.5	29.1	29.6	30.2
Pig meat	kg/person	9.5	9.9	8.5	8.8	9.1	9.3	9.5	9.8	10.0	10.2	10.5	10.8
Poultry meat	kg/person	26.0	28.7	29.3	29.9	30.7	31.9	33.0	33.9	34.8	36.1	36.9	38.1
CHINA													
Total meat[b]	kg/person	36.3	38.5	38.8	39.3	39.7	40.0	40.9	41.7	42.1	42.4	43.4	44.4
Beef and veal	kg/person	2.9	3.3	3.4	3.5	3.5	3.6	3.6	3.7	3.8	3.9	4.1	4.2
Pig meat	kg/person	25.1	26.3	26.4	26.8	26.9	27.1	27.7	28.3	28.5	28.5	29.3	30.2
Poultry meat	kg/person	6.4	6.9	6.9	6.9	7.0	7.1	7.2	7.3	7.4	7.4	7.5	7.4
RUSSIA													
Total meat[b]	kg/person	33.7	34.0	34.0	34.7	35.7	36.9	37.8	38.4	39.3	40.2	41.2	42.2
Beef and veal	kg/person	11.9	11.3	11.2	11.3	11.5	12.1	12.4	12.5	12.9	13.2	13.6	13.9
Pig meat	kg/person	11.3	11.9	11.9	12.1	12.6	13.0	13.3	13.4	13.6	13.8	14.1	14.3
Poultry meat	kg/person	9.7	10.1	10.1	10.6	10.9	11.2	11.5	11.8	12.1	12.5	12.9	13.3

a) Year ended 30 September for New Zealand. Consumption expressed in retail weight. Carcass weight to retail weight conversion factors of 0.7 for beef and veal, 0.78 for pig meat and 0.88 for sheep meat. Rtc to retail weight conversion factor 0.88 for poultry meat.
b) Excludes sheep meat.
c) Includes goat meat.
d) Includes Norway, Switzerland and Turkey. Excludes Iceland.
est.: estimate.
Source: OECD Secretariat.

Annex Table 24. **MILK PROJECTIONS**

Calendar year[a]		Average 1998-02	2003 est.	2004	2005	2006	2007	2008	2009	2010	2011	2012	2013
AUSTRALIA													
Production	mt pw	10.8	10.3	10.3	10.3	10.4	10.6	10.8	11.1	11.4	11.6	11.8	11.9
Liquid sales	mt pw	2.0	2.0	2.0	2.0	2.0	2.0	2.0	2.0	2.0	2.0	2.0	2.0
Industrial use	mt pw	8.8	8.4	8.4	8.3	8.4	8.6	8.9	9.1	9.4	9.6	9.7	9.9
Price[b]	AUDc/litre	29.2	29.6	27.9	26.4	28.3	30.2	31.1	31.6	32.0	32.4	32.8	33.4
Price[c]	AUDc/litre	25.6	29.6	27.9	26.4	28.3	30.2	31.1	31.6	32.0	32.4	32.8	33.4
CANADA													
Production	mt pw	8.2	8.1	8.1	8.1	8.2	8.3	8.3	8.4	8.5	8.5	8.6	8.6
Liquid sales	mt pw	2.9	2.9	2.9	2.9	3.0	3.0	3.0	3.0	3.0	3.0	3.1	3.1
Industrial use	mt pw	4.8	4.6	4.6	4.7	4.7	4.8	4.8	4.9	4.9	5.0	5.0	5.1
Price[d]	CADc/litre	64.3	71.0	74.4	75.3	77.1	77.9	79.3	80.4	81.8	83.0	84.2	85.6
Price[e]	CADc/litre	57.1	61.4	64.0	64.6	65.8	66.4	67.3	68.0	68.9	69.7	70.4	71.3
EU-25													
Production	mt pw	..	143.4	143.5	143.4	144.1	144.5	144.8	144.6	144.4	144.3	144.1	143.9
EU-15	mt pw	121.5	121.6	121.5	121.4	121.8	122.2	122.6	122.5	122.4	122.3	122.1	122.0
EU-10	mt pw	..	21.8	22.0	22.0	22.3	22.3	22.2	22.1	22.1	22.0	22.0	21.9
Liquid sales	mt pw	29.1	28.3	28.7	28.8	28.9	29.0	29.1	29.2	29.4	29.5	29.6	29.7
Industrial use	mt pw	85.4	86.2	85.9	85.8	86.2	86.5	86.9	86.7	86.5	86.4	86.2	86.0
Price[f]	EUR/litre	0.314	0.310	0.291	0.284	0.272	0.258	0.257	0.259	0.264	0.267	0.268	0.269
JAPAN													
Production	mt pw	8.4	8.4	8.4	8.4	8.4	8.4	8.4	8.4	8.4	8.4	8.4	8.4
Liquid sales	mt pw	5.0	5.1	5.1	5.2	5.2	5.2	5.2	5.2	5.2	5.3	5.3	5.3
Industrial use	mt pw	3.4	3.2	3.2	3.1	3.1	3.1	3.1	3.1	3.0	3.0	3.0	3.0
Price[g]	JPY/litre	85.0	86.5	84.7	84.1	84.2	84.0	83.1	82.7	82.1	81.4	80.8	80.1
KOREA													
Production	mt pw	2.2	2.6	2.8	3.0	3.2	3.3	3.4	3.5	3.6	3.7	3.8	3.9
Liquid sales	mt pw	1.3	1.4	1.4	1.4	1.4	1.4	1.4	1.4	1.4	1.5	1.4	1.5
Industrial use	mt pw	0.9	1.2	1.4	1.6	1.8	1.9	2.0	2.1	2.2	2.3	2.4	2.5
Price[h]	KRW/litre	607.5	644.0	652.8	664.4	661.7	675.8	685.5	693.5	702.8	714.0	719.7	733.2
MEXICO													
Production	mt pw	9.3	10.4	10.6	10.6	11.3	11.4	11.6	11.7	11.7	11.8	11.8	11.8
On farm use	mt pw	2.8	3.3	3.2	3.0	3.4	3.2	3.1	2.9	2.7	2.5	2.2	2.0
Liquid sales	mt pw	3.1	3.5	3.7	3.8	4.0	4.1	4.3	4.4	4.6	4.7	4.9	5.0
Industrial use	mt pw	3.4	3.6	3.7	3.8	3.9	4.0	4.2	4.3	4.5	4.6	4.7	4.8
Price[i]	MXN/litre	3.1	3.0	3.1	3.3	3.4	3.6	3.7	3.8	3.9	4.0	4.0	4.1
NEW ZEALAND													
Production	mt pw	12.3	15.3	15.9	16.7	17.3	17.8	18.3	18.9	19.4	19.9	20.5	21.1
Liquid sales	mt pw	0.3	0.3	0.3	0.3	0.3	0.3	0.3	0.3	0.3	0.3	0.3	0.3
Industrial use	mt pw	12.0	14.9	15.5	16.3	16.9	17.4	17.9	18.5	19.0	19.5	20.1	20.7
Price[j]	NZDc/litre	35.7	40.0	42.9	42.8	45.6	46.5	46.8	46.2	45.5	44.6	43.7	43.1

For notes, see end of the table.

Annex Table 24. **MILK PROJECTIONS** (cont.)

Calendar year[a]		Average 1998-02	2003 est.	2004	2005	2006	2007	2008	2009	2010	2011	2012	2013
UNITED STATES													
Production	mt pw	74.7	77.0	77.7	79.0	80.1	81.4	82.8	83.9	85.1	86.5	87.9	89.0
Liquid sales	mt pw	24.9	26.1	25.6	25.6	25.6	25.6	25.6	25.5	25.5	25.4	25.4	25.3
Industrial use	mt pw	48.0	50.4	51.7	52.9	54.0	55.4	56.8	57.9	59.2	60.7	62.1	63.2
Price[k]	USDc/litre	28.2	26.6	27.4	28.4	28.8	29.0	28.8	28.8	29.0	28.9	29.1	29.4
Price[l]	USDc/litre	31.6	28.6	29.0	29.9	30.3	30.5	30.2	30.3	30.4	30.3	30.5	30.8
OTHER OECD[m]													
Production	mt pw	12.3	13.6	13.8	14.0	14.2	14.4	14.5	14.7	14.9	15.1	15.3	15.5
On farm use	mt pw	2.8	3.2	3.3	3.3	3.3	3.4	3.4	3.5	3.5	3.6	3.6	3.6
Liquid sales	mt pw	2.8	3.2	3.2	3.3	3.3	3.3	3.3	3.4	3.4	3.4	3.4	3.5
Industrial use	mt pw	6.6	7.2	7.3	7.4	7.5	7.7	7.8	7.9	8.0	8.1	8.3	8.4
ARGENTINA													
Production	mt pw	9.5	7.7	7.9	8.3	8.7	9.0	9.3	9.6	10.0	10.4	10.8	11.3
Liquid sales	mt pw	1.6	1.3	1.4	1.4	1.5	1.5	1.6	1.6	1.7	1.7	1.8	1.8
Industrial use	mt pw	7.9	6.4	6.5	6.9	7.2	7.5	7.7	8.0	8.3	8.7	9.1	9.5
Price[n]	ARSc/litre	18.5	45.1	46.3	47.3	48.0	48.9	49.5	49.7	49.9	50.3	50.8	51.3
BRAZIL													
Production	mt pw	19.8	22.1	22.7	23.3	23.8	24.4	25.1	25.8	26.5	27.3	28.1	28.9
CHINA													
Production	mt pw	11.0	14.8	15.1	16.0	17.4	18.3	19.5	20.7	22.0	23.3	24.8	26.3
Industrial use	mt pw	3.5	4.5	4.6	4.9	5.3	5.6	5.9	6.3	6.6	7.0	7.4	7.9
Other uses	mt pw	5.5	10.2	10.4	11.1	12.1	12.7	13.6	14.5	15.3	16.3	17.4	18.5
Price[o]	CNY/litre	2584.1	1625.9	1759.5	1859.4	1933.6	2062.5	2187.2	2314.9	2454.7	2600.3	2747.3	2898.9
RUSSIA													
Production	mt pw	32.5	32.7	32.7	32.8	33.2	33.5	33.9	34.3	34.7	35.1	35.6	36.1
Price	RUR/100 kg	346.1	430.9	462.7	485.1	507.7	524.5	539.8	554.7	561.9	572.2	583.9	597.9
REST OF WORLD[p]													
Production	mt pw	..	193.2	199.5	205.9	212.3	218.7	225.5	232.6	239.9	247.5	255.4	263.6
Industrial use	mt pw	..	96.4	100.6	105.0	109.3	113.0	117.4	121.8	126.4	131.3	136.4	141.8
Other uses	mt pw	..	96.8	98.9	101.0	103.1	105.7	108.2	110.8	113.5	116.2	119.0	121.8
WORLD													
Production	mt pw	559.5	573.9	583.3	594.2	607.0	618.3	630.9	642.9	655.2	668.1	681.6	695.3

a) Year ended 30 June for Australia and 31 May for New Zealand.
b) Weighted average farm price, market and manufacturing milk.
c) Average price, manufacturing milk.
d) Fluid milk price, class 1, Ontario.
e) Industrial milk target return.
f) Weighted average farm price, raw cow's milk.
g) Average producer price, all milk.
h) Producer price, 4th grade raw milk.
i) Average producer price.
j) Average farm price, all milk, milk to milkfat conversion factor 0.043.
k) Average farm price, manufacturing milk, 3.5% fat, Minnesota-Wisconsin.
l) Average received by farmers for all milk.
m) Includes Norway, Switzerland and Turkey. Excludes Iceland.
n) Price of milk to producers.
o) Producer price.
p) Excluding OIS.
est.: estimate.

Source: OECD Secretariat.

Annex Table 25. **BUTTER PROJECTIONS**

Calendar year[a]		Average 1998-02	2003 est.	2004	2005	2006	2007	2008	2009	2010	2011	2012	2013
AUSTRALIA													
Production	kt pw	177	138	131	130	129	131	133	135	136	136	136	135
Consumption	kt pw	81	46	60	72	66	67	68	68	69	70	72	73
Exports	kt pw	108	100	77	64	69	70	72	73	73	72	71	69
Price[b]	AUD/100 kg	263	224	182	215	230	243	253	261	269	277	286	296
CANADA													
Production	kt pw	83	81	79	79	79	79	79	79	80	80	81	81
Consumption	kt pw	86	90	81	82	83	83	82	83	83	83	84	84
Exports	kt pw	3	2	3	3	3	3	3	3	3	3	3	3
Price[c]	CAD/100 kg	560	610	625	635	644	654	663	673	683	693	704	714
EU-25													
Production	kt pw	..	2 174	2 163	2 032	2 002	1 998	1 998	1 994	1 992	1 981	1 980	1 970
EU-15	kt pw	1 857	1 867	1 850	1 715	1 681	1 674	1 670	1 663	1 657	1 643	1 639	1 626
EU-10	kt pw	..	308	314	317	321	324	327	331	334	338	341	344
Consumption	kt pw	..	1 937	1 959	1 999	2 032	2 025	2 024	2 015	2 005	1 995	1 986	1 977
EU-15	kt pw	1 741	1 679	1 702	1 738	1 769	1 761	1 759	1 750	1 739	1 729	1 718	1 708
EU-10	kt pw	..	257	257	262	263	264	265	265	266	267	268	268
Net trade[d]	kt pw	..	191	194	123	54	47	43	32	–11	–25	–31	–41
Closing stocks	kt pw	296	429	439	348	263	190	121	68	65	76	100	135
Intervention stocks	kt pw	133	290	289	271	233	205	175	175	175	175	175	175
Price[e]	EUR/100 kg	332	317	299	292	277	260	254	256	264	267	268	270
JAPAN													
Production	kt pw	86	74	76	79	81	82	83	83	84	85	85	86
Consumption	kt pw	86	86	84	82	84	84	84	84	85	86	86	87
Imports	kt pw	1	9	7	3	3	2	1	1	1	1	1	1
Price[f]	'000 JPY/100 kg	96	96	95	96	96	96	96	96	95	95	95	94
KOREA													
Production	kt pw	4	4	4	4	5	5	5	5	6	6	6	6
Consumption	kt pw	5	5	5	5	6	6	6	6	7	7	7	7
Imports	kt pw	1	1	1	1	1	1	1	1	1	1	1	1
MEXICO													
Production	kt pw	15	16	16	16	16	16	16	16	17	17	17	17
Consumption	kt pw	48	52	54	58	60	63	69	73	77	81	84	88
Imports	kt pw	33	36	39	43	44	48	53	57	60	64	67	71
Price[g]	MXN/100 kg	2 552	2 690	2 638	2 719	2 719	2 733	2 732	2 749	2 769	2 785	2 805	2 827
NEW ZEALAND													
Production[h]	kt pw	369	485	499	531	548	555	573	585	601	612	621	632
Consumption	kt pw	31	32	31	31	31	31	31	31	31	31	31	31
Exports[i, j]	kt pw	317	395	467	498	517	523	541	554	569	581	589	599
Price[b, j]	NZD/100 kg	313	253	248	270	290	295	299	305	308	312	314	318

For notes, see end of the table.

Annex Table 25. **BUTTER PROJECTIONS** (cont.)

Calendar year[a]		Average 1998-02	2003 est.	2004	2005	2006	2007	2008	2009	2010	2011	2012	2013
UNITED STATES													
Production	kt pw	571	570	564	572	572	578	578	578	582	587	592	591
Consumption	kt pw	579	587	599	577	572	573	573	573	577	582	587	586
Exports	kt pw	2	10	10	20	20	20	20	20	20	20	20	20
Imports	kt pw	23	15	15	15	15	15	15	15	15	15	15	15
Closing stocks	kt pw	26	60	30	20	15	15	15	15	15	15	15	15
Price[k]	USD/100 kg	307	249	255	276	286	290	291	289	281	271	270	268
OTHER OECD[l]													
Production	kt pw	184	165	167	169	171	173	175	177	178	180	182	183
Consumption	kt pw	189	175	176	178	180	182	184	185	187	189	191	193
Net trade	kt pw	−5	−9	−9	−9	−9	−9	−9	−9	−9	−9	−9	−10
ARGENTINA													
Production	kt pw	47	42	40	42	44	44	46	47	49	50	52	54
Consumption	kt pw	42	39	41	43	46	46	48	49	51	53	55	57
Net trade	kt pw	4	−1	−1	−1	−2	−2	−2	−2	−3	−3	−3	−4
Price[m]	ARS/100 kg	429	558	578	600	608	617	629	631	634	640	649	656
BRAZIL													
Production	kt pw	79	86	92	92	93	94	96	98	101	104	106	108
Consumption	kt pw	85	85	97	100	104	107	110	113	116	119	121	126
Net trade	kt pw	−7	1	−4	−8	−11	−12	−14	−14	−15	−15	−16	−18
CHINA													
Production	kt pw	82	107	109	115	125	131	139	148	156	165	175	185
Consumption	kt pw	100	126	128	135	145	151	160	169	177	187	197	208
Imports	kt pw	18	19	19	20	20	20	21	21	21	22	22	22
RUSSIA													
Production	kt pw	268	252	259	259	262	263	269	273	278	282	288	295
Consumption	kt pw	369	401	408	412	418	424	430	436	441	444	446	449
Imports	kt pw	99	149	156	160	162	166	166	167	168	167	163	159
OIS													
Production	kt pw	127	130	130	131	131	127	125	121	118	115	112	109
Consumption	kt pw	101	101	101	102	102	102	102	102	103	103	103	103
Net trade	kt pw	24	29	29	29	29	25	22	19	16	12	9	6
REST OF WORLD													
Production	kt pw	..	3 805	3 977	4 156	4 330	4 483	4 662	4 842	5 031	5 229	5 436	5 658
Consumption	kt pw	..	4 286	4 498	4 628	4 748	4 889	5 077	5 249	5 400	5 589	5 793	6 008
Net trade	kt pw	..	−481	−521	−472	−418	−406	−415	−407	−369	−360	−357	−350

a) Year ending 30 June for Australia and 31 May for New Zealand.
b) Average export price, f.o.b.
c) Wholesale support price.
d) Excludes intra-EU-25.
e) Producer price.
f) Average wholesale price for major users.
g) Value of production divided by volume of production.
h) Includes AMF measured in butter equivalent.
i) Includes AMF measured in product weight.
j) Year ended 30 June.
k) Average wholesale price, grade A butter, Chicago.
l) Includes Norway, Switzerland and Turkey. Excludes Iceland.
m) Wholesale price (precios mayoristas).
est.: estimate.

Source: OECD Secretariat.

Annex Table 26. **CHEESE PROJECTIONS**

Calendar year[a]		Average 1998-02	2003 est.	2004	2005	2006	2007	2008	2009	2010	2011	2012	2013
AUSTRALIA													
Production	kt pw	363	352	372	361	375	378	388	400	411	422	432	440
Consumption	kt pw	204	187	225	186	195	188	189	191	194	197	199	201
Exports	kt pw	197	209	192	222	228	240	251	262	273	283	293	301
Price[b]	AUD/100 kg	413	385	379	406	434	464	485	503	522	542	564	587
CANADA													
Production	kt pw	335	358	352	362	367	373	379	384	389	394	400	404
Consumption	kt pw	336	367	374	379	384	390	396	401	406	411	417	421
Imports	kt pw	22	20	21	21	21	21	21	21	21	21	21	21
Exports	kt pw	20	10	5	5	5	5	5	5	5	5	5	5
Price[c]	CAD/100 kg	729	802	827	852	876	899	921	942	965	988	1 011	1 036
EU-25													
Production	kt pw	..	8 289	8 410	8 509	8 624	8 680	8 764	8 798	8 851	8 892	8 935	8 979
EU-15	kt pw	6 960	7 315	7 426	7 514	7 618	7 664	7 736	7 760	7 802	7 832	7 864	7 897
EU-10	kt pw	..	974	984	995	1 006	1 017	1 028	1 038	1 049	1 060	1 071	1 082
Consumption	kt pw	..	7 874	8 024	8 129	8 293	8 368	8 493	8 555	8 633	8 702	8 773	8 848
EU-15	kt pw	6 663	7 012	7 145	7 220	7 355	7 398	7 493	7 520	7 565	7 598	7 633	7 669
EU-10	kt pw	..	862	879	909	938	970	1 000	1 035	1 068	1 105	1 140	1 179
Net trade[d]	kt pw	..	381	345	312	281	263	243	228	211	194	177	160
Price[e]	EUR/100 kg	413	411	418	415	403	387	370	376	377	379	380	381
JAPAN													
Production[f]	kt pw	123	128	129	134	137	139	142	145	148	151	154	157
of which: domestic	kt pw	35	37	37	37	36	36	36	36	36	36	36	36
Consumption	kt pw	232	241	244	250	255	258	263	268	273	277	282	286
Imports[g]	kt pw	197	205	207	214	219	222	227	232	237	241	246	250
Price[h]	'000 JPY/100 kg	33	33	33	33	33	33	33	33	33	33	33	34
KOREA													
Production	kt pw	16	24	27	30	33	36	39	41	44	47	50	53
Consumption	kt pw	42	60	66	73	80	83	87	91	94	98	101	105
Imports	kt pw	26	36	39	43	47	47	48	49	50	50	51	52
MEXICO													
Production	kt pw	130	140	150	153	155	163	173	181	186	192	198	204
Consumption	kt pw	182	209	221	225	228	237	249	258	265	271	279	287
Imports	kt pw	52	69	71	72	73	75	76	77	78	80	81	82
Price[i]	MXN/100 kg	4 216	4 221	4 289	4 507	4 677	4 794	4 885	4 946	4 999	5 034	5 075	5 112
NEW ZEALAND													
Production	kt pw	278	292	303	319	338	356	360	371	382	393	407	420
Consumption	kt pw	28	28	28	28	28	28	28	28	28	28	28	28
Exports[j]	kt pw	253	308	273	290	308	326	331	341	353	364	378	389
Price[b, j]	NZD/100 kg	435	325	314	332	355	365	366	376	381	387	393	399

For notes, see end of the table.

Annex Table 26. **CHEESE PROJECTIONS** (cont.)

Calendar year[a]		Average 1998-02	2003 est.	2004	2005	2006	2007	2008	2009	2010	2011	2012	2013
UNITED STATES													
Production	kt pw	3 674	3 930	4 052	4 174	4 282	4 395	4 537	4 664	4 790	4 929	5 055	5 164
Consumption	kt pw	3 795	4 090	4 215	4 331	4 438	4 549	4 692	4 818	4 945	5 083	5 210	5 318
Imports	kt pw	190	210	210	210	210	212	214	216	218	220	222	224
Exports	kt pw	46	52	52	53	54	55	57	59	61	63	65	67
Price[k]	USD/100 kg	298	287	274	285	298	296	296	296	296	296	296	298
OTHER OECD[l]													
Production	kt pw	497	503	513	523	533	543	552	563	573	583	594	602
Consumption	kt pw	453	462	472	481	491	501	511	521	532	543	554	565
Net trade	kt pw	45	41	41	42	42	43	41	42	41	40	40	37
ARGENTINA													
Production	kt pw	433	420	426	450	477	497	515	533	553	573	593	614
Consumption	kt pw	416	392	419	443	459	476	490	504	517	533	549	566
Net trade	kt pw	16	19	7	7	18	21	25	30	36	41	45	49
Price[m]	ARS/100 kg	307	421	437	449	466	472	478	480	481	487	492	497
BRAZIL													
Production	kt pw	446	481	517	523	535	546	556	568	583	599	612	626
Consumption	kt pw	460	484	526	531	543	555	566	578	596	616	632	653
Net trade	kt pw	−14	−4	−9	−7	−9	−8	−10	−10	−12	−17	−20	−27
CHINA													
Production	kt pw	200	264	269	284	309	324	344	365	386	408	432	456
Consumption	kt pw	211	276	281	297	323	337	358	379	400	422	446	471
Imports	kt pw	11	13	13	13	13	13	14	14	14	14	15	15
RUSSIA													
Production	kt pw	227	309	316	305	304	306	311	315	320	326	332	339
Consumption	kt pw	330	457	473	489	514	533	557	579	601	622	643	664
Imports	kt pw	106	161	167	194	219	237	257	274	291	306	321	335
OIS													
Production	kt pw	80	129	127	131	136	140	144	148	152	157	161	165
Consumption	kt pw	64	92	92	93	93	93	93	93	93	94	94	94
Net trade	kt pw	16	37	35	39	43	47	51	55	59	63	67	71
REST OF WORLD													
Production	kt pw	..	1 996	2 072	2 147	2 228	2 298	2 375	2 457	2 544	2 636	2 733	2 835
Consumption	kt pw	..	2 307	2 250	2 303	2 356	2 421	2 469	2 540	2 613	2 689	2 772	2 853
Net trade	kt pw	..	−305	−179	−157	−129	−123	−94	−84	−69	−54	−40	−19

a) Year ended 30 June for Australia and 31 May for New Zealand.
b) Average export price, f.o.b.
c) Industry price of cheddar cheese.
d) Excludes intra-EU-25 trade.
e) Producer price.
f) Includes cheese produced from natural cheese imports.
g) Includes natural cheese imports.
h) Average import price, natural cheese, c.i.f.
i) Value of production divided by volume of production.
j) Year ended 30 June.
k) Average wholesale price, American cheese, 40 lb blocks, f.o.b., Winconsin.
l) Includes Norway, Switzerland and Turkey. Excludes Iceland.
m) Wholesale price (precios mayoristas).
est.: estimate.

Source: OECD Secretariat.

Annex Table 27. **SKIM MILK POWDER PROJECTIONS**

Calendar year[a]		Average 1998-02	2003 est.	2004	2005	2006	2007	2008	2009	2010	2011	2012	2013
AUSTRALIA													
Production	kt pw	242	192	183	206	217	224	233	241	246	251	256	261
Consumption	kt pw	42	21	43	55	55	56	58	59	61	62	64	65
Exports	kt pw	206	180	141	154	165	171	179	185	188	192	195	198
Price[b]	AUD/100 kg	274	225	256	282	297	322	332	338	343	349	355	361
CANADA													
Production	kt pw	79	85	78	77	79	78	78	79	81	83	85	87
Consumption	kt pw	42	49	68	67	70	69	69	70	72	73	75	77
Exports	kt pw	40	32	11	10	10	10	10	9	9	9	9	9
Price[c]	CAD/100 kg	467	515	543	546	557	561	568	572	579	585	590	596
EU-25													
Production	kt pw	..	1 347	1 269	1 162	1 107	1 108	1 065	1 055	1 041	1 030	1 018	1 006
EU-15	kt pw	1 084	1 074	993	887	834	836	794	784	771	761	749	739
EU-10	kt pw	..	273	276	274	273	272	272	271	270	269	268	267
Consumption	kt pw	..	992	999	1 012	1 018	1 017	1 017	1 014	995	980	965	954
EU-15	kt pw	916	896	903	915	920	918	918	914	895	879	863	852
EU-10	kt pw	..	96	97	97	98	99	99	100	100	101	102	102
Net trade[d]	kt pw	..	184	176	136	91	83	66	59	55	47	45	45
Closing stocks	kt pw	124	367	461	475	473	481	464	446	437	441	449	456
Intervention stocks	kt pw	106	185	184	146	101	69	25	0	0	0	0	0
Price[e]	EUR/100 kg	222	203	194	191	187	183	185	186	187	188	188	188
JAPAN													
Production	kt pw	188	176	177	177	176	175	175	174	174	173	173	173
Consumption	kt pw	232	213	224	219	218	220	221	227	229	233	239	246
Imports	kt pw	49	44	45	46	47	48	50	52	56	60	66	73
Price[f]	'000 JPY/100 kg	18	15	16	17	18	18	18	19	19	20	20	21
KOREA													
Production	kt pw	22	36	36	36	36	36	36	37	37	37	37	37
Consumption	kt pw	25	41	41	41	41	39	39	39	39	39	39	39
Imports	kt pw	4	4	4	4	4	4	4	3	3	3	3	3
MEXICO													
Production	kt pw	27	31	31	32	33	33	34	35	35	36	37	37
Consumption	kt pw	138	145	147	151	155	159	164	169	173	178	183	187
Imports	kt pw	110	116	118	121	124	128	132	136	140	144	148	152
Price[g]	MXN/100 kg	1 849	2 518	2 615	2 576	2 508	2 562	2 587	2 614	2 625	2 637	2 633	2 633
NEW ZEALAND													
Production	kt pw	273	438	459	496	515	520	534	551	565	581	595	609
Consumption	kt pw	30	35	35	35	34	34	34	34	34	34	34	34
Exports[h]	kt pw	238	379	423	461	479	484	498	515	530	545	559	571
Price[b, h]	NZD/100 kg	360	326	324	341	362	367	369	373	376	377	378	381

For notes, see end of the table.

OECD AGRICULTURAL OUTLOOK: 2004-2013 – ISBN 92-64-02008-X – © OECD 2004

Annex Table 27. **SKIM MILK POWDER PROJECTIONS** (cont.)

Calendar year[a]		Average 1998-02	2003 est.	2004	2005	2006	2007	2008	2009	2010	2011	2012	2013
UNITED STATES													
Production	kt pw	629	670	638	647	650	652	648	628	618	607	592	575
Consumption[i]	kt pw	403	595	601	589	593	574	535	504	479	491	495	499
Exports	kt pw	137	152	152	153	152	143	143	144	144	122	101	81
Closing stocks	kt pw	280	450	340	250	160	100	75	60	60	60	60	60
Price[j]	USD/100 kg	223	187	187	183	182	181	181	181	183	190	193	196
OTHER OECD[k]													
Production	kt pw	29	29	24	22	19	21	21	20	20	20	19	19
Consumption	kt pw	25	23	23	22	22	22	21	21	21	20	20	20
Net trade	kt pw	5	5	1	−1	−3	−1	−1	−1	−1	−1	−1	−1
ARGENTINA													
Production	kt pw	42	44	36	38	39	40	42	43	45	47	49	51
Consumption	kt pw	23	22	23	25	27	27	28	30	31	32	33	34
Net trade	kt pw	20	12	13	13	12	13	13	14	14	15	16	17
Price[l]	ARS/100 kg	300	468	478	483	491	501	505	508	510	513	517	522
BRAZIL													
Production	kt pw	61	69	75	69	66	65	65	65	68	68	67	68
Consumption	kt pw	93	78	88	94	97	100	103	106	107	111	115	118
Net trade	kt pw	−30	−9	−14	−24	−31	−35	−38	−40	−39	−43	−47	−51
CHINA													
Consumption	kt pw	78	111	118	122	126	133	144	150	156	164	171	179
Imports	kt pw	19	27	31	33	33	36	41	43	45	47	50	52
RUSSIA													
Production	kt pw	139	126	128	128	130	131	133	135	137	139	141	144
Consumption	kt pw	167	165	168	171	173	175	178	180	182	185	187	190
Imports	kt pw	52	63	66	68	68	70	70	70	71	71	71	71
Exports	kt pw	25	25	25	25	25	25	25	25	25	25	25	25
OIS													
Production	kt pw	56	58	66	73	81	89	99	108	118	127	136	146
Consumption	kt pw	17	25	25	25	25	25	25	25	25	25	26	26
Net trade	kt pw	39	33	41	48	56	63	74	83	92	101	111	120
REST OF WORLD													
Production	kt pw	..	253	285	329	352	386	405	426	453	484	517	564
Consumption	kt pw	..	1 001	980	1 022	1 022	1 048	1 068	1 105	1 147	1 162	1 183	1 218
Net trade	kt pw	..	−726	−694	−694	−670	−662	−664	−679	−694	−678	−666	−655

a) Year ended 30 June for Australia and 31 May for New Zealand.
b) Average export price, f.o.b.
c) Average wholesale price.
d) Excludes intra-EU-25 trade.
e) Producer price.
f) Unit import price for feed use.
g) Average import price c.i.f., SMP and WMP.
h) Year ended 31 June.
i) Excludes domestic feed use.
j) Average wholesale price, non-fat dry milk, f.o.b., Central States.
k) Includes Norway, Switzerland and Turkey. Excludes Iceland.
l) Wholesale price (precios mayoristas).
est.: estimate.
Source: OECD Secretariat.

Annex Table 28. **WHEY POWDER AND CASEIN PROJECTIONS**

Calendar year		Average 1998-02	2003 est.	2004	2005	2006	2007	2008	2009	2010	2011	2012	2013
AUSTRALIA													
Net trade, whey	kt pw	37.6	76.4	42.4	39.5	42.2	42.1	43.5	45.2	46.8	48.2	49.7	50.8
Exports, casein	kt pw	10.8	8.5	9.9	10.2	11.6	12.6	13.9	15.1	16.2	17.2	18.2	19.2
CANADA													
Net trade, whey	kt pw	−6.8	−7.0	−7.0	−7.0	−7.0	−7.0	−7.0	−7.0	−7.0	−7.0	−7.0	−7.0
EU-25													
Net trade, whey	kt pw	199.3	235.1	233.9	239.9	235.3	235.3	232.3	231.2	229.7	228.2	226.6	225.1
Casein EU-15													
production	kt pw	154.3	170.1	176.7	181.9	188.0	194.8	200.8	207.1	213.5	219.8	226.5	232.8
consumption	kt pw	146.3	158.1	163.5	169.5	174.5	179.4	185.5	190.7	196.2	201.6	206.8	212.4
net trade	kt pw	8.5	12.0	13.2	12.4	13.6	15.5	15.3	16.4	17.3	18.2	19.6	20.5
JAPAN													
Net trade, whey	kt pw	−35.3	−42.3	−46.5	−51.0	−56.1	−61.6	−67.1	−73.6	−80.5	−88.1	−96.4	−105.5
Imports, casein	kt pw	9.1	6.3	6.1	6.0	5.7	5.5	5.1	4.8	4.5	4.2	3.9	3.7
Import price, casein	'000 JPY/100 kg	56.6	53.8	54.2	51.9	50.8	49.8	46.5	45.0	43.2	41.5	40.1	38.4
KOREA													
Net trade, whey	kt pw	−33.5	−42.7	−41.3	−40.0	−38.6	−38.6	−38.6	−38.6	−38.6	−38.6	−38.6	−38.6
MEXICO													
Net trade, whey	kt pw	−62.2	−70.7	−71.0	−71.4	−72.3	−73.3	−74.3	−75.2	−76.1	−77.1	−78.0	−78.8
NEW ZEALAND													
Net trade, whey	kt pw	7.0	9.6	9.6	9.6	9.6	9.6	9.6	9.6	9.6	9.6	9.6	9.6
Exports, casein	kt pw	104.3	146.4	143.6	151.3	156.3	158.8	164.7	169.5	174.8	180.1	185.0	190.4
Export price, casein	USD/100 kg	434.4	359.6	396.3	411.6	425.9	440.1	436.5	437.1	435.1	434.1	435.3	433.0
UNITED STATES													
Whey													
production	kt pw	516.4	524.4	551.6	580.1	602.6	616.9	632.2	646.8	661.6	676.4	691.2	705.9
consumption	kt pw	363.7	359.8	400.4	442.6	470.6	483.9	497.5	510.9	524.4	537.9	551.4	564.9
exports	kt pw	160.5	173.2	159.8	146.2	140.6	141.6	143.2	144.4	145.7	147.0	148.3	149.5
price[a]	USD/100 kg	47.0	45.8	47.1	45.9	45.8	46.0	47.1	47.7	47.7	47.5	47.5	47.6
Imports, casein	kt pw	70.2	69.4	69.4	70.0	70.0	69.7	70.2	70.3	70.3	70.3	70.0	69.9
ARGENTINA													
Net trade, whey	kt pw	13.3	30.2	31.2	32.2	33.2	34.2	35.2	36.2	37.2	38.2	39.2	40.2
BRAZIL													
Net trade, whey	kt pw	−35.4	−34.8	−34.8	−34.8	−34.8	−34.8	−34.8	−34.8	−34.8	−34.8	−34.8	−34.8
CHINA													
Net trade, whey	kt pw	−113.4	−142.6	−148.8	−153.8	−156.3	−162.6	−166.9	−171.9	−177.8	−183.4	−189.7	−196.4
RUSSIA													
Net trade, whey	kt pw	−6.0	−9.1	−9.1	−9.1	−9.1	−9.1	−9.1	−9.1	−9.1	−9.1	−9.1	−9.1
OIS													
Net trade, whey	kt pw	1.3	1.6	1.6	1.6	1.6	1.6	1.6	1.6	1.6	1.6	1.6	1.6
REST OF WORLD[b]													
Net trade, whey	kt pw	−127.7	−201.8	−146.7	−123.1	−115.5	−104.6	−97.1	−86.8	−75.2	−63.4	−49.9	−35.4

a) Wholesale price, edible dry whey, Wisconsin, plant.
b) Excluding OIS.
est.: estimate.

Source: OECD Secretariat.

Annex Table 29. **WHOLE MILK POWDER PROJECTIONS**

Calendar year[a]		Average 1998-02	2003 est.	2004	2005	2006	2007	2008	2009	2010	2011	2012	2013
AUSTRALIA													
Production	kt pw	181	170	156	186	191	198	204	209	213	216	218	220
Consumption	kt pw	44	44	29	38	41	41	41	41	41	41	41	42
Exports	kt pw	142	142	125	154	156	163	169	174	178	181	183	185
EU-25													
Production	kt pw	933	898	902	906	908	912	914	917	920	923	926	929
EU-15	kt pw	857	814	815	816	817	817	818	819	820	821	821	822
EU-10	kt pw	76	84	87	90	92	95	96	98	100	102	104	106
Consumption	kt pw	330	329	317	313	323	337	344	356	367	377	388	399
EU-15	kt pw	288	282	267	263	271	284	291	302	312	322	331	341
EU-10	kt pw	42	47	49	51	52	53	53	54	55	56	56	57
Exports[b]	kt pw	616	581	597	605	598	587	582	574	566	558	550	542
JAPAN													
Production	kt pw	52	55	53	51	49	47	45	43	41	40	38	37
Consumption	kt pw	52	55	53	51	49	47	45	43	41	40	38	37
MEXICO													
Production	kt pw	140	143	152	159	167	175	183	191	199	208	216	224
Consumption	kt pw	181	181	187	192	197	203	209	215	222	228	234	241
Imports	kt pw	43	40	37	34	32	29	27	25	23	21	19	17
Price[c]	MXN/100 kg	4 728	5 401	5 478	5 765	6 007	6 197	6 365	6 508	6 650	6 777	6 921	7 067
NEW ZEALAND													
Production	kt pw	405	573	593	638	656	678	705	723	740	758	781	811
Consumption	kt pw	9	4	4	4	4	4	4	4	4	4	4	4
Exports[d]	kt pw	398	569	589	634	652	675	701	719	736	754	777	807
Price[e]	NZD/100 kg	377	296	298	314	330	340	342	341	339	337	335	336
UNITED STATES													
Production	kt pw	42	17	16	16	16	16	16	16	16	16	16	16
Consumption	kt pw	39	19	18	18	18	18	18	18	18	18	18	18
Exports	kt pw	6	0	0	0	0	0	0	0	0	0	0	0
Closing stocks	kt pw	2	1	1	1	1	1	1	1	1	1	1	1
OTHER OECD[f]													
Production	kt pw	19	29	26	26	29	30	32	33	34	35	36	38
Consumption	kt pw	20	28	30	31	32	33	34	36	37	38	40	41
Net trade	kt pw	−2	1	−3	−4	−3	−3	−3	−3	−3	−3	−3	−4
ARGENTINA													
Production	kt pw	211	140	148	154	160	165	171	178	186	194	202	211
Consumption	kt pw	94	61	64	68	72	74	77	80	83	86	89	92
Net trade	kt pw	117	79	84	86	88	90	94	98	103	108	114	119
Price[g]	ARS/100 kg	362	488	501	513	522	531	535	540	542	546	551	555
BRAZIL													
Production	kt pw	271	296	318	318	322	327	333	340	350	359	366	374
Consumption	kt pw	364	331	370	387	402	416	430	441	453	467	479	496
Net trade	kt pw	−92	−35	−52	−68	−80	−89	−97	−101	−103	−108	−114	−122
CHINA													
Consumption	kt pw	545	670	702	733	759	791	826	864	899	935	973	1 012
Exports	kt pw	14	10	10	10	10	10	10	10	10	10	10	10
Imports	kt pw	54	73	76	75	78	82	85	90	94	98	102	104

For notes, see end of the table.

Annex Table 29. **WHOLE MILK POWDER PROJECTIONS** (cont.)

Calendar year[a]		Average 1998-02	2003 est.	2004	2005	2006	2007	2008	2009	2010	2011	2012	2013
RUSSIA													
Production	kt pw	83	92	94	92	93	93	95	96	98	100	101	103
Consumption	kt pw	100	103	105	107	108	109	111	112	113	114	115	115
Imports	kt pw	20	14	14	17	18	19	19	19	18	17	16	15
OIS													
Production	kt pw	12	13	17	21	25	26	27	28	29	31	32	33
Consumption	kt pw	7	8	7	7	7	7	6	6	6	6	5	5
Net trade	kt pw	5	5	9	14	18	20	21	22	24	25	26	28
REST OF WORLD													
Production	kt pw	..	511	528	545	563	578	594	611	629	648	667	687
Consumption	kt pw	..	1 711	1 732	1 821	1 846	1 873	1 912	1 940	1 974	2 006	2 042	2 086
Net trade	kt pw	..	−1 200	−1 205	−1 276	−1 284	−1 294	−1 318	−1 329	−1 345	−1 359	−1 376	−1 398

a) Year ended 30 June for Australia and 31 May for New Zealand.
b) Excludes intra-EU-25 trade.
c) Value of production divided by volume of production.
d) Including exports of other dairy products made from WMP.
e) Export price.
f) Includes Korea, Norway, Switzerland and Turkey. Excludes Iceland.
g) Wholesale price (precios mayoristas).
est.: estimate.

Source: OECD Secretariat.

Annex Table 30. **DAIRY PER CAPITA CONSUMPTION PROJECTIONS**[a]

Calendar year[b]		Average 1998-02	2003 est.	2004	2005	2006	2007	2008	2009	2010	2011	2012	2013
AUSTRALIA													
Milk	L/person	100.2	96.5	95.2	95.1	95.9	94.1	93.2	92.9	92.6	92.4	92.2	91.8
Butter	kg/person	4.2	2.3	3.0	3.6	3.2	3.3	3.3	3.3	3.3	3.3	3.4	3.4
Cheese	kg/person	10.6	9.4	11.2	9.2	9.6	9.2	9.1	9.2	9.3	9.3	9.4	9.4
SMP	kg/person	2.2	1.1	2.2	2.7	2.7	2.7	2.8	2.8	2.9	3.0	3.0	3.1
WMP	kg/person	2.3	2.2	1.4	1.9	2.0	2.0	2.0	2.0	2.0	2.0	2.0	1.9
CANADA													
Milk	L/person	91.4	90.2	89.5	88.8	88.9	89.2	89.2	89.3	89.2	89.2	89.3	89.4
Butter	kg/person	2.8	2.8	2.5	2.6	2.6	2.5	2.5	2.5	2.5	2.5	2.5	2.5
Cheese	kg/person	10.9	11.6	11.7	11.8	11.9	12.0	12.1	12.2	12.3	12.4	12.5	12.6
SMP	kg/person	1.4	1.6	2.1	2.1	2.2	2.1	2.1	2.1	2.2	2.2	2.3	2.3
EU-25													
Milk	L/person	..	71.8	72.3	72.5	72.5	72.7	72.9	73.1	73.3	73.5	73.7	73.9
Butter	kg/person	..	4.3	4.3	4.4	4.4	4.4	4.4	4.4	4.4	4.3	4.3	4.3
Cheese	kg/person	..	17.3	17.6	17.8	18.1	18.3	18.5	18.6	18.8	18.9	19.1	19.2
SMP	kg/person	..	2.2	2.2	2.2	2.2	2.2	2.2	2.2	2.2	2.1	2.1	2.1
WMP	kg/person	0.7	0.7	0.7	0.7	0.7	0.7	0.8	0.8	0.8	0.8	0.8	0.9
EU-15													
Milk	L/person	74.8	72.0	72.6	72.7	72.8	72.8	72.9	73.0	73.1	73.1	73.1	73.1
Butter	kg/person	4.6	4.4	4.4	4.5	4.6	4.6	4.5	4.5	4.5	4.4	4.4	4.3
Cheese	kg/person	17.7	18.4	18.6	18.8	19.1	19.1	19.3	19.3	19.4	19.4	19.4	19.4
SMP	kg/person	2.4	2.3	2.4	2.4	2.4	2.4	2.4	2.3	2.3	2.2	2.2	2.2
WMP	kg/person	0.8	0.7	0.7	0.7	0.7	0.7	0.7	0.8	0.8	0.8	0.8	0.9
EU-10													
Milk	L/person	..	71.2	70.4	71.2	71.2	72.0	72.7	73.5	74.5	75.7	77.1	78.6
Butter	kg/person	..	3.5	3.5	3.6	3.7	3.7	3.8	3.8	3.8	3.9	4.0	4.1
Cheese	kg/person	..	11.7	12.1	12.6	13.1	13.6	14.2	14.8	15.4	16.2	16.9	17.8
SMP	kg/person	..	1.3	1.3	1.3	1.4	1.4	1.4	1.4	1.4	1.5	1.5	1.5
WMP	kg/person	..	0.6	0.7	0.7	0.7	0.7	0.8	0.8	0.8	0.8	0.8	0.9
JAPAN													
Milk	L/person	38.1	39.0	39.2	39.2	39.4	39.4	39.7	39.8	40.1	40.3	40.6	40.8
Butter	kg/person	0.7	0.7	0.7	0.6	0.7	0.7	0.7	0.7	0.7	0.7	0.7	0.7
Cheese	kg/person	1.8	1.9	1.9	2.0	2.0	2.0	2.1	2.1	2.2	2.2	2.2	2.3
SMP	kg/person	1.8	1.7	1.8	1.7	1.7	1.7	1.7	1.8	1.8	1.8	1.9	2.0
WMP	kg/person	0.4	0.4	0.4	0.4	0.4	0.4	0.4	0.3	0.3	0.3	0.3	0.3
KOREA													
Milk	L/person	27.5	27.6	27.6	27.9	27.5	27.9	27.5	27.9	27.5	28.2	27.6	28.3
Butter	kg/person	0.1	0.1	0.1	0.1	0.1	0.1	0.1	0.1	0.1	0.1	0.1	0.1
Cheese	kg/person	0.9	1.2	1.4	1.5	1.6	1.7	1.8	1.8	1.9	2.0	2.0	2.1
SMP	kg/person	0.5	0.9	0.8	0.8	0.8	0.8	0.8	0.8	0.8	0.8	0.8	0.8
MEXICO													
Milk[c]	L/person	31.0	33.4	34.2	35.1	36.0	36.9	37.9	38.7	39.6	40.4	41.2	42.0
Butter	kg/person	0.5	0.5	0.5	0.6	0.6	0.6	0.6	0.7	0.7	0.7	0.7	0.8
Cheese	kg/person	1.8	2.0	2.1	2.1	2.1	2.2	2.3	2.3	2.4	2.4	2.4	2.5
SMP	kg/person	1.4	1.4	1.4	1.4	1.4	1.5	1.5	1.5	1.5	1.6	1.6	1.6
WMP	kg/person	1.8	1.8	1.8	1.8	1.8	1.9	1.9	1.9	2.0	2.0	2.0	2.1
NEW ZEALAND													
Milk	L/person	85.4	83.4	82.5	81.8	81.3	80.9	80.4	79.9	79.5	79.1	78.7	78.3
Butter	kg/person	8.1	8.0	7.9	7.8	7.7	7.7	7.6	7.6	7.6	7.5	7.5	7.5
Cheese	kg/person	7.2	7.1	7.0	7.0	6.9	6.9	6.9	6.8	6.8	6.8	6.7	6.7
SMP	kg/person	7.8	8.8	8.7	8.7	8.6	8.6	8.5	8.5	8.4	8.4	8.3	8.3
WMP	kg/person	2.3	0.9	0.9	0.9	0.9	0.9	0.9	0.9	0.9	0.9	0.9	0.9

For notes, see end of the table.

Annex Table 30. **DAIRY PER CAPITA CONSUMPTION PROJECTIONS**[a] *(cont.)*

Calendar year[b]		Average 1998-02	2003 est.	2004	2005	2006	2007	2008	2009	2010	2011	2012	2013
UNITED STATES													
Milk	L/person	89.8	87.0	84.7	84.2	83.6	83.0	82.3	81.6	80.8	80.2	79.5	78.9
Butter	kg/person	2.0	2.0	2.0	2.0	1.9	1.9	1.9	1.9	1.9	1.9	1.9	1.9
Cheese	kg/person	13.4	14.0	14.4	14.6	14.9	15.2	15.6	15.9	16.2	16.5	16.8	17.1
SMP[d]	kg/person	1.4	2.0	2.0	2.0	2.0	1.9	1.8	1.7	1.6	1.6	1.6	1.6
WMP	kg/person	0.1	0.1	0.1	0.1	0.1	0.1	0.1	0.1	0.1	0.1	0.1	0.1
OTHER OECD[e]													
Milk[c]	L/person	35.0	38.0	37.8	37.7	37.6	37.4	37.4	37.3	37.2	37.1	37.1	37.0
Butter	kg/person	2.4	2.1	2.1	2.1	2.1	2.1	2.1	2.1	2.1	2.1	2.1	2.1
Cheese	kg/person	5.7	5.6	5.7	5.7	5.8	5.8	5.9	6.0	6.0	6.1	6.2	6.2
SMP	kg/person	0.3	0.3	0.3	0.3	0.3	0.3	0.2	0.2	0.2	0.2	0.2	0.2
WMP	kg/person	0.2	0.2	0.2	0.2	0.2	0.2	0.3	0.3	0.3	0.3	0.3	0.3
ARGENTINE													
Milk	L/person	40.8	32.3	34.5	35.5	36.5	36.9	37.6	38.5	39.3	40.2	41.1	42.0
Butter	kg/person	1.1	1.0	1.1	1.1	1.2	1.2	1.2	1.2	1.3	1.3	1.3	1.4
Cheese	kg/person	11.2	10.2	10.8	11.3	11.6	11.9	12.2	12.4	12.6	12.9	13.2	13.5
SMP	kg/person	0.6	0.6	0.6	0.6	0.7	0.7	0.7	0.7	0.7	0.8	0.8	0.8
WMP	kg/person	2.5	1.6	1.7	1.7	1.8	1.9	1.9	2.0	2.0	2.1	2.1	2.2
BRAZIL													
Butter	kg/person	0.5	0.5	0.5	0.6	0.6	0.6	0.6	0.6	0.6	0.6	0.6	0.6
Cheese	kg/person	2.7	2.7	2.9	2.9	3.0	3.0	3.0	3.1	3.1	3.2	3.3	3.3
SMP	kg/person	0.5	0.4	0.5	0.5	0.5	0.5	0.6	0.6	0.6	0.6	0.6	0.6
WMP	kg/person	2.1	1.9	2.1	2.1	2.2	2.3	2.3	2.3	2.4	2.4	2.5	2.5
CHINA													
Milk[f]	L/person	4.3	7.7	7.8	8.2	8.9	9.3	9.9	10.5	11.1	11.7	12.4	13.1
Butter	kg/person	0.1	0.1	0.1	0.1	0.1	0.1	0.1	0.1	0.1	0.1	0.1	0.2
Cheese	kg/person	0.2	0.2	0.2	0.2	0.2	0.3	0.3	0.3	0.3	0.3	0.3	0.3
SMP	kg/person	0.1	0.1	0.1	0.1	0.1	0.1	0.1	0.1	0.1	0.1	0.1	0.1
WMP	kg/person	0.4	0.5	0.5	0.6	0.6	0.6	0.6	0.6	0.7	0.7	0.7	0.7
RUSSIA													
Butter	kg/person	2.5	2.8	2.8	2.9	2.9	3.0	3.0	3.1	3.1	3.2	3.2	3.2
Cheese	kg/person	2.3	3.2	3.3	3.4	3.6	3.7	3.9	4.1	4.3	4.4	4.6	4.8
SMP	kg/person	1.1	1.1	1.2	1.2	1.2	1.2	1.3	1.3	1.3	1.3	1.3	1.4
WMP	kg/person	0.7	0.7	0.7	0.7	0.8	0.8	0.8	0.8	0.8	0.8	0.8	0.8
OIS													
Butter	kg/person	0.7	0.7	0.7	0.7	0.7	0.7	0.7	0.7	0.7	0.7	0.7	0.7
Cheese	kg/person	0.5	0.7	0.7	0.7	0.7	0.7	0.7	0.7	0.7	0.7	0.7	0.7
SMP	kg/person	0.1	0.2	0.2	0.2	0.2	0.2	0.2	0.2	0.2	0.2	0.2	0.2
WMP	kg/person	0.0	0.1	0.1	0.1	0.0	0.0	0.0	0.0	0.0	0.0	0.0	0.0
REST OF WORLD[g]													
Milk[f]	L/person	..	28.1	28.2	28.3	28.5	28.7	29.0	29.2	29.5	29.7	30.0	30.3
Butter	kg/person	..	1.3	1.3	1.3	1.4	1.4	1.4	1.4	1.4	1.5	1.5	1.5
Cheese	kg/person	..	0.7	0.7	0.7	0.7	0.7	0.7	0.7	0.7	0.7	0.7	0.7
SMP	kg/person	..	0.3	0.3	0.3	0.3	0.3	0.3	0.3	0.3	0.3	0.3	0.3
WMP	kg/person	0.0	0.5	0.5	0.5	0.5	0.5	0.5	0.5	0.5	0.5	0.5	0.5

a) Milk excludes on farm use.
b) Year ended 30 June for Australia and 31 May for New Zealand.
c) In Mexico, Switzerland and Turkey on farm use is large.
d) Excludes feed use.
e) Includes Norway, Switzerland and Turkey (and Korea for WMP). Excludes Iceland.
f) Fluid milk and other dairy products not specified.
g) Excludes OIS.
est.: estimate.

Source: OECD Secretariat.

Annex Table 31. **OTHER SELECTED COUNTRIES' PROJECTIONS: BUTTER AND SMP**

Calendar year		Average 1998-02	2003 est.	2004	2005	2006	2007	2008	2009	2010	2011	2012	2013
BUTTER													
NORWAY[b]													
Production	kt	15	13	13	13	13	13	13	13	13	13	12	12
Consumption	kt	12	11	11	11	11	12	12	12	12	12	12	12
Exports	kt	3	2	2	2	2	2	2	2	2	1	1	1
Imports	kt	0	0	0	0	0	0	1	1	1	1	1	1
SWITZERLAND[b]													
Production	kt	40	32	32	32	32	32	32	32	32	31	31	31
Consumption	kt	44	39	38	38	38	37	37	37	37	36	36	36
Exports	kt	0	0	0	0	0	0	0	0	0	0	0	0
Imports	kt	5	6	6	6	6	5	5	5	5	5	5	5
TURKEY[a]													
Production	kt	129	120	122	124	126	128	130	132	134	136	138	140
Consumption	kt	133	125	127	129	131	133	135	137	139	141	143	145
Exports	kt	0	0	0	0	0	0	0	0	0	0	0	0
Imports	kt	4	5	5	5	5	5	5	5	5	5	5	5
SMP													
NORWAY[b]													
Production	kt	4	4	4	3	3	3	3	3	3	2	2	2
Consumption	kt	4	4	4	3	3	3	3	3	3	2	2	2
Exports	kt	0	0	0	0	0	0	0	0	0	0	0	0
Imports	kt	0	0	0	0	0	0	0	0	0	0	0	0
SWITZERLAND[b]													
Production	kt	25	25	20	18	16	18	18	18	18	17	17	17
Consumption	kt	16	15	15	14	14	14	14	14	14	13	13	13
Exports	kt	9	10	6	4	2	4	4	4	4	4	4	4
Imports	kt	0	0	0	0	0	0	0	0	0	0	0	0
TURKEY[a]													
Production	kt	0	0	0	0	0	0	0	0	0	0	0	0
Consumption	kt	5	5	5	5	5	5	5	5	5	5	5	5
Exports	kt	0	0	0	0	0	0	0	0	0	0	0	0
Imports	kt	5	5	5	5	5	5	5	5	5	5	5	5

a) Questionnaire response, summer 2003.
b) OECD estimates.
est.: estimate.

Source: OECD Secretariat.

Annex Table 32. **OTHER SELECTED COUNTRIES' PROJECTIONS: WHEAT**

Crop year[a]		Average 1998-02	03/04 est.	04/05	05/06	06/07	07/08	08/09	09/10	10/11	11/12	12/13	13/14
NORWAY[c]													
Production	kt	280	280	280	281	282	283	284	285	286	287	288	289
Consumption	kt	549	550	551	554	555	556	557	558	560	561	562	563
Feed use	kt	80	80	80	80	80	80	80	80	80	80	80	80
Food	kt	469	470	471	474	475	476	477	478	480	481	482	483
Exports	kt	0	0	0	0	0	0	0	0	0	0	0	0
Imports	kt	259	270	271	273	273	273	273	274	274	274	274	274
Ending stocks	kt	259	200	200	200	200	200	200	200	200	200	200	200
SWITZERLAND[c]													
Production	kt	539	427	507	509	511	512	514	515	517	519	520	522
Consumption	kt	805	769	772	779	785	791	798	804	811	817	824	830
Feed use	kt	205	290	220	220	220	220	220	220	220	220	220	220
Food	kt	600	479	552	559	565	571	578	584	591	597	604	610
Exports	kt	38	38	38	38	38	38	38	38	38	38	38	38
Imports	kt	290	333	318	298	306	311	315	320	325	330	335	339
Ending stocks	kt	717	634	649	640	633	626	620	613	606	600	593	586
TURKEY[b]													
Production	kt	16 288	16 000	16 289	16 578	16 871	17 170	17 473	17 773	18 073	18 374	18 675	18 975
Consumption	kt	16 587	16 973	17 211	17 452	17 696	17 944	18 195	18 295	18 592	18 820	19 049	19 277
Feed use	kt	1 230	1 086	1 086	1 086	1 086	1 086	1 086	1 086	1 086	1 086	1 086	1 086
Food	kt	15 357	15 887	16 125	16 366	16 610	16 858	17 109	17 209	17 505	17 734	17 962	18 191
Exports	kt	1 845	4 222	4 289	4 357	4 427	4 498	4 571	4 642	4 714	4 787	4 859	4 931
Imports	kt	1 929	5 195	3 420	3 420	3 420	3 420	3 420	3 420	3 420	3 420	3 420	3 420
Ending stocks	kt	818	500	480	440	380	299	197	223	181	139	97	55

a) Beginning crop marketing year – see the Glossary of Terms for definitions.
b) Questionnaire response, summer 2003.
c) OECD estimates.
est.: estimate.

Source: OECD Secretariat.

OECD AGRICULTURAL OUTLOOK: 2004-2013 – ISBN 92-64-02008-X – © OECD 2004

Annex Table 33. **OTHER SELECTED COUNTRIES' PROJECTIONS: COARSE GRAINS**

Crop year[a]		Average 1998-02	03/04 est.	04/05	05/06	06/07	07/08	08/09	09/10	10/11	11/12	12/13	13/14
NORWAY[c]													
Production	kt	1 004	1 061	1 061	1 061	1 061	1 061	1 061	1 061	1 061	1 061	1 061	1 061
Consumption	kt	1 167	1 216	1 217	1 217	1 218	1 219	1 219	1 220	1 221	1 221	1 222	1 223
Feed use	kt	1 047	1 100	1 100	1 100	1 100	1 100	1 100	1 100	1 100	1 100	1 100	1 100
Food	kt	119	116	117	117	118	119	119	120	121	121	122	123
Exports	kt	0	0	0	0	0	0	0	0	0	0	0	0
Imports	kt	130	130	156	156	157	158	158	159	160	160	161	162
Ending stocks	kt	255	228	228	228	228	228	228	228	228	228	228	228
SWITZERLAND[c]													
Production	kt	589	366	539	537	533	525	527	526	524	518	516	516
Consumption	kt	787	760	687	682	682	688	687	677	678	678	678	678
Feed use	kt	597	644	593	591	591	599	600	590	591	590	590	590
Food	kt	190	116	94	92	91	89	87	87	87	88	88	88
Exports	kt	9	7	7	7	7	7	7	7	7	7	7	7
Imports	kt	213	353	208	152	152	165	162	153	155	161	163	164
Ending stocks	kt	310	285	337	338	334	329	323	318	313	307	302	297
TURKEY[b]													
Production	kt	10 112	10 220	10 346	10 474	10 604	10 735	10 867	10 999	11 131	11 263	11 395	11 526
Consumption	kt	10 277	10 714	10 825	10 938	11 053	11 168	11 285	11 401	11 517	11 633	11 749	11 865
Feed use	kt	9 073	9 265	9 363	9 462	9 562	9 663	9 765	9 866	9 968	10 070	10 171	10 273
Food	kt	1 204	1 449	1 463	1 477	1 491	1 505	1 520	1 534	1 549	1 563	1 577	1 592
Exports	kt	584	719	719	719	719	719	719	719	719	719	719	719
Imports	kt	879	1 213	1 198	1 183	1 168	1 152	1 136	1 120	1 105	1 089	1 073	1 057
Ending stocks	kt	2 176	2 487	2 487	2 487	2 487	2 487	2 487	2 487	2 487	2 487	2 487	2 487

a) Beginning crop marketing year – see the Glossary of Terms for definitions.
b) Questionnaire response, Summer 2003.
c) OECD estimates.
est.: estimate.

Source: OECD Secretariat.

Annex Table 34. **OTHER SELECTED COUNTRIES' PROJECTIONS: RICE**

Crop year[a]		Average 1998-02	03/04 est.	04/05	05/06	06/07	07/08	08/09	09/10	10/11	11/12	12/13	13/14
NORWAY[c]													
Production	kt	0	0	0	0	0	0	0	0	0	0	0	0
Consumption	kt	0	0	0	0	0	0	0	0	0	0	0	0
Exports	kt	0	0	0	0	0	0	0	0	0	0	0	0
Imports	kt	0	0	0	0	0	0	0	0	0	0	0	0
Ending stocks	kt	0	0	0	0	0	0	0	0	0	0	0	0
SWITZERLAND[c]													
Production	kt	0	0	0	0	0	0	0	0	0	0	0	0
Consumption	kt	55	75	76	77	78	80	81	82	83	84	86	87
Exports	kt	0	0	0	0	0	0	0	0	0	0	0	0
Imports	kt	63	75	76	77	78	80	81	82	83	84	86	87
Ending stocks	kt	18	45	45	45	45	45	45	45	45	45	45	45
TURKEY[b]													
Production	kt	212	220	208	208	208	208	208	208	208	208	208	208
Consumption	kt	540	623	635	648	661	674	687	701	715	729	744	759
Exports	kt	3	0	0	0	0	0	0	0	0	0	0	0
Imports	kt	331	403	427	440	453	466	479	493	507	521	536	551
Ending stocks	kt	45	45	45	45	45	45	45	45	45	45	45	45

a) Beginning crop marketing year – see the Glossary of Terms for definitions.
b) Questionnaire response, Summer 2003.
c) OECD estimates.
est.: estimate.

Source: OECD Secretariat.

Annex Table 35. **OTHER SELECTED COUNTRIES' PROJECTIONS: OILSEEDS**

Crop year[a]		Average 1998-02	03/04 est.	04/05	05/06	06/07	07/08	08/09	09/10	10/11	11/12	12/13	13/14
NORWAY[c]													
Production	kt	11	10	10	11	11	11	11	11	11	12	12	12
Consumption	kt	413	413	415	417	418	420	422	424	426	428	430	432
Crush	kt	407	410	412	413	414	416	417	419	420	422	423	425
Feed	kt	6	2	3	3	4	4	5	5	6	6	7	7
Exports	kt	0	0	0	0	0	0	0	0	0	0	0	0
Imports	kt	403	402	403	405	407	409	411	412	414	416	418	419
Ending stocks	kt	8	9	8	8	7	6	5	4	3	2	2	1
SWITZERLAND[c]													
Production	kt	55	59	68	68	69	70	71	72	73	73	74	75
Consumption	kt	154	141	162	164	166	168	170	172	175	177	179	181
Crush	kt	141	127	148	150	152	154	156	158	160	163	165	167
Feed	kt	13	14	14	14	14	14	14	14	14	14	14	14
Exports	kt	0	0	0	0	0	0	0	0	0	0	0	0
Imports	kt	101	79	98	96	97	98	100	101	102	104	105	107
Ending stocks	kt	5	5	9	9	9	9	9	9	9	9	10	10
TURKEY[b]													
Production	kt	882	926	940	954	968	983	998	1 013	1 028	1 043	1 059	1 075
Consumption	kt	1 661	1 954	1 983	2 013	2 043	2 074	2 105	2 137	2 169	2 201	2 234	2 268
Crush	kt	1 650	1 624	1 648	1 673	1 698	1 724	1 750	1 776	1 802	1 829	1 857	1 885
Feed	kt	10	330	335	340	345	350	356	361	366	372	377	383
Exports	kt	2	2	2	2	2	2	2	2	2	2	2	2
Imports	kt	798	1 030	1 048	1 064	1 080	1 096	1 112	1 129	1 146	1 163	1 181	1 198
Ending stocks	kt	288	187	190	193	196	198	201	204	208	211	214	217

a) Beginning crop marketing year – see the Glossary of Terms for definitions.
b) Questionnaire response, Summer 2003.
c) OECD estimates.
est.: estimate.

Source: OECD Secretariat.

Annex Table 36. **OTHER SELECTED COUNTRIES' PROJECTIONS: OILSEED MEALS**

Marketing year[a]		Average 1998-02	03/04 est.	04/05	05/06	06/07	07/08	08/09	09/10	10/11	11/12	12/13	13/14
NORWAY[c]													
Production	kt	317	319	321	322	323	325	326	328	329	330	332	333
Consumption	kt	230	259	258	256	255	253	251	250	248	246	245	243
Exports	kt	135	129	132	135	138	141	144	147	150	153	156	159
Imports	kt	42	70	70	70	70	70	70	70	70	70	70	70
Ending stocks	kt	24	14	16	17	18	20	21	22	24	25	26	28
SWITZERLAND[c]													
Production	kt	98	85	99	100	102	103	104	106	107	109	110	112
Consumption	kt	232	325	327	330	332	335	338	340	343	346	349	351
Exports	kt	1	2	2	2	2	2	2	2	2	2	2	2
Imports	kt	137	235	230	231	232	234	235	236	237	239	240	241
Ending stocks	kt	8	9	9	9	9	9	9	9	9	9	9	9
TURKEY[b]													
Production	kt	908	934	948	962	976	991	1 006	1 021	1 036	1 052	1 068	1 084
Consumption	kt	1 587	1 877	1 914	1 953	1 992	2 031	2 072	2 114	2 156	2 199	2 243	2 288
Exports	kt	9	20	20	20	20	20	20	20	20	20	20	20
Imports	kt	699	963	987	1 011	1 036	1 061	1 087	1 113	1 140	1 168	1 196	1 225
Ending stocks	kt	43	55	56	57	58	58	59	60	61	62	63	64

a) Beginning crop marketing year – see the Glossary of Terms for definitions.
b) Questionnaire response, Summer 2003.
c) OECD estimates.
est.: estimate.

Source: OECD Secretariat.

Annex Table 37. **OTHER SELECTED COUNTRIES' PROJECTIONS: VEGETABLE OILS**

Marketing year[a]		Average 1998-02	03/04 est.	04/05	05/06	06/07	07/08	08/09	09/10	10/11	11/12	12/13	13/14
NORWAY[c]													
Production	kt	77	77	77	78	78	78	79	79	80	80	81	81
Consumption	kt	86	86	85	85	85	85	85	84	84	84	84	83
Exports	kt	23	25	26	27	27	28	28	29	30	30	31	32
Imports	kt	29	35	35	35	35	35	35	35	35	35	35	35
Ending stocks	kt	9	6	7	8	8	9	10	11	12	13	13	14
SWITZERLAND[c]													
Production	kt	36	34	39	40	40	41	41	42	42	43	44	44
Consumption	kt	76	102	103	104	105	106	107	108	109	110	111	112
Exports	kt	10	1	1	1	1	1	1	1	1	1	1	1
Imports	kt	52	68	64	65	65	66	66	67	67	68	68	69
Ending stocks	kt	7	10	10	10	10	10	10	10	10	10	10	10
TURKEY[b]													
Production	kt	566	514	521	529	537	545	553	562	570	578	587	596
Consumption	kt	1 064	1 101	1 118	1 135	1 152	1 169	1 186	1 204	1 222	1 241	1 259	1 278
Exports	kt	39	16	16	16	16	16	16	16	16	16	16	16
Imports	kt	528	604	613	622	631	640	650	659	669	679	689	699
Ending stocks	kt	39	25	25	26	26	27	27	27	28	28	29	29

a) Beginning crop marketing year – see the Glossary of Terms for definitions.
b) Questionnaire response, Summer 2003.
c) OECD estimates.
est.: estimate.
Source: OECD Secretariat.

Annex Table 38. **MAIN POLICY ASSUMPTIONS FOR SUGAR MARKETS**

Crop year[a]		Average 98/99-02/03	03/04 est.	04/05	05/06	06/07	07/08	08/09	09/10	10/11	11/12	12/13	13/14
ARGENTINA													
Tariff, sugar	ARS/t	35.0	35.0	35.0	35.0	35.0	35.0	35.0	35.0	35.0	35.0	35.0	35.0
BRAZIL													
Tariff, raw sugar	%	43.9	37.2	35.0	35.0	35.0	35.0	35.0	35.0	35.0	35.0	35.0	35.0
Tariff, white sugar	%	56.0	40.0	35.0	35.0	35.0	35.0	35.0	35.0	35.0	35.0	35.0	35.0
CANADA													
Tariff, raw sugar	CAD/t	24.5	24.1	24.1	24.1	24.1	24.1	24.1	24.1	24.1	24.1	24.1	24.1
Tariff, white sugar	CAD/t	36.0	35.4	35.4	35.4	35.4	35.4	35.4	35.4	35.4	35.4	35.4	35.4
CHINA													
TRQ sugar	kt	1 697	1 852	1 954	1 954	1 954	1 954	1 954	1 954	1 954	1 954	1 954	1 954
Tariff, in-quota, raw sugar	%	20.0	20.0	15.0	15.0	15.0	15.0	15.0	15.0	15.0	15.0	15.0	15.0
Tariff, in-quota, white sugar	%	30.0	30.0	20.0	20.0	20.0	20.0	20.0	20.0	20.0	20.0	20.0	20.0
Tariff, over-quota	%	75.0	75.0	50.0	50.0	50.0	50.0	50.0	50.0	50.0	50.0	50.0	50.0
EU													
Intervention price, white sugar	EUR/t	632	632	632	632	632	632	632	632	632	632	632	632
Total quota, white sugar[c]	kt rse	..	15 415	18 957	18 957	18 957	18 957	18 957	18 957	18 957	18 957	18 957	18 957
from A quota	kt rse	..	12 846	16 004	16 004	16 004	16 004	16 004	16 004	16 004	16 004	16 004	16 004
from B quota	kt rse	..	2 569	2 954	2 954	2 954	2 954	2 954	2 954	2 954	2 954	2 954	2 954
Subsidised export limits													
EU-15	kt rse	1 307	1 274	1 274	1 274	1 274	1 274	1 274	1 274	1 274	1 274	1 274	1 274
EU-15	'000 EUR	527 180	499 100	499 100	499 100	499 100	499 100	499 100	499 100	499 100	499 100	499 100	499 100
EU-10	kt rse	178	157	157	157	157	157	157	157	157	157	157	157
EU-10	'000 EUR	35 879	32 560	32 560	32 560	32 560	32 560	32 560	32 560	32 560	32 560	32 560	32 560
Tariff, raw sugar	EUR/t	348	339	339	339	339	339	339	339	339	339	339	339
Tariff, white sugar[b]	EUR/t	430	419	419	419	419	419	419	420	421	422	423	424
INDIA													
Intervention price, sugar cane	INR/t	590	750	750	750	750	750	750	750	750	750	750	750
INDONESIA													
Tariff, white sugar	%	15	25	25	25	25	25	25	25	25	25	25	25
JAPAN													
Minimum stabilisation price, raw sugar	JPY/kg	150	152	152	152	152	152	152	152	152	152	152	152
Tariff, raw sugar	JPY/kg	73.1	71.8	71.8	71.8	71.8	71.8	71.8	71.8	71.8	71.8	71.8	71.8
Tariff, white sugar	JPY/kg	104.9	103.1	103.1	103.1	103.1	103.1	103.1	71.8	71.8	71.8	71.8	71.8
KOREA													
Tariff	%	20.3	18.6	18.0	18.0	18.0	18.0	18.0	18.0	18.0	18.0	18.0	18.0
MEXICO													
Mexico common external tariff, raw sugar	MXN/t	3 509	4 247	4 341	4 341	4 341	4 341	4 341	4 341	4 341	4 341	4 341	4 341
Mexico common external tariff, white sugar	MXN/t	3 598	4 247	4 341	4 341	4 341	4 341	4 341	4 341	4 341	4 341	4 341	4 341
RUSSIA													
Tariff, raw sugar[d]	%	30.2	114.3	153.6	108.9	147.9	147.0	168.6	154.1	123.7	160.0	143.1	133.4
Tariff, white sugar	%	34.8	108.5	144.2	106.4	140.0	139.4	157.7	146.2	120.2	151.0	137.3	129.7
TRQ, raw sugar	kt rse	3 650	3 950	0	0	0	0	0	0	0	0	0	0

For notes, see end of table.

Annex Table 38. **MAIN POLICY ASSUMPTIONS FOR SUGAR MARKETS** (cont.)

Crop year[a]		Average 98/99-02/03	03/04 est.	04/05	05/06	06/07	07/08	08/09	09/10	10/11	11/12	12/13	13/14
UNITED STATES[b]													
Loan rate, cane sugar	USD/t	397	397	397	397	397	397	397	397	397	397	397	397
Loan rate, white sugar	USD/t	504.9	504.9	504.9	504.9	504.9	504.9	504.9	504.9	504.9	504.9	504.9	504.9
TRQ, raw sugar	kt rse	1 122	1 117	1 117	1 117	1 117	1 117	1 117	1 117	1 117	1 117	1 117	1 117
TRQ, refined sugar	kt rse	22	22	22	22	22	22	22	22	22	22	22	22
Raw sugar high tier tariff, over quota	USD/t	345	339	339	339	339	339	339	339	339	339	339	339
White sugar high tier tariff, over quota	USD/t	364	357	357	357	357	357	357	357	357	357	357	357

a) Beginning crop marketing year – see the Glossary of Terms for definitions.
b) Price based special safeguard actions apply.
c) Includes the 10 new member countries from May 2004.
d) Assumes a wholesale price target of USD 470 per tonne as the basis for setting the floating tariff duty.
Note: The source for tariffs (except United States and Russia) is AMAD. The source for Russia and United States tariffs is ERS, USDA.
rse: raw sugar equivalent.
est.: estimate.
Source: OECD Secretariat.

Annex Table 39. **WORLD SUGAR PROJECTIONS (in raw sugar equivalent)**

Crop year[a]		Average 98/99-02/03	03/04 est.	04/05	05/06	06/07	07/08	08/09	09/10	10/11	11/12	12/13	13/14
OECD													
Production	kt rse	41 213	41 595	42 992	42 570	42 789	43 071	43 042	43 140	43 214	43 400	43 617	43 747
Consumption	kt rse	38 540	39 212	39 402	39 356	39 859	40 284	40 040	40 285	40 743	40 906	41 166	41 408
Closing stocks	kt rse	12 182	9 967	10 370	10 647	10 597	10 419	10 498	10 425	10 302	10 389	10 474	10 566
NON-OECD													
Production	kt rse	95 305	103 291	104 035	106 560	111 147	112 104	114 928	117 042	120 603	123 452	125 913	128 905
Consumption	kt rse	94 170	105 878	107 219	110 819	112 856	114 726	117 327	120 372	122 818	125 769	128 114	131 196
Net trade[b]	kt rse	–3 206	–2 323	–3 185	–2 905	–3 044	–2 882	–2 779	–2 819	–2 518	–2 280	–2 255	–2 134
Closing stocks	kt rse	49 946	59 031	59 033	57 679	59 014	59 274	59 654	59 142	59 445	59 407	59 461	59 304
WORLD													
Production	kt rse	137 613	145 951	148 100	150 181	154 945	156 200	159 000	161 213	164 852	167 891	170 574	173 699
Consumption	kt rse	133 827	146 237	147 810	151 330	153 850	156 140	158 510	161 810	164 720	167 840	170 451	173 780
Closing stocks	kt rse	62 568	69 391	69 681	68 532	69 627	69 687	70 177	69 580	69 712	69 763	69 886	69 805
Price, raw sugar[c]	USD/t	176.7	149.9	160.3	200.0	164.6	165.3	150.0	160.0	185.1	155.8	168.4	176.4
Price, white sugar[d]	USD/t	224.9	196.2	190.0	229.3	193.8	194.3	178.8	188.3	213.4	184.1	196.2	203.6

a) Beginning crop marketing year – see the Glossary of Terms for definitions.
b) Non-OECD net exports (imports) equal OECD net imports (exports).
c) Raw sugar world price, New York No. 11, f.o.b. stowed Caribbean port (including Brazil), bulk spot price, September/August.
d) Refined sugar price, London No. 5, f.o.b. Europe, spot, September/August.
est.: estimate.

Source: OECD Secretariat.

Annex Table 40. **SUGAR PROJECTIONS (in raw sugar equivalent)**

Crop year[a]		Average 98/99-02/03	03/04 est.	04/05	05/06	06/07	07/08	08/09	09/10	10/11	11/12	12/13	13/14
AUSTRALIA													
Production	kt rse	4 971	5 106	5 334	5 334	5 441	5 549	5 660	5 540	5 510	5 600	5 729	5 746
Consumption	kt rse	1 098	1 201	1 208	1 217	1 221	1 229	1 235	1 243	1 251	1 258	1 261	1 272
Exports, raw sugar	kt rse	3 698	3 727	4 101	4 080	4 172	4 261	4 194	4 214	4 223	4 133	4 215	4 227
Exports, white sugar	kt rse	105	115	122	129	137	145	154	163	173	183	194	206
Closing stocks	kt rse	1 852	2 271	2 180	2 093	2 009	1 929	2 012	1 938	1 807	1 838	1 903	1 950
Price, raw sugar[b]	AUD/t	314.6	239.9	233.4	288.2	239.0	240.1	218.4	232.5	267.5	226.7	244.0	255.1
CANADA													
Production	kt rse	97	98	105	105	99	106	104	103	103	103	103	103
Imports, raw sugar	kt rse	1 149	1 139	1 151	1 158	1 190	1 204	1 232	1 241	1 262	1 283	1 304	1 325
Imports, white sugar	kt rse	43	44	44	48	40	42	34	34	34	34	34	34
Consumption	kt rse	1 249	1 275	1 283	1 294	1 306	1 318	1 321	1 329	1 334	1 340	1 346	1 354
Closing stocks	kt rse	220	289	257	252	258	258	268	280	281	279	285	278
EU-25													
Production		..	21 068	21 401	21 318	21 224	21 196	20 940	20 949	20 849	20 754	20 671	20 613
EU-15	kt rse	17 765	17 808	18 098	18 030	17 953	17 899	17 608	17 585	17 473	17 362	17 251	17 140
EU-10	kt rse	..	3 261	3 303	3 289	3 271	3 297	3 332	3 365	3 376	3 392	3 420	3 472
Imports, raw sugar[c]	kt rse	1 743	1 750	1 753	1 753	1 753	1 753	1 753	1 753	2 071	2 151	2 250	2 374
Consumption		..	17 526	17 225	17 059	16 871	17 036	16 789	16 749	16 896	16 888	16 872	16 881
EU-15	kt rse	14 343	14 133	14 020	13 908	13 796	13 951	13 690	13 677	13 833	13 840	13 838	13 861
EU-10	kt rse	..	3 393	3 205	3 151	3 074	3 085	3 099	3 073	3 062	3 048	3 034	3 020
Exports, white sugar[c]	kt rse	6 463	6 471	6 296	6 290	6 389	6 220	6 175	6 268	6 314	6 282	6 333	6 389
Closing stocks	kt rse	4 743	2 290	2 154	2 129	2 121	2 095	2 110	2 079	2 073	2 090	2 088	2 085
JAPAN													
Production	kt rse	873	939	935	949	950	942	937	942	941	941	941	941
Imports, raw sugar	kt rse	1 553	1 558	1 578	1 590	1 609	1 637	1 660	1 675	1 695	1 715	1 734	1 753
Consumption	kt rse	2 426	2 492	2 508	2 533	2 554	2 573	2 592	2 611	2 631	2 651	2 670	2 689
Price, white[d]	'000 JPY/t	125.8	115.9	114.3	113.4	112.9	113.2	113.2	113.2	113.2	113.2	113.2	113.2
KOREA													
Imports, raw sugar	kt rse	1 479	1 694	1 641	1 688	1 729	1 776	1 820	1 882	1 925	1 956	1 998	2 067
Consumption	kt rse	1 177	1 340	1 360	1 386	1 419	1 451	1 495	1 535	1 570	1 590	1 620	1 678
Exports, white sugar	kt rse	329	304	292	306	315	326	338	349	360	371	382	394
Closing stocks	kt rse	342	243	232	229	224	223	210	209	204	199	194	189
MEXICO													
Production	kt rse	5 124	5 460	5 507	5 619	5 756	5 882	5 998	6 129	6 250	6 367	6 481	6 595
Consumption	kt rse	4 744	5 286	5 404	5 501	5 599	5 672	5 510	5 621	5 757	5 768	5 881	5 912
Exports, raw sugar	kt rse	125	123	140	168	211	136	555	522	556	590	665	707
Closing stocks	kt rse	1 712	1 725	1 700	1 668	1 656	1 759	1 725	1 748	1 723	1 772	1 749	1 767
UNITED STATES													
Production	kt rse	7 692	8 094	8 594	8 082	8 113	8 168	8 160	8 217	8 283	8 341	8 381	8 423
Imports, raw sugar	kt rse	1 400	1 368	1 259	1 419	1 378	1 349	1 672	1 639	1 673	1 707	1 782	1 825
Consumption	kt rse	9 103	8 972	9 163	9 050	9 520	9 617	9 681	9 778	9 870	9 958	10 043	10 130
Exports, white sugar	kt rse	94	149	169	158	158	158	158	158	158	158	158	158
Closing stocks	kt rse	1 963	2 003	2 588	2 945	2 822	2 628	2 685	2 670	2 663	2 659	2 685	2 709
Price, raw[e]	USD/t	458.5	479.7	467.6	452.2	455.9	467.8	456.8	459.9	462.1	464.1	463.2	462.5
Price, white[f]	USD/t	545.6	543.9	541.0	508.2	509.7	536.1	523.5	524.5	525.4	527.7	526.4	524.5

For notes, see end of the table.

Annex Table 40. **SUGAR PROJECTIONS (in raw sugar equivalent)** (cont.)

Crop year[a]		Average 98/99-02/03	03/04 est.	04/05	05/06	06/07	07/08	08/09	09/10	10/11	11/12	12/13	13/14
OTHER OECD[g]													
Production	kt rse	2 640	2 130	2 431	2 455	2 452	2 491	2 508	2 524	2 543	2 562	2 580	2 599
Consumption	kt rse	2 674	2 777	2 826	2 850	2 879	2 893	2 926	2 935	2 958	2 980	3 003	3 026
Net trade, raw sugar	kt rse	−227	−239	−239	−244	−264	−259	−269	−278	−286	−293	−301	−308
Net trade, white sugar	kt rse	115	−157	−152	−150	−145	−141	−137	−133	−130	−126	−122	−118
ARGENTINA													
Production	kt rse	1 670	1 690	1 708	1 727	1 745	1 764	1 782	1 801	1 821	1 840	1 860	1 879
Consumption	kt rse	1 502	1 441	1 454	1 475	1 491	1 512	1 537	1 566	1 597	1 627	1 658	1 732
Exports, raw sugar	kt rse	73	67	64	66	70	70	70	71	72	73	74	75
Exports, white sugar	kt rse	117	124	129	130	126	127	130	131	132	134	135	136
Closing stocks	kt rse	1 173	1 136	1 197	1 255	1 314	1 369	1 415	1 449	1 470	1 477	1 471	1 409
BRAZIL													
Production													
Sugar	kt rse	20 881	24 530	25 282	26 504	27 600	27 800	28 200	28 900	29 300	31 400	32 370	33 990
Ethanol	mn L	12 344	14 208	17 610	18 295	19 100	19 064	19 340	19 822	20 099	21 541	22 209	23 323
Consumption	kt rse	9 444	10 147	10 407	10 669	10 935	11 204	11 475	11 748	12 024	12 303	12 583	12 956
Exports, raw sugar	kt rse	6 712	7 933	8 547	8 691	9 722	10 119	11 670	12 037	12 147	13 433	13 884	14 801
Exports, white sugar	kt rse	4 433	6 041	6 289	6 928	6 498	6 077	4 861	5 014	5 060	5 596	5 783	6 166
Closing stocks	kt rse	5 165	5 992	6 032	6 248	6 692	7 092	7 286	7 387	7 456	7 525	7 645	7 712
Price, raw sugar[b]	BRL/t	358.8	463.6	479.5	625.5	537.6	559.5	523.6	574.9	689.1	599.5	669.6	724.9
Price, white sugar[b]	BRL/t	455.0	606.8	568.3	717.2	632.9	657.6	623.9	676.4	794.5	708.5	780.3	836.9
CHINA													
Production	kt rse	8 949	10 948	10 679	10 415	10 520	10 625	10 703	10 799	10 840	10 782	10 874	10 965
Imports, raw sugar	kt rse	745	703	750	809	925	1 015	1 141	1 242	1 350	1 458	1 566	1 673
Imports, white sugar	kt rse	115	97	134	153	187	215	259	269	279	289	299	309
Consumption	kt rse	9 278	11 035	11 183	11 296	11 486	11 643	11 790	12 014	12 151	12 331	12 517	12 660
Exports, white sugar	kt rse	331	347	338	289	371	269	326	277	254	231	209	221
Closing stocks	kt rse	2 221	3 460	3 501	3 292	3 064	3 003	2 986	3 002	3 063	3 026	3 036	3 099
CUBA													
Production	kt rse	3 504	2 200	2 198	2 193	2 190	2 187	2 190	2 194	2 198	2 200	2 215	2 220
Consumption	kt rse	684	695	710	728	745	756	768	790	801	820	837	850
Exports, raw sugar	kt rse	2 538	1 196	1 417	1 411	1 343	1 339	1 321	1 314	1 305	1 287	1 289	1 286
Exports, white sugar	kt rse	368	354	227	224	223	222	220	216	217	217	213	208
Closing stocks	kt rse	349	348	322	282	291	291	302	306	311	317	323	328
INDIA													
Production	kt rse	19 755	19 050	22 154	23 361	24 157	23 189	22 665	23 360	24 945	25 362	25 501	25 362
Consumption	kt rse	17 641	20 709	21 228	21 748	22 268	22 787	23 305	23 667	24 054	24 436	24 818	25 198
Exports, white sugar	kt rse	875	1 050	1 074	857	991	897	793	789	768	756	738	702
Closing stocks	kt rse	12 731	12 254	12 077	12 804	13 672	13 146	11 685	10 559	10 653	10 793	10 709	10 141
INDONESIA													
Production	kt rse	1 768	2 114	2 131	2 156	2 190	2 199	2 210	2 248	2 341	2 387	2 580	2 690
Imports, raw sugar	kt rse	489	563	356	367	378	389	401	412	423	434	445	456
Imports, white sugar	kt rse	1 369	1 314	1 305	1 505	1 452	1 648	1 619	1 732	1 758	1 747	1 758	1 747
Consumption	kt rse	3 595	3 817	3 899	3 999	4 078	4 189	4 278	4 401	4 524	4 640	4 767	4 880
Closing stocks	kt rse	2 061	2 156	2 048	2 075	2 015	2 061	2 012	2 002	1 998	1 925	1 941	1 954
OIS													
Production	kt rse	1 771	1 735	1 758	1 774	1 793	1 804	1 811	1 825	1 877	1 896	1 902	1 934
Consumption	kt rse	2 178	2 259	2 252	2 270	2 287	2 305	2 324	2 213	2 232	2 252	2 272	2 294
Net trade	kt rse	−757	−1 060	−1 107	−1 189	−861	−948	−877	−987	−955	−938	−963	−943

For notes, see end of the table.

Annex Table 40. **SUGAR PROJECTIONS (in raw sugar equivalent)** *(cont.)*

Crop year[a]		Average 98/99-02/03	03/04 est.	04/05	05/06	06/07	07/08	08/09	09/10	10/11	11/12	12/13	13/14
RUSSIA													
Production	kt rse	1 651	1 902	1 927	1 945	1 966	1 978	1 985	2 001	2 058	2 078	2 085	2 120
Imports, raw sugar	kt rse	4 919	4 896	4 976	5 203	5 179	5 313	5 417	5 602	5 627	5 725	5 836	5 921
Imports, white sugar	kt rse	190	204	216	196	156	178	175	175	175	175	175	175
Consumption	kt rse	6 416	6 800	6 912	7 024	7 138	7 252	7 367	7 482	7 598	7 716	7 835	7 955
Closing stocks	kt rse	2 386	2 504	2 511	2 666	2 623	2 653	2 623	2 678	2 699	2 720	2 741	2 763
SOUTH AFRICA													
Production	kt rse	2 562	2 538	2 452	2 576	2 690	2 620	2 735	2 678	2 650	2 720	2 790	2 910
Consumption	kt rse	1 555	1 616	1 662	1 710	1 733	1 768	1 799	1 834	1 868	1 901	1 934	1 968
Exports, raw sugar	kt rse	1 014	897	953	1 151	978	942	838	918	1 012	908	992	1 024
Exports, white sugar	kt rse	282	283	208	117	231	125	186	163	159	155	152	148
Closing stocks	kt rse	1 126	964	851	705	711	753	922	942	810	823	791	819
Price, raw sugar[b]	ZAR/t	1 359.3	1 208.3	1 317.3	1 602.4	1 357.8	1 370.2	1 257.3	1 354.9	1 605.0	1 370.4	1 505.3	1 602.3
Price, white sugar[b]	ZAR/t	1 732.3	1 581.5	1 561.3	1 837.3	1 598.4	1 610.3	1 498.3	1 594.2	1 850.4	1 619.7	1 754.2	1 849.7
THAILAND													
Production	kt rse	6 182	7 500	7 616	7 471	8 099	8 354	8 374	8 250	8 456	8 624	8 991	9 201
Consumption	kt rse	1 915	2 108	2 142	2 208	2 251	2 316	2 383	2 445	2 499	2 531	2 560	2 598
Exports, raw sugar	kt rse	2 237	3 096	3 370	3 415	3 503	3 657	3 793	3 643	3 576	3 782	3 861	3 820
Exports, white sugar	kt rse	1 779	2 238	2 131	2 189	2 122	2 102	2 202	2 215	2 321	2 332	2 416	2 623
Closing stocks	kt rse	1 501	2 157	2 130	1 789	2 012	2 291	2 287	2 234	2 294	2 274	2 428	2 588
Price, raw sugar[b]	THB/t	7 305	6 337	6 722	8 384	6 917	6 945	6 285	6 691	7 751	6 768	7 404	7 850
Price, white sugar[b]	THB/t	9 318	8 295	7 967	9 613	8 142	8 163	7 489	7 872	8 937	7 999	8 628	9 063
REST OF WORLD													
Production	kt rse	22 851	25 179	22 224	22 529	24 265	25 650	28 327	29 040	30 160	30 908	31 544	32 486
Consumption	kt rse	37 872	42 991	43 193	45 463	46 156	46 640	47 910	49 767	50 964	52 654	53 723	55 440
Net trade, raw sugar	kt rse	−4 367	−4 662	−6 449	−6 206	−7 085	−7 301	−8 513	−8 350	−7 867	−8 742	−8 953	−9 379
Net trade, white sugar	kt rse	−12 612	−14 435	−14 810	−14 831	−14 934	−13 742	−12 704	−12 736	−12 920	−13 445	−13 736	−14 410
Closing stocks	kt rse	19 487	25 995	26 285	24 389	24 517	24 569	26 203	26 562	26 545	26 306	26 067	26 094

a) Beginning crop marketing year – see the Glossary of Terms for definitions.
b) Export price, f.o.b.
c) Excludes intra-EU trade.
d) White sugar, refined, lower price; Tokyo market.
e) Raw sugar price, September-August New York No. 14.
f) Refined beet sugar price (Midwest), September-August.
g) Includes New Zealand, Norway, Switzerland and Turkey.
est.: estimate.

Source: OECD Secretariat.

Annex Table 41. SUGAR PER CAPITA CONSUMPTION PROJECTIONS
(in raw sugar equivalent)

Crop year[a]		Average 98/99-02/03	03/04 est.	04/05	05/06	06/07	07/08	08/09	09/10	10/11	11/12	12/13	13/14
Australia	kg/person	57.1	60.5	60.3	60.3	60.0	59.9	59.8	59.7	59.7	59.7	59.5	59.6
Canada	kg/person	40.5	40.2	40.2	40.3	40.4	40.5	40.4	40.5	40.4	40.5	40.5	40.6
EU-25	kg/person	39.0	38.5	37.8	37.4	36.9	37.2	36.6	36.5	36.8	36.7	36.6	36.6
Japan	kg/person	19.1	19.5	19.7	19.9	20.0	20.2	20.4	20.6	20.8	21.0	21.2	21.4
Korea	kg/person	25.1	27.9	28.2	28.5	29.0	29.5	30.2	30.9	31.5	31.7	32.2	33.3
Mexico	kg/person	48.3	51.5	51.9	52.1	52.4	52.5	50.4	50.8	51.4	50.9	51.3	50.9
United States	kg/person	33.0	31.5	31.9	31.3	32.7	32.9	32.9	33.2	33.5	33.8	34.1	34.4
Argentina	kg/person	40.5	37.5	37.5	37.6	37.7	37.9	38.2	38.6	39.0	39.4	39.8	41.2
Brazil	kg/person	55.5	57.5	58.3	59.1	59.9	60.7	61.5	62.3	63.1	63.9	64.7	66.0
China	kg/person	7.4	8.6	8.6	8.7	8.7	8.8	8.9	9.0	9.0	9.1	9.2	9.2
Cuba	kg/person	61.1	61.2	62.3	63.6	64.8	65.4	66.2	67.8	68.5	69.8	70.9	71.7
India	kg/person	17.4	19.5	19.7	19.9	20.1	20.3	20.5	20.6	20.7	20.8	20.9	21.0
Indonesia	kg/person	17.1	17.4	17.6	17.8	18.0	18.2	18.4	18.7	19.0	19.3	19.6	19.9
Russia	kg/person	43.9	47.0	47.9	48.9	49.8	50.8	51.9	52.9	54.0	55.0	56.1	57.3
South Africa	kg/person	36.4	36.4	37.1	37.8	38.0	38.5	38.8	39.2	39.6	40.0	40.3	40.7
Thailand	kg/person	30.8	32.9	33.1	33.8	34.1	34.8	35.5	36.1	36.6	36.8	36.9	37.1
Rest of World	kg/person	20.7	22.0	21.6	22.2	22.1	21.8	21.9	22.2	22.2	22.5	22.4	22.7

a) Beginning crop marketing year – see the Glossary of Terms for definitions.
est.: estimate.

Source: OECD Secretariat.

References

Argentina

Wheat production, export, price	SAGPYA, Reply to OECD medium term
Coarse grains production, export, stocks and price	questionnaire (*October 2003*),
Oilseed prices	Buenos Aires, Argentina.
Oilseeds production, import export crush	USDA (*January 2004*), *PS&D Database*,
Vegetable oils production, import export	Washington DC.
Oilseed meals production, import export	
Rice production, exports, stocks and price	
Milk production, liquid sales, industrial use	SAGPYA, Reply to OECD medium term
Milk, butter, cheese, SMP and WMP prices	questionnaire (*October 2003*),
Butter production, export	Buenos Aires, Argentina.
Cheese production, export	
SMP production, export	FAO, FAOSTAT PC database, Rome (*2003*).
WMP production, import export	
Whey powder, net trade	
Beef balance	SAGPYA, Reply to OECD medium term
Poultry balance	questionnaire (*October 2003*),
Pork balance	Buenos Aires, Argentina.
Egg balance	
Pigmeat, poultry and beef meat price	EAP, Buenos Aires, Argentina.
Consumption of all products	Calculated as production + imports – exports – change in stocks.

Australia

Wheat production, feed use, trade, price Coarse grain production, feed use, trade, price Oilseed production, crush, trade, price Oilseed meal trade Vegetable oils price Beef production, trade, price Pig meat production, trade, prices Poultry meat production, trade, prices Sheep meat production, trade, prices Milk production, liquid sales, industrial use, prices Butter production, trade, price Cheese production, trade, price SMP production, trade, price WMP production, trade Casein production, trade	ABARE, *Australian Commodity Statistics 2003*, Canberra.
Whey powder, net trade Egg production, price	ABARE, Reply to OECD medium term questionnaire, Canberra (*October 2003*).
Oilseed meals production, trade, feed use Vegetable oils production, imports Rice, production, exports	USDA (*December 2003*), *PS&D Database*, Washington DC.
Consumption of all products	Calculated as production + imports – exports – change in stocks.

Brazil

Wheat utilisation, supply, price Coarse grains (except buckwheat, rye and other cereals) utilisation, supply, price Cotton, supply, price Soyabean seed, meal and oil, utilisation, supply, Sunflower, utilisation, consumption Beef utilisation, supply, price Pig meat utilisation, supply, prices Poultry meat utilisation, supply, prices Sheep meat utilisation, supply, prices Milk utilisation, liquid sales, industrial use, prices Butter utilisation, supply, price Cheese utilisation, supply, price SMP utilisation, supply, price WMP utilisation, supply, price	Ministry of Agriculture, Reply to OECD medium term questionnaire, Brasilia (*November 2003*).

Buckwheat utilisation, supply	FAO, FAOSTAT PC database, Rome (2003).
Other cereals utilisation, supply Oilseeds, meal and oil prices Rapeseed , production, supply Sunflower, trade Palm oil, utilisation, supply	
Rye, utilisation, supply	USDA (2003),*PS&D Database,* Washington DC.

Canada

Wheat production, exports, stocks, price Coarse grain production, exports, stocks, price Oilseed production, crush, exports, feed use, price Oilseed meal production, imports, exports, price Vegetable oils production, imports, exports, price Beef production, imports, exports, price Pig meat production, exports, price Poultry meat production, imports, price Sheep meat production, imports, price Milk production, liquid sales, industrial use, prices, target return Dairy subsidy Butter production, exports, price, support price Cheese production, imports SMP production, exports, price	Agriculture and Agri-Food Canada (*January 2004*), *CANSIM Database,* Ottawa.
Whey powder net trade	FAO, FAOSTAT PC database, Rome (2003).
Consumption of all products	Calculated as production + imports – exports – change in stocks.

China

Wheat balance, price Coarse grains price Rice balance, price Oilseed balance, price Beef balance, price Pig meat balance, price Poultry balance, price Milk price Coarse grains production, imports, exports, stocks Soyabean oil balance Rapeseed meal balance Rapeseed oil balance	USDA China team, Washington DC.

Palm oil balance	USDA *(January 2003)*, *PS&D Database*, Washington DC.
Milk production, industrial use, other use Whey powder net trade Butter production, imports, exports Cheese production, imports, exports SMP imports WMP imports, exports	FAO, FAOSTAT PC database, Rome (2003).
Consumption of all products	Calculated as production + imports – exports – change in stocks.

EU

Wheat price Coarse grain price Rice price Poultry meat price Sheep meat price Milk price	EUROSTAT (2003), OECD PSE database (2003), Meat and Livestock Commission, *European Market survey* (2003).
Pig meat price	Meat and Livestock Commission, *European Market survey* (2003).
Oilseed price Oilseed meal price Vegetable oil price	ISTA Mielke BmbH, *Oil World Annual* (2003), Hamburg.
Wheat production, exports, stocks Coarse grains production, exports, stocks Rice production, imports, stocks Oilseeds production Beef and Veal production, exports, imports, stocks, male bovine premium Pig meat production, exports, imports, stocks Poultry meat production, exports, imports, stocks Sheep meat production, imports Butter production, imports, exports, stocks Cheese production, imports, exports, stocks SMP production, imports, exports, stocks	EU Commission, Reply to OECD medium term questionnaire, Brussels *(January 2004)*, April 2004 for the accession countries.
Oilseed crush, imports, stocks Oilseed meals production, imports, exports, stocks Vegetable oils production, imports, exports, stocks	ISTA Mielke BmbH, *Oil World Statistics* (2003), Hamburg.

Butter price Cheese price SMP price	Agra Europe (2003), *Milk Products*, London.
Consumption of all products	Calculated as production + imports – exports – change in stocks.

Japan

Wheat price Coarse grain price Oilseed price Oilseed meal price Oilseed meal imports	MAFF, *Monthly Statistics of Agriculture Forestry and Fisheries (various issues)* – Japan, Tokyo.
Wheat production, imports, stocks Coarse grain production, imports, stocks Rice production, imports, stocks Oilseed production, crush, imports, stocks Oilseed meal production Vegetable oil production, imports, stocks	MAFF, *Food balance sheet,* Japan, Tokyo. USDA *PS&D Database*, Washington DC. MAFF, Monthly Statistics of Agriculture Forestry and Fisheries (various issues) – Japan, Tokyo.
Beef production, imports, price Pig meat production, imports, price Sheepmeat imports Poultry meat production, imports, price Milk production, fluid sales, industrial use, price, support price, transaction price, deficiency payment Butter production, imports, price, stabilisation price Cheese production, imports, price SMP production, imports, price, stabilisation price WMP production	MAFF, *Monthly Statistics of Agriculture Forestry and Fisheries (various issues)* – Japan, Tokyo. ALIC, *Monthly Statistics (various issues)*, Japan, Tokyo. USDA *PS&D Database*, Washington DC. United States.
Consumption of all products	Calculated as production + imports – exports – change in stocks.

Korea

Wheat price Coarse grains price Rice price Oilseed price	Replies to OECD medium term questionnaire, Seoul (*September 2003*).

Wheat imports Coarse grains production, imports, stocks Rice production, imports, stocks Oilseed production, crush, imports Oilseed meals production, imports Vegetable oils production, imports	MAFF, Statistical Yearbook (2003) and FAO, FAOSTAT PC database, Rome (2003).
Beef production, imports, price Pig meat production, net trade, price Poultry meat production, imports, price Milk production, liquid sales, industrial use Butter production, imports Cheese production, imports SMP production, imports	Replies to OECD medium term questionnaire, Seoul (*September* 2003).
Whey Powder net trade	FAO, FAOSTAT PC database, Rome (2003).
Consumption of all products	Calculated as production + imports – exports – change in stocks.

Mexico

Wheat production, price Coarse grains production, price Oilseed production, price Beef production, price Pig meat production, price Poultry meat production, price Sheep meat production, price Rice production, export, stocks and price	SAGAR, Reply to OECD medium term questionnaire (*August* 2003), Mexico City. CEA (Centro de Esta distica Agropecuaria), SAGAR, Mexico City. USDA (*September* 2003), *PS&D Database,* Washington DC and FAS reports.
Butter production SMP production	FAO, FAOSTAT PC database, Rome (2003).
Wheat support price Maize support price Cereal income payment Oilseed support price Soyabean income payment	SAGAR (2003), Reply to OECD medium term questionnaire (*August* 2003), Mexico City.

Milk production, price Milk liquid sales, industrial use Butter price Cheese price SMP price WMP price	SAGAR, *Medium Term Questionnaire Reply* *(August 2003)*, Mexico City.
Consumption of all products	Calculated as production + imports – exports – change in stocks.

New Zealand

Wheat production, imports, price Coarse grain production, price Beef production, exports, price Pig meat production, imports, price Poultry meat production, price Sheep meat production, exports, prices Milk production, liquid sales, industrial use, prices Butter production, exports, price Cheese production, exports, price SMP production, exports, price WMP production, consumption, exports, price Casein price	MAF, Reply to OECD Questionnaire *(September 2003)*, Wellington. MAF SONZAF *(December 2003)*, Wellington. New Zealand Meat and Wool Innovation Economic Service *(December 2003)*.
Wheat feed use Coarse grain imports, feed use Butter consumption SMP consumption	
Whey powder net trade	FAO, FAOSTAT PC database, Rome (2003).
Casein, exports	USDA *(January 2004)*, PS&D FAO.
Consumption of all products	Calculated as production + imports – exports – change in stocks.

Russia

Wheat production, imports, exports, ending stocks Coarse grains production, imports, exports, ending stocks Oilseed production, crush, imports, exports Oilseed meals production, imports, exports Vegetable oils production, imports, exports Rice production, imports, exports	USDA *(January 2004)*, PS&D Database, Washington DC.

Beef production, imports Pig meat production, imports	FAO, FAOSTAT PC database (2003), Rome.
Poultry meat production, imports	USDA *(January 2004), PS&D Database,* Washington DC.
Milk production	FAO, FAOSTAT PC database (2003), Rome.
Butter production, imports Cheese production, imports SMP production, imports, exports WMP production, imports	USDA *(January 2004), PS&D Database,* Washington DC.
Consumption of wheat, coarse grain, rice, oilseeds, oilseed meals, vegetable oils, beef, pig meat, poultry meat, sheep meat, butter, cheese, SMP and WMP	Calculated as production – imports + exports – change in stocks.
Prices	OECD PSE database (2004)

United States

Wheat production, imports, exports, stocks, price, EEP payment	USDA, *Wheat Outlook (January 2004),* *Wheat S&O Yearbook (March 2003),* Washington DC.
Coarse grains production, exports and price Rice production, imports, exports, stocks and price	USDA, *Feed Outlook (January 2004),* *Feed S&O Yearbook (April 2003),* Washington DC. USDA, *Rice Outlook (January 2004),* *Rice S&O Yearbook (December 2003),* Washington DC.
Beef production, imports, exports, price Pig meat production, imports, exports, price Poultry meat production, exports, price Sheep meat production, imports, price	USDA, *Livestock, Dairy and Poultry Outlook* *(December 2003 and January 2004),* Washington DC.
Milk production, liquid sales, industrial use, support price, prices Butter production, exports, stocks, price Cheese production, imports, exports, price SMP production, exports, stocks, price WMP production, exports, stocks	USDA, *Livestock, Dairy and Poultry* *(January 2004),* Foreign Agricultural Service *Dairy Production & Trade Summary* *(December 2003),* USDA *Dairy Yearbook* tables (2003), Washington DC.
Whey powder production, exports, price	USDA, *Dairy Yearbook (2003),* Washington DC. FAO, FAOSTAT PC database, Rome (2003).

226

Casein imports	USDA (*January 2004*), PS&D database, Washington DC.
Oilseed production, crush, exports, and price Oilseed meals production, imports, exports and price Vegetable oils production, imports, exports, stocks and price	USDA, *Oil Crops Outlook (January 2004)*, *Oil Crops S&O Yearbook (October 2003)*, Washington DC.
Wheat target price, loan rate, LDP, DP and CCP payments, rates and areas, other land idled Coarse grains, target price, loan rate, LDP, DP and CCP payments, rates and areas, other land idled Maize target price, loan rate, LDP, DP and CCP payments, rates and areas Soyabean target price, loan rate, LDP, DP and CCP payments, rates and areas	USDA, *Agricultural Outlook (December 2003)*, Farm Service Agency (*November 2003/January 2004*), Washington DC .
Consumption of all products	Calculated as production + imports – exports – change in stocks.

Other OECD

Wheat production, consumption Coarse grains production, consumption Oilseed production, crush, consumption Oilseed meals production, consumption Vegetable oils production, consumption Rice production, consumption	Replies to OECD Questionnaires (*September 2003*). USDA (*September 2003*), *PS&D Database*, Washington DC.
Beef production, consumption Pig meat production, consumption Poultry meat production, consumption Sheep meat production, consumption	Replies to OECD Questionnaires (*September 2003*). USDA (*September 2003*), *PS&D Database*, Washington DC.
Milk production, on farm use, liquid sales, industrial use Butter production, consumption Cheese production, consumption SMP production, consumption WMP production, consumption	Replies to OECD Questionnaires (*September 2003*).
Net trade in wheat, coarse grain, rice, oilseeds, oilseed meals, vegetable oils, beef, pig meat, poultry meat, sheep meat, butter, cheese, SMP and WMP	Calculated as production – consumption – change in stocks.

OECD

Production of wheat, coarse grains, rice, oilseeds, oilseed meals, vegetable oils, butter, cheese, SMP, WMP Consumption of wheat, coarse grains, rice, oilseeds, oilseed meals, vegetable oils, butter, cheese, SMP, whole milk powder Imports of butter, cheese, SMP, WMP Exports of butter, cheese, SMP, WMP Stocks of wheat, coarse grains, rice, oilseeds, oilseed meals, vegetable oils, butter, cheese, SMP Feed use of wheat, coarse grains Oilseed crush	Calculated as Australia + Canada + EU-25 + Japan + New Zealand + United States + Mexico + Korea + other OECD. Therefore the OECD total excludes Iceland but includes Cyprus, Estonia, Latvia, Lithuania, Malta and Slovenia.

Rest of world

Wheat production, stocks Coarse grains production, stocks Rice production, stocks Oilseed production, crush, stocks Oilseed meals production, stocks Vegetable oils production, stocks	USDA (*November 2003*), *PS&D Database*, Washington DC.
Net trade of wheat, coarse grains, rice, oilseeds, oilseed meals, vegetable oils, butter, cheese, SMP, WMP, whey powder	Calculated as – net trade of (OECD + RUS + Other independent states + Brazil + China + Argentina).
Milk production, industrial use, other uses Butter production Cheese production SMP production WMP production	Calculated as World – (OECD + RUS + Other independent states + Argentina + Brazil + China).
Consumption of all products	Calculated as production – net trade – change in stocks.

Chinese Tapei, India
Rice production, stocks

Indonesia
Rice production, imports, stocks

Thailand
Rice production, exports, stocks

USDA (*November 2003*) *PS&D Database*, Washington, DC.

Chinese Taipei, India, Indonesia, Thailand

Rice price	University of Arkansas rice database (2003), Fayetteville, USA. USDA, *FAS reports (various* issues), Washington, DC. IRRI *World Rice Statistics(various issues)*, Makati, Philippines.
Consumption of all products	Calculated as production – net trade – change in stocks.

OIS (Other Independent States)

Wheat production, net trade, ending stocks Coarse grains production, net trade, ending stocks Rice production, net trade Oilseed production, crush, net trade, ending stocks Oilseed meals production, net trade Vegetable oils production, net trade	USDA *(September 2003), PS&D Database,* Washington DC for FSU. Calculated as FSU-RUS.
Butter production, net trade Cheese production, net trade SMP production, net trade WMP production, net trade	USDA *(September 2003), PS&D Database,* Washington DC for FSU. Calculated as FSU-RUS.
Consumption of wheat, coarse grain, rice, oilseeds, oilseed meals, vegetable oils, beef, pig meat, poultry meat, sheep meat, butter, cheese, SMP and WMP	Calculated as production – net trade – change in stocks.

World

Wheat production, feed use, stocks, Coarse grains production, feed use, stocks Rice production, stocks Oilseed production, crush, stocks Oilseed meals production, stocks Vegetable oils production, stocks Butter, cheese, skim milk powder, stocks	Calculated as Rest of world + OECD + Argentina + Brazil + China + OIS + Russia.
Production of butter, cheese, skim milk powder, whole milk powder	FAO, FAOSTAT PC database, Rome (2003).
Wheat price	USDA, *Wheat Outlook (January 2004),* Washington DC.

Coarse grains price Rice price	USDA, *Feed Outlook (January 2004)*, Washington DC. USDA, *Rice Outlook (January 2004)*, Washington DC.
Oilseed price Oilseed meals price Oilseed oils price Palm oil price	ISTA Mielke GmbH, *Oil World Annual 2003*, Hamburg.
Butter price SMP price	USDA, *Dairy World Markets and Trade (December 2003)*, Washington DC.
Cheese price	USDA, *Dairy World Markets and Trade (December 2003)*, Washington DC.
WMP price	USDA, *Dairy World Markets and Trade (December 2003)*, Washington DC.
Whey powder price	USDA, *Livestock, Dairy and Poultry (December 2003)*, Washington DC.
Casein price	New Zealand Dairy Board, *International Market Update*, Wellington.
Tariffs, tariff-quotas and subsidised export limits for OECD countries unless otherwise specified	GATT (1996), *Uruguay Round GATT Schedules*, Geneva.
Consumption of all products	Calculated as production – net trade – change in stocks.

Sugar

Sugar production, raw and white exports, raw and white imports, consumption, stocks	FO Licht World Sugar Balances, 2003

ISBN 92-64-02008-X
OECD Agricultural Outlook: 2004-2013
© OECD 2004

ANNEX II

Glossary of Terms

Agenda 2000

A CAP reform package proposed by the European Commission in 1998. After a number of modifications, the European Union Heads of State agreed to a package of reforms in March 1999. Beginning in 2000, the package reduces price supports and increases direct payments for cereals and beef, while lowering oilseed direct payments (by harmonising them with cereals) and raising the milk quota. Dairy support price reductions and the introduction of new dairy direct payments are delayed until 2005, along with a second round of milk quota increase. The package is sometimes referred to as the "Berlin Agreement" to distinguish the agreement from the initial European Commission proposals.

AMAD

Agricultural Market Access database. A co-operative effort between Agriculture and Agri-food Canada, EU Commission-Agriculture Directorate-General, FAO, OECD, The World Bank, UNCTAD and the United States Department of Agriculture, Economic Research Service. Data in the database is obtained from countries' schedules and notifications submitted to the WTO.

Avian virus

Avian influenza is an infectious disease of birds caused by type A strains of the influenza virus. The disease, which was first identified in Italy more than 100 years ago, occurs worldwide. The quarantining of infected farms and destruction of infected or potentially exposed flocks are standard control measures.

Atlantic beef market

World beef market excluding the Pacific Rim beef trade.

Baseline

The set of market projections used for the Outlook analysis in this report and as a benchmark for the analysis of the impact of different economic and policy scenarios. A detailed description of the generation of the baseline is provided in the chapter on Methodology in this report.

Berlin Agreement

The CAP reform package to which European Union Heads of State agreed in March 1999. Beginning in 2000, the package reduces price supports and increases direct payments for cereals and beef, while lowering oilseed direct payments (by harmonising them with cereal payments) and raising the milk quota. Dairy support price reductions and the introduction of new dairy direct payments are delayed until 2005, along with a second round of milk quota increase. Like the initial proposal by the European Commission which was not accepted, the agreement is often referred to as "Agenda 2000".

Biomass

Biomass is defined as any plant matter used directly as fuel or converted into other forms before combustion. Included are wood, vegetal waste (including wood waste and crops used for energy production), animal materials/wastes and other solid biomass.

Bovine Spongiform Encephalopathy (BSE)

A fatal disease of the central nervous system of cattle, first identified in the United Kingdom in 1986. On 20 March 1996 the UK Spongiform Encephalopathy Advisory Committee (SEAC) announced the discovery of a new variant of Creutzfeldt-Jacob Disease (vCJD), a fatal disease of the central nervous system in humans, which might be linked to consumption of beef affected by exposure to BSE.

Cereals

Defined as wheat, coarse grains and rice.

Common Agricultural Policy (CAP)

The European Union's agricultural policy, first defined in Article 39 of the Treaty of Rome signed in 1957.

CAP reform

The EU Commission has published a Communication on the Mid-Term Review on the Common Agricultural Policy in July 2002, in January 2003 the Commission adopted a formal proposal. A formal decision on the "CAP reform – a long-term perspective for sustainable agriculture" was taken by the EU farm ministers. The reform includes far-reaching amendments of current policies, including further reductions in support prices, partly offset by direct payments, and a further decoupling of most direct payments from current production.

Coarse grains

Defined as barley, maize, oats, sorghum and other coarse grains in all countries except Australia, where it includes triticale and in the European Union where it includes rye and other mixed grains.

Conservation Reserve Program (CRP)

A major provision of the United States' Food SecurityActof1985 and extended under the Food, Agriculture, Conservation and Trade Act of 1990, and the Food, Agriculture, Improvement and Reform Act of 1996, designed to reduce erosion on 40 to 45 million acres (16 to 18 million hectares) of farm land. Under the programme, producers who sign contracts agree to convert erodable crop land to approved conservation uses for ten years. Participating producers receive annual rental payments and cash or payment in kind to share up to 50% of the cost of establishing permanent vegetative cover. The CRP is part of the *Environmental Conservation Acreage Reserve Program*. The 1996 FAIR Act authorised a 36.4 million acre (14.7 million hectares) maximum under CRP, its 1995 level.

Commonwealth of Independent States (CIS)

The heads of twelve sovereign states (except the Baltic states) have signed the Treaty on establishment of the Economic Union, in which they stressed that the Azerbaijan Republic, Republic of Armenia, Republic of Belarus, Republic of Georgia, Republic of Kazakhstan, Kyrgyz Republic, Republic of Moldova, Russian Federation, Republic of Tajikistan, Turkmenistan, Republic of Uzbekistan and Ukraine on equality basis established the Commonwealth of Independent States. OIS is used in the Outlook tables to refer to all these states excluding the Russian Federation

Common Market Organisation (CMO) for sugar

The common organisation of the sugar market (CMO) in the European Union was established in 1968 to ensure a fair income to community sugar producers and self-supply of the Community market. At present the CMO is governed by Council Regulation (EC) No. 1260/2001 (the basic regulation) which are applicable until 30 June 2006.

Country-of-origin labelling (CoOL)

Refers to the 2002 United States Farm Act which requires that by September 30, 2004, United States Department of Agriculture will implement a mandatory regulation for country-of-origin labelling.

Crop year, coarse grains

Refers to the crop marketing year beginning 1 April for Japan, 1 July for the European Union and New Zealand, 1 August for Canada and 1 October for Australia. The US crop year begins 1 June for barley and oats and 1 September for maize and sorghum.

Crop year, oilseeds

Refers to the crop marketing year beginning 1 April for Japan, 1 July for the European Union and New Zealand, 1 August for Canada and 1 October for Australia. The US crop year begins 1 June for rapeseed, 1 September for soyabeans and for sunflower seed.

Crop year, rice

Refers to the crop marketing year beginning 1 April for Japan, Australia, 1 August for the United States, 1 September for the European Union, 1 October for Mexico, 1 November for Korea and 1 January for other countries.

Crop year, wheat

Refers to the crop marketing year beginning 1 April for Japan, 1 June for the United States, 1 July for the European Union and New Zealand, 1 August for Canada and 1 October for Australia.

De-coupled payments

Budgetary payments paid to eligible recipients who are not linked to current production of specific commodities or livestock numbers or the use of specific factors of production.

Direct payments

Payments made directly by governments to producers.

Doha Development Agenda

Mulitlateral trade negotiations in the World Trade Organisation that were initiated in November 2001, in Doha, Qatar

Domestic support

Refers to the annual level of support, expressed in monetary terms, provided to agricultural production. It is one of the three pillars of the Uruguay Round Agreement on Agriculture targeted for reduction.

Double-zero agreement

In the double zero agreements, the EU and the CEEC typically agree to offer duty free quotas for a specific quantity of a given agricultural product, while anything above the quota is subject to duty. Further the EU and the CEEC agree not to use any export subsidies for the given agricultural product. Every agreement has been concluded bilaterally between the EU and each CEEC country and that, consequently, the contents of the agreements vary from one case to another.

Everything-But-Arms (EBA)

The Everything-But-Arms (EBA) Initiative of the European Union eliminates EU import tariffs and restrictions for numerous goods, including agricultural products, from the least developed countries. The tariff elimination is scheduled in four steps from 2006/07 to 2009/10.

Export credits (with official support)

Government financial support, direct financing, guarantees, insurance or interest rate support provided to foreign buyers to assist in the financing of the purchase of goods from national exporters.

Export Enhancement Program (EEP)

A US programme initiated in May 1985 under a *Commodity Credit Corporation* charter to subsidise the export of certain products to specified countries. The programme was formally authorised by the Food Security Act of 1985 and has been extended since under the Farm Act of 1990 and the FAIR Act of 1996. Under the EEP, exporters are awarded generic commodity certificates which are redeemable for commodities held in CCC stores, thus enabling them to sell commodities to designated countries at prices below those on the US market.

Export restitutions (refunds)

EU export subsidies provided to cover the difference between internal prices and world market prices for particular commodities.

Export subsidies

Subsidies given to traders to cover the difference between internal market prices and world market prices, such as for example the EU *export restitutions* and the US *Export Enhancement Program* (see above). Export subsidies are now subject to value and volume restrictions under the *Uruguay Round Agreement on Agriculture*.

FAIR Act, 1996

Officially known as the Federal Agriculture Improvement and Reform Act of 1996. This US legislation replaces the 1990 Farm Act and governs almost all aspects of food and agriculture policy during the period 1996-2002.

Foot and Mouth Disease (FMD)

Foot and mouth disease is a highly contagious disease, which chiefly affects cloven-hoofed animal species (cattle, sheep, goats and pigs). Its symptoms are the appearance of vesicles (aphthae) on the animals' mouths (with a consequent reduction in appetite) and feet. It is caused by a virus which may be found in the animals' blood, saliva and milk. The virus is transmitted in a number of ways, via humans, insects, most meat products, urine and faeces, feed, water or soil. Although the mortality rate in adult animals from this disease is generally low and the disease presents no risk for humans, because it is highly contagious, infected animals in a given country are generally put down and other countries place an embargo on imports of live animals and fresh, chilled or frozen meat from the country of infection; in that case, only smoked, salted or dried meat and meat preserves may be imported from the country concerned. In addition, given the possibility of contagion between different species of cloven-hoofed animals, when foot and mouth disease breaks out in one species in a given country, exports of meat from all four types of animal are suspended

FSRI Act, 2002

Officially known as the Farm Security and Rural Investment Act of 2002. This US legislation replaces the FAIR Act of 1996, covering a wide range of policy aspects for the period 2002-2007.

Hazard Analysis and Critical Control Point (HACCP)

HACCP describes a system of control for assuring food safety and provides a more structured and critical approach to the control of identified hazards than that achievable by traditional inspection and quality control procedures. It has the potential to identify areas of concern where failure has not yet been experienced, making it particularly useful for new operations.

Industrial oilseeds

A category of oilseed production in the European Union for industrial use (*i.e.* biofuels) that is subject to subsidy limits.

Intervention purchases

Purchases by the EC Commission of certain commodities to support internal market prices.

Intervention purchase price

Price at which the European Commission will purchase produce to support internal market prices. It usually is below 100% of the intervention price, which is an annually decided policy price.

Intervention stocks

Stocks held by national intervention agencies in the European Union as a result of *intervention* buying of commodities subject to market price support. Intervention stocks may be released onto the internal markets if internal prices exceed intervention prices; otherwise, they may be sold on the world market with the aid of *export restitutions*.

International Organization for Standardization (ISO)

An international organisation, of which the object is "to facilitate the international coordination and unification of industrial standards". The new organisation, ISO, officially began operations on 23 February 1947.

Inulin

Inulin syrups are extracted from chicory through a process commercially developed in the 1980s. They usually contain 83 per cent fructose. Inulin syrup production in the European Union is covered by the sugar regime and subject to a production quota.

Isoglucose

Isoglucose is a starch-based fructose sweetener, produced by the action of glucose isomerase enzyme on dextrose. This isomerisation process can be used to produce glucose/fructose blends containing up to 42 per cent fructose. Application of a further process can raise the fructose content to 55 per cent. Where the fructose content is 42 per cent, isoglucose is equivalent in sweetness to sugar. Isoglucose production in the European Union is covered by the sugar regime and subject to a production quota.

OECD AGRICULTURAL OUTLOOK: 2004-2013 – ISBN 92-64-02008-X – © OECD 2004

Loan deficiency payments (United States)

Loan deficiency payments are a type of non-recourse loan whereby, for wheat, feed grain, upland cotton, rice and oilseeds, a producer may agree to forgo loan eligibility and receive an output subsidy, the rate of payment of which is the amount by which the applicable county's loan rate exceeds the marketing loan repayment rate. Producers may elect to apply for this payment during the loan availability period on a quantity of the programme crop not exceeding their loan-eligible production. This, combined with marketing loan gains, represent the benefits made available to US farmers when commodity prices fall relative to loan rates.

Loan rate

The commodity price at which the *Commodity Credit Corporation* (CCC) offers *non-recourse loans* to participating farmers. The crops covered by the programme are used as collateral for these loans. The loan rate serves as a floor price, with the effective level lying somewhat above the announced rate, for participating farmers in the sense that they can default on their loan and forfeit their crop to the CCC rather than sell it in the open market at a lower price.

Luxembourg agreement

A formal decision on further "CAP reform – a long-term perspective for sustainable agriculture" was taken by the EU Council of farm ministers meeting in Luxembourg on 26 June 2003. The reform includes far-reaching amendments of current policies, including further reductions in support prices, partly offset by direct payments and a further decoupling of most direct payments, such as the new single farm payment from current production. The different elements of the reform will enter into force in 2004 and 2005. A single farm payment will enter into force in 2005. If a member State needs a transitional period due to its specific agricultural conditions, it may apply the single farm payment from 2007 at the latest.

Market access

Governed by provisions of the *Uruguay Round Agreement* on *Agriculture* which refer to concessions contained in the country schedules with respect to bindings and reductions of tariffs and to other minimum import commitments.

Marketing Assistance Loan Programme

US loan programme since 1986 designed to provide producers of certain crops with financial assistance when prices are low while avoiding a disadvantage of the traditional loan programme (*see loan rate*) – the accumulation of government stocks that depress prices when disposed of. The programme effectively guarantees farmers a minimum price. Farmers can obtain payments in two ways. They can sell the crop and repay the loan at the posted county price (a USDA estimate of the local market price) and keep the difference known as "marketing gain". They can also obtain a payment without taking out a loan – see *loan deficiency payments*.

Marketing year, oil meal

Refers to the production year beginning 1 October for the United States.

Marketing year, oilseed oil

Refers to the production year beginning 1 October for the United States.

MERCOSUR

A multilateral agreement on trade, including agricultural trade between Argentina, Brazil, Paraguay and Uruguay. The agreement was signed in 1991 and came into effect on 1 January 1995. Its main goal is to create a customs union between the four countries by 2006.

Market Price Support (MPS) Payment

Indicator of the annual monetary value of gross transfers from consumers and taxpayers to agricultural producers arising from policy measures creating a gap between domestic market prices and *border prices* of a specific agricultural commodity, measured at the farm gate level. Conditional on the production of a specific commodity, MPS includes the transfer to producers associated with both production for domestic use and exports, and is measured by the price gap applied to current production. The MPS isnet of financial contributions from individual producers through producer levies on sales of the specific commodity or penalties for not respecting regulations such as production quotas (*Price levies*), and in the case of livestock production is net of the market price support on domestically produced coarse grains and oilseeds used as animal feed (*Excess feed cost*).

Mid-Term Review

See Luxembourg agreement on CAP reform.

Milk quota scheme

A supply control measure to limit the volume of milk produced or supplied. Quantities up to a specified quota amount benefit from full *market price support*. Over-quota volumes may be penalised by a levy (as in the European Union, where the "super levy" is 115% of the target price) or may receive a lower price. Allocations are usually fixed at individual producer level. Other features, including arrangements for quota reallocation, differ according to scheme.

Modulation

A partial transfer of support from the first (support to agriculture) to the second pillar (support to other rural activities) of the EU Common Agricultural Policy (CAP). With the latest reform of the CAP, modulation was made compulsory, resulting in a gradual reduction of payments directly to farmers with the aim of boosting rural development.

Non-member economies

Countries outside of the OECD area or membership.

North American Free Trade Agreement (NAFTA)

A trilateral agreement on trade, including agricultural trade, between Canada, Mexico and the United States, phasing out tariffs and revising other trade rules between the three countries over a 15-year period. The agreement was signed in December 1992 and came into effect on 1 January 1994.

Oilmeal

Defined as rapeseed meal (canola), soyabean meal, and sunflower meal in all countries, except in Japan where it excludes sunflower meal.

Oilseeds

Defined as rapeseed (canola), soyabeans, and sunflower seed in all countries, except in Japan where it excludes sunflower seed.

Over Thirty Months Scheme (OTMS)

The purpose of the Scheme is to provide cattle producers with an alternative market for cattle aged over thirty months which have come to the end of their productive lives and can no longer be entered into the human or animal food chain as a result of the Over Thirty Month Rule. The OTM Rule bans meat from most cattle aged over 30 months at slaughter from being sold for human consumption. This is to remove older animals, which are more likely to have developed a significant amount of BSE agent in any tissue, from the human food chain. It applies equally to home-produced and imported meat.

Pacific beef market

Beef trade between countries in the Pacific Rim where foot and mouth disease is not endemic.

PROCAMPO

A programme of direct support to farmers in Mexico. It provides for direct payments per hectare on a historical basis.

Producer Support Estimate (PSE)

Indicator of the annual monetary value of gross transfers from consumers and taxpayers to agricultural producers, measured at farm gate level, arising from policy measure, regardless of their nature, objectives or impacts on farm production or income. The PSE measure support arising from policies targeted to agriculture relative to a situation without such policies, i.e. when producers are subject only to general policies (including economic, social, environmental and tax policies) of the country. The PSE is a gross notion implying that any costs associated with those policies and incurred by individual producers are not deducted. It is also a nominal assistance notion meaning that increased costs associated with import duties on inputs are not deducted. But it is an indicator net of producer contributions to help finance the policy measure (e.g. producer levies) providing a given transfer to producers. The PSE includes implicit and explicit payments. The percentage PSE is the ration of the PSE to the value of total gross farm receipts, measured by the value of total production (at farm gate prices), plus budgetary support. The nomenclature and definitions of this indicator replaced the former Producer Subsidy Equivalent in 1999.

Purchasing Power Parity (PPP)

Purchasing power parities (PPPs) are the rates of currency conversion that eliminate the differences in price levels between countries. The PPPs are given in national currency units per US dollar.

Recourse loan programme

Programme to be implemented under the US FAIR Act of 1996 for butter, non-fat dry milk and cheese after 1999 in which loans must be repaid with interest to processors to assist them in the management of dairy product inventories.

SARS

Severe Acute Respiratory Syndrome is an atypical pneumonia caused by SARS coronavirus (SARS-CoV). First cases were reported in February 2003, and the subsequent four month epidemic was considered a health threat of global proportions, because of limited knowledge concerning the transmission of the virus.

Scenario

A model-generated set of market projections based on alternative assumptions than those used in the baseline. Used to provide quantitative information on the impact of changes in assumptions on the outlook.

Set-aside programme

European Union programme for cereal, oilseed and protein crops that both requires and allows producers to set-aside a portion of their historical base acreage from current production. Mandatory set-aside rates for commercial producers are set at 10% until 2006.

Single Farm Payment

With the 2003 CAP reform, the EU introduced a farm-based payment largely independent of current production decisions and market developments, but based on the level of former payments received by farmers. To facilitate land transfers, entitlements are calculated by dividing the reference amount of payment by the number of eligible hectares (incl. forage area) in the reference year. Farmers receiving the new SFP are obliged to keep their land in good agricultural and environmental condition and have the flexibility to produce any commodity on their land except fruits, vegetables and table potatoes.

SPS Agreement

WTO Agreement on Sanitary and Phyto-sanitary measures, including standards used to protect human, animal or plant life and health.

Support price

Prices fixed by government policy makers in order to determine, directly or indirectly, domestic market or producer prices. All administered price schemes set a minimum guaranteed support price or a target price for the commodity, which is maintained by associated policy measures, such as quantitative restrictions on production and imports; taxes, levies and tariffs on imports; export subsidies; and public stockholding.

Tariff-rate quota (TRQ)

Resulted from the Uruguay Round Agreement on Agriculture. Certain countries agreed to provide minimum import opportunities for products previously protected by non-tariff barriers. This import system established a quota and a two-tier tariff regime for affected commodities. Imports within the quota enter at a lower (in-quota) tariff rate while a higher (out-of-quota) tariff rate is used for imports above the concessionary access level.

Uruguay Round Agreement on Agriculture (URAA)

The terms of the URAA are contained in the section entitled the "Agreement on Agriculture" of the Final Act Embodying the Results of the Uruguay Round of Multilateral Trade Negotiations. This text contains commitments in the areas of *market access*, domestic support (see AMS), and *export subsidies*, and general provisions concerning monitoring and continuation. In addition, each country's schedule is an integral part of its contractual commitment under the URAA. There is a separate agreement entitled the Agreement on the Application of Sanitary and Phyto-sanitary Measures. This agreement seeks establishing a multilateral framework of rules and disciplines to guide the adoption, development and the enforcement of sanitary and phyto-sanitary measures in order to minimise their negative effects on trade. See also *Phyto-sanitary regulations* and *Sanitary regulations*.

Vegetable oil

Defined as rapeseed oil (canola), soyabean oil, sunflower seed oil and palm oil, except in Japan where it excludes sunflower seed oil.

WTO

World Trade Organisation created by the Uruguay Round agreement.

OECD PUBLICATIONS, 2, rue André-Pascal, 75775 PARIS CEDEX 16
PRINTED IN FRANCE
(51 2004 03 1 P) ISBN 92-64-02008-X – No. 53339 2004